METHODS IN MOLECULAR BIOLOGY™

Series Editor
John M. Walker
School of Life Sciences
University of Hertfordshire
Hatfield, Hertfordshire, AL10 9AB, UK

For further volumes:
http://www.springer.com/series/7651

Ribosome Display and Related Technologies

Methods and Protocols

Edited by

Julie A. Douthwaite and Ronald H. Jackson

*Department of Antibody Discovery and Protein Engineering,
MedImmune Limited, Cambridge, UK*

Editors
Julie A. Douthwaite, Ph.D
Department of Antibody Discovery
and Protein Engineering
MedImmune Limited
Cambridge, UK
douthwaitej@medimmune.com

Ronald H. Jackson, Ph.D
Department of Antibody Discovery
and Protein Engineering
MedImmune Limited
Cambridge, UK
jacksonr@medimmune.com

ISSN 1064-3745   e-ISSN 1940-6029
ISBN 978-1-4939-5872-6   ISBN 978-1-61779-379-0 (eBook)
DOI 10.1007/978-1-61779-379-0
Springer New York Dordrecht Heidelberg London

© Springer Science+Business Media, LLC 2012
Softcover re-print of the Hardcover 1st edition 2011
All rights reserved. This work may not be translated or copied in whole or in part without the written permission of the publisher (Humana Press, c/o Springer Science+Business Media, LLC, 233 Spring Street, New York, NY 10013, USA), except for brief excerpts in connection with reviews or scholarly analysis. Use in connection with any form of information storage and retrieval, electronic adaptation, computer software, or by similar or dissimilar methodology now known or hereafter developed is forbidden.
The use in this publication of trade names, trademarks, service marks, and similar terms, even if they are not identified as such, is not to be taken as an expression of opinion as to whether or not they are subject to proprietary rights.

Printed on acid-free paper

Humana Press is part of Springer Science+Business Media (www.springer.com)

Preface

Over the last 20 years, display technologies have become a very powerful way of generating therapeutic lead molecules and specific reagents for increasing our understanding of biology. Despite first being described shortly after phage display, the use of ribosome display and related methods has been much less widespread. Since this is in part due to the complexity of the methods, it is our hope that the availability of this volume of *Methods in Molecular Biology*™ will allow their extended use. The protocols described range from well-established methods that have been used for a decade to generate high affinity antibodies which are in the clinic to methods that are more at their early stages of application, such as display of peptides incorporating noncanonical amino acids.

At the core of ribosome display and related technologies, such as mRNA display, "in vitro virus," and cDNA display, is the in vitro generation of a library of diverse molecules in which a peptide or protein is associated with the nucleic acid encoding it. From the large libraries of over 10^{12} variants that can be made by these purely in vitro methods, molecules with desired properties can be selected by binding to a partner molecule. In all of these methods, the displayed protein is generated by a transcription/translation process which may use semi-purified extracts, such as the *E. coli* S30 system or purified components of the transcription and translation systems.

In ribosome display, a non-covalent ternary complex is formed between the ribosome, mRNA, and the encoded translated protein, and the complex is selected by binding to the target protein. The complex is stabilized by magnesium ions and can be readily disassociated by the addition of EDTA. In mRNA display, a covalent complex is formed between the mRNA and the encoded protein, and this complex is then selected. In both cases, the mRNA recovered from the selected complexes is reverse transcribed to DNA, followed by amplification to generate DNA that can be assembled and used for subsequent rounds of translation and selection. The non-covalent ribosome display method is most widely used and is described in a number of the methods described (Chapters 3–5, 9–12, 14, 15, 17, and 18), including modifications such as the introduction of a protein–RNA interaction to stabilize ternary complexes (Hara et al., Chapter 4) or in situ DNA recovery (He et al., Chapter 5). An mRNA display protocol is described in Chapter 6 by Wang et al. and applications of mRNA display are described by Cotten et al. in Chapter 16.

The covalent coupling in mRNA display is achieved by the use of puromycin, an antibiotic that mimics the aminoacyl moiety of the tRNA, entering the ribosome A site and accepting the nascent polypeptide forming a peptide bond. Puromycin is attached to the mRNA at its 3' end by a short DNA linker. During in vitro translation, the ribosome reaches the RNA–DNA junction, translation pauses, and covalent coupling of the mRNA to the translated polypeptide takes place. In Chapter 8, Ueno and Nemoto describe an adaptation of mRNA display in which the mRNA–protein fusion, derived using puromycin, is converted into a more stable mRNA/cDNA protein fusion, wherein the cDNA is covalently linked to its encoded protein. The application of disulfide shuffling reactions to a library of disulfide-rich peptides displayed in this format is described by Mochizuki and Nemoto in Chapter 13. A further related technique is SNAP display, described by Kaltenbach and Hollfelder in Chapter 7, in which the translated molecule is covalently coupled to the DNA encoding it following translation in microdroplets in an emulsion. The complex is subsequently selected after breaking the emulsion. Methods involving stable covalent coupling of the nucleic acid and protein allow chemical modification of displayed molecules or their selection under harsher conditions. They may be most applicable to smaller molecules, such as peptides or proteins that refold readily, as in Chapter 13 (Mochizuki and Nemoto), Chapter 16 (Cotten et al.), and Chapter 21 (Hartman et al.).

In his perspective in Chapter 1, Andreas Plueckthun examines critically the features and advantages of ribosome display and related technologies and gives guidance on how to use methods most effectively, for instance when performing selections to enrich for binding molecules with slower off-rates. We recommend that all readers planning to use the methods described in this book read this chapter to help their overall understanding and experimental design.

At the core of all technologies related to ribosome display are efficient, high quality translation extracts. The "workhorse" for most display technology has been the *E. coli* S30 in vitro transcription/translation extract. Its preparation is described in the chapter by Zawada (Chapter 2), and its application is described in several chapters (e.g., Lewis and Lloyd, Chapter 9, for antibodies and Dreier and Plueckthun, Chapter 15, for Darpins). Use of the S30 extract has been particularly successful for the selection of variant molecules from libraries having fixed frameworks that provide consistent translation yields. In our laboratories at MedImmune, we routinely use an in-house, purified translation system reconstituted from *E. coli* components, which we find gives more consistent results in our generation of antibodies by affinity maturation or selection from naive antibody ribosome display libraries. These purified systems have been used for instance in protocols from Ravn (Chapter 12) and Ohashi (Chapter 14). Purified recombinant *E. coli* systems are now available, for example PURE Express™ from New England Biolabs and PURESYSTEM from Wako. Eukaryotic translation systems, such as rabbit reticulocyte lysate, are also available commercially but require adjustments to the protocols compared to the *E. coli* systems. Display and selection protocols for eukaryotic systems are described in the chapters by Douthwaite (Chapter 3) and He et al. (Chapter 5).

Ribosome display has been most widely used for the generation of high affinity antibodies. This has most usually involved site directed or error prone mutagenesis of lead antibody molecules as described by Lewis and Lloyd in Chapter 9. Ribosome display has also been used to select antibodies with desired characteristics from more diverse populations, either sub-cloned from initial phage display selections (affinity maturation of pools, Chapter 10, Groves and Nickson) or directly from naive antibody ribosome display libraries (Ravn, Chapter 12). The intrinsic error rate of the PCR step between rounds of selection allows the generation of increased diversity and the selection of higher affinity variants. Selection from naive RD libraries generates a diverse range of antibodies with a different bias to those from phage display. Thus, although it is easier to generate a diverse range of antibodies from naive libraries by phage display, ribosome display can be a useful supplement or alternative to widen the diversity of initial lead antibodies. For maturation of antibodies for affinity, ribosome display can offer significant advantages compared to methods involving steps in cells, which arises from its increased library diversity, monovalent nature, and disruption of complexes under mild conditions. To give a fuller picture of how ribosome display has fitted into the development of antibodies, two case studies are given. In Chapter 22, Thom and Minter describe the use of ribosome display to generate an antibody, CAT-354, directed against interleukin-13, that is currently in clinical trial. In Chapter 23, Hufton describes the use of ribosome display to affinity mature a humanized antibody directed against RAGE.

Ribosome display has also been used extensively for selection of molecules from novel scaffold libraries, for example ankyrin repeat domains (Darpins) (Chapter 15) and Sac7d scaffolds (Mouratou et al., Chapter 18). Selection from synthetic peptide libraries has also been performed by ribosome and mRNA display, for the purpose of epitope mapping of antibodies and for derivation of peptides interacting with specific proteins (Chapters 14 and 16). Natural sources of variants have been used to generate libraries for selection in ribosome display or mRNA display format and are described here for isolation of calmodulin-binding proteins from cDNA libraries by Cotten et al. (Chapter 16) and for the identification of vaccine genes from pathogenic bacteria by Lei (Chapter 17).

Ribosome display can not only be used to derive molecules with improved affinity or specificity, but also for derivation of molecules with improved stability by taking advantage of the flexible conditions employable in a pure in vitro method. This is described by Buchanan in Chapter 11, where stress conditions can be used during the translation step (such as the presence of dithiothreitol) or during the selection step (such as high temperature) using binding to a hydrophobic interaction chromatography matrix to discriminate between unfolded and more stable, folded molecules.

A significant advantage of in vitro translation methods is the ability to modify the genetic code to allow the incorporation of non-canonical, unnatural amino acids, to give molecules with novel properties, such as cyclic peptides with increased serum, stability. Three chapters in this book illustrate different methods that are being developed to incorporate unnatural amino acids. Reid et al. (Chapter 19) describe the use of flexizyme, a ribozyme that allows the charging of tRNA, with essentially any amino acid. This technology has been applied to the selection of peptide molecules incorporating thioether linkages. Watts and Forster (Chapter 20) describe pure translational display in which codons are reassigned by chemical synthesis of aminoacyl-tRNAs and unnatural amino acids then incorporated into the chain. In contrast, Ma and Hartman (Chapter 21) use the natural broad substrate specificity of aminoacyl tRNA synthetases to charge tRNAs with unnatural amino acids. This is followed by chemical derivatization to generate cyclic peptide libraries.

We hope that this book will be of value to those with general molecular biology or protein engineering experience who wish to select peptides or proteins by display, those with phage display experience who would benefit from the application of ribosome display, and those with some ribosome display experience who would like to expand the range of applications to which they are applying the technology. We would like to thank all the contributors for sharing their knowledge, the Series Editor, John Walker, for his advice and the invitation to edit the book, and our colleagues at MedImmune for their support.

Cambridge, UK *Julie A. Douthwaite*
Ronald H. Jackson

Contents

Preface... *v*
Contributors... *xi*

PART I REVIEW

1 Ribosome Display: A Perspective 3
   Andreas Plückthun

PART II TRANSLATION EXTRACT PREPARATION

2 Preparation and Testing of *E. coli* S30 In Vitro Transcription
   Translation Extracts 31
   James F. Zawada

PART III BASIC RIBOSOME DISPLAY AND RELATED SELECTION METHODS

3 Eukaryotic Ribosome Display Selection Using Rabbit
   Reticulocyte Lysate.. 45
   Julie A. Douthwaite

4 Stabilized Ribosome Display for In Vitro Selection 59
   *Shuta Hara, Mingzhe Liu, Wei Wang, Muye Xu, Zha Li,
   and Yoshihiro Ito*

5 Eukaryotic Ribosome Display with In Situ DNA Recovery 75
   *Mingyue He, Bryan M. Edwards, Damjana Kastelic,
   and Michael J. Taussig*

6 mRNA Display Using Covalent Coupling of mRNA
   to Translated Proteins... 87
   Rong Wang, Steve W. Cotten, and Rihe Liu

7 SNAP Display: In Vitro Protein Evolution in Microdroplets 101
   Miriam Kaltenbach and Florian Hollfelder

8 cDNA Display: Rapid Stabilization of mRNA Display 113
   Shingo Ueno and Naoto Nemoto

PART IV APPLICATIONS OF RIBOSOME DISPLAY METHODS USING NATURAL AMINO ACIDS

9 Optimisation of Antibody Affinity by Ribosome Display
   Using Error-Prone or Site-Directed Mutagenesis 139
   Leeanne Lewis and Chris Lloyd

10 Affinity Maturation of Phage Display Antibody Populations
   Using Ribosome Display... 163
   Maria A. Groves and Adrian A. Nickson

11 Evolution of Protein Stability Using Ribosome Display................ 191
   Andrew Buchanan

12 Selection of Lead Antibodies from Naive Ribosome
   Display Antibody Libraries............................... 213
   Peter Ravn

13 Evolution of Disulfide-Rich Peptide Aptamers Using cDNA Display........... 237
   Yuki Mochizuki and Naoto Nemoto

14 Peptide Screening Using PURE Ribosome Display................ 251
   *Hiroyuki Ohashi, Takashi Kanamori, Eriko Osada,
   Bintang K. Akbar, and Takuya Ueda*

15 Rapid Selection of High-Affinity Binders Using Ribosome Display............. 261
   Birgit Dreier and Andreas Plückthun

16 mRNA Display-Based Selections Using Synthetic Peptide
   and Natural Protein Libraries............................. 287
   *Steve W. Cotten, Jianwei Zou, Rong Wang, Bao-cheng Huang,
   and Rihe Liu*

17 Identification of Candidate Vaccine Genes Using Ribosome Display............ 299
   Liancheng Lei

18 Ribosome Display for the Selection of Sac7d Scaffolds................ 315
   *Barbara Mouratou, Ghislaine Béhar, Lauranne Paillard-Laurance,
   Stéphane Colinet, and Frédéric Pecorari*

PART V INCORPORATION OF NON-NATURAL AMINO ACIDS FOR SELECTION
        BY RIBOSOME DISPLAY AND RELATED METHODS

19 Charging of tRNAs Using Ribozymes and Selection
   of Cyclic Peptides Containing Thioethers...................... 335
   Patrick C. Reid, Yuki Goto, Takayuki Katoh, and Hiroaki Suga

20 Update on Pure Translation Display with Unnatural Amino
   Acid Incorporation................................... 349
   R. Edward Watts and Anthony C. Forster

21 In Vitro Selection of Unnatural Cyclic Peptide Libraries
   via mRNA Display.................................... 367
   Zhong Ma and Matthew C.T. Hartman

PART VI CASE STUDIES

22 Optimization of CAT-354, a Therapeutic Antibody Directed
   Against Interleukin-13, Using Ribosome Display.................. 393
   George Thom and Ralph Minter

23 Affinity Maturation and Functional Dissection of a Humanised
   Anti-RAGE Monoclonal Antibody by Ribosome Display.............. 403
   Simon E. Hufton

Index.. *423*

Contributors

BINTANG K. AKBAR • *Department of Medical Genome Sciences, Graduate School of Frontier Sciences, University of Tokyo, Kashiwa, Chiba Prefecture, Japan*
GHISLAINE BÉHAR • *Unité de Biotechnologie, Biocatalyse et Biorégulation, UMR6204 CNRS, Université de Nantes, Nantes, France*
ANDREW BUCHANAN • *Department of Antibody Discovery and Protein Engineering, MedImmune Limited, Cambridge, UK*
STÉPHANE COLINET • *Unité de Biotechnologie, Biocatalyse et Biorégulation, UMR6204 CNRS, Université de Nantes, Nantes, France*
STEVE W. COTTEN • *Eshelman School of Pharmacy and Carolina Center for Genome Sciences, University of North Carolina at Chapel Hill, Chapel Hill, NC, USA*
JULIE A. DOUTHWAITE • *Department of Antibody Discovery and Protein Engineering, MedImmune Limited, Cambridge, UK*
BIRGIT DREIER • *Department of Biochemistry, University of Zurich, Zurich, Switzerland*
BRYAN M. EDWARDS • *Crescendo Biologics Limited, Cambridge, UK*
ANTHONY C. FORSTER • *Department of Cell and Molecular Biology, Uppsala University, Uppsala, Sweden*
YUKI GOTO • *Research Center for Advanced Science and Technology, The University of Tokyo, Tokyo, Japan*
MARIA A. GROVES • *Department of Antibody Discovery and Protein Engineering, MedImmune Limited, Cambridge, UK*
SHUTA HARA • *Nano Medical Engineering Laboratory, RIKEN Advanced Science Institute, Wako, Saitama, Japan; Department of Materials and Applied Chemistry, Graduate School of Science and Technology, Nihon University, Tokyo, Japan*
MATTHEW C. T. HARTMAN • *Department of Chemistry, Massey Cancer Center, Virginia Commonwealth University, Richmond, VA, USA*
MINGYUE HE • *The Inositide Laboratory, The Babraham Institute, Cambridge, UK*
FLORIAN HOLLFELDER • *Department of Biochemistry, University of Cambridge, Cambridge, UK*
BAO-CHENG HUANG • *Eshelman School of Pharmacy and Carolina Center for Genome Sciences, University of North Carolina at Chapel Hill, Chapel Hill, NC, USA*
SIMON E. HUFTON • *Biotherapeutics Division, National Institute for Biological Standards and Controls, South Mimms, Potters Bar, UK*
YOSHIHIRO ITO • *Nano Medical Engineering Laboratory, RIKEN Advanced Science Institute, Wako, Saitama, Japan; Department of Biological Sciences, Tokyo Metropolitan University, Tokyo, Japan; Department of Biomolecular Engineering, Graduate School of Bioscience and Technology, Tokyo Institute of Technology, Kanagawa, Japan*

MIRIAM KALTENBACH • *Department of Biochemistry, University of Cambridge, Cambridge, UK*
TAKASHI KANAMORI • *Department of Medical Genome Sciences, Graduate School of Frontier Sciences, University of Tokyo, Kashiwa, Chiba Prefecture, Japan*
DAMJANA KASTELIC • *Medical Centre for Molecular Biology, Institute of Biochemistry, University of Ljubljana, Ljubljana, Slovenia; Babraham Bioscience Technologies, Cambridge, UK*
TAKAYUKI KATOH • *Research Center for Advanced Science and Technology, The University of Tokyo, Tokyo, Japan*
LIANCHENG LEI • *College of Animal Science and Veterinary Medicine, Jinlin University, Changchun, China*
LEEANNE LEWIS • *Department of Antibody Discovery and Protein Engineering, MedImmune Limited, Cambridge, UK*
ZHA LI • *Nano Medical Engineering Laboratory, RIKEN Advanced Science Institute, Wako, Saitama, Japan; Department of Biomolecular Engineering, Graduate School of Bioscience and Technology, Tokyo Institute of Technology, Kanagawa, Japan*
MINGZHE LIU • *Nano Medical Engineering Laboratory, RIKEN Advanced Science Institute, Wako, Saitama, Japan*
RIHE LIU • *Eshelman School of Pharmacy and Carolina Center for Genome Sciences, University of North Carolina at Chapel Hill, Chapel Hill, NC, USA*
CHRIS LLOYD • *Department of Antibody Discovery and Protein Engineering, MedImmune Limited, Cambridge, UK*
ZHONG MA • *Department of Chemistry, Virginia Commonwealth University, Richmond, VA, USA; Massey Cancer Center, Virginia Commonwealth University, Richmond, VA, USA*
RALPH MINTER • *Department of Antibody Discovery and Protein Engineering, MedImmune, Cambridge, UK*
YUKI MOCHIZUKI • *Graduate School of Science and Engineering, Saitama University, Saitama, Japan*
BARBARA MOURATOU • *Unité de Biotechnologie, Biocatalyse et Biorégulation, UMR6204 CNRS, Université de Nantes, Nantes, France*
NAOTO NEMOTO • *Graduate School of Science and Engineering, Saitama University, Saitama, Japan*
ADRIAN A. NICKSON • *Department of Antibody Discovery and Protein Engineering, MedImmune Limited, Cambridge, UK*
HIROYUKI OHASHI • *Department of Medical Genome Sciences, Graduate School of Frontier Sciences, University of Tokyo, Kashiwa, Chiba Prefecture, Japan*
ERIKO OSADA • *Department of Medical Genome Sciences, Graduate School of Frontier Sciences, University of Tokyo, Kashiwa, Chiba Prefecture, Japan*
LAURANNE PAILLARD-LAURANCE • *Unité de Biotechnologie, Biocatalyse et Biorégulation, UMR6204 CNRS, Université de Nantes, Nantes, France*
FRÉDÉRIC PECORARI • *Unité de Biotechnologie, Biocatalyse et Biorégulation, UMR6204 CNRS, Université de Nantes, Nantes, France*
ANDREAS PLÜCKTHUN • *Department of Biochemistry, University of Zurich, Zurich, Switzerland*

PETER RAVN • *Department of Antibody Discovery and Protein Engineering, MedImmune Limited, Cambridge, UK*

PATRICK C. REID • *PeptiDream Inc., Tokyo, Japan*

HIROAKI SUGA • *Research Center for Advanced Science and Technology, The University of Tokyo, Tokyo, Japan; Department of Chemistry, School of Science, The University of Tokyo, Tokyo, Japan*

MICHAEL J. TAUSSIG • *Babraham Bioscience Technologies, Cambridge, UK*

GEORGE THOM • *Department of Antibody Discovery and Protein Engineering, MedImmune, Cambridge, UK*

TAKUYA UEDA • *Department of Medical Genome Sciences, Graduate School of Frontier Sciences, University of Tokyo, Kashiwa, Chiba Prefecture, Japan*

SHINGO UENO • *Graduate School of Science and Engineering, Saitama University, Saitama, Japan*

RONG WANG • *Eshelman School of Pharmacy and Carolina Center for Genome Sciences, University of North Carolina at Chapel Hill, Chapel Hill, NC, USA*

WEI WANG • *Nano Medical Engineering Laboratory, RIKEN Advanced Science Institute, Wako, Saitama, Japan; Department of Biological Sciences, Tokyo Metropolitan University, Tokyo, Japan*

R. EDWARD WATTS • *Department of Pharmacology, Vanderbilt Institute of Chemical Biology, Vanderbilt University Medical Center, Nashville, TN, USA*

MUYE XU • *Nano Medical Engineering Laboratory, RIKEN Advanced Science Institute, Wako, Saitama, Japan; Department of Biological Sciences, Tokyo Metropolitan University, Tokyo, Japan*

JAMES F. ZAWADA • *Sutro Biopharma, Inc., South San Francisco, CA, USA*

JIANWEI ZOU • *Eshelman School of Pharmacy and Carolina Center for Genome Sciences, University of North Carolina at Chapel Hill, Chapel Hill, NC, USA*

Part I

Review

Chapter 1

Ribosome Display: A Perspective

Andreas Plückthun

Abstract

Ribosome display is an in vitro evolution technology for proteins. It is based on in vitro translation, but prevents the newly synthesized protein and the mRNA encoding it from leaving the ribosome. It thereby couples phenotype and genotype. Since no cells need to be transformed, very large libraries can be used directly in selections, and the in vitro amplification provides a very convenient integration of random mutagenesis that can be incorporated into the procedure. This review highlights concepts, mechanisms, and different variations of ribosome display and compares it to related methods. Applications of ribosome display are summarized, e.g., the directed evolution of proteins for higher binding affinity, for higher stability or other improved biophysical parameters and enzymatic properties. Ribosome display has developed into a robust technology used in academia and industry alike, and it has made the cell-free Darwinian evolution of proteins over multiple generations a reality.

Key words: Directed evolution, Cell-free translation, Ribosome display, Protein engineering, Antibody engineering, DARPins, Designed ankyrin repeat proteins, Affinity maturation

1. Introduction: Ribosome Display in Context

All technologies of molecular evolution must couple phenotype and genotype. There are two fundamental possibilities for achieving this. The first one is compartmentalization. Nature's compartments are cells: they secure that the superior phenotype expressed by one cell's mutant genotype can be replicated, without the gene products from the wild type interfering. All selections based on microbial phenotypes use this principle. The second possibility is a direct physical coupling of genetic material to the protein product. Nature's example would be viruses: the virus coat and its receptor-binding properties are the phenotype whose genetic information is encoded on the viral genome, inside the virion.

In order to use selection methods in biotechnology, i.e., for the enrichment of binders from libraries and their evolutionary improvement, one can exploit both compartmentalization and physical linkages. The most frequently used compartments are microbial cells, where intracellular interactions can be detected by genetic means. As exemplified by the yeast two hybrid system (1) or by the protein fragment complementation assay (2), only those cells grow in which a desired molecular interaction restores the critical factor, which has previously been split in two pieces. Alternatively, one can use enzymatic or optical means for detecting interactions (3–5). These examples are only meant to be illustrative – a large research field has developed around exploiting such phenomena.

Most popular, however, have been systems where the interaction itself occurs outside of the cell, even though the polypeptides are still being produced by cells. Thus, expression on the surface of bacteria (6) or yeast (7) has been used successfully, as has been the display of peptides (8–11) and polypeptides (12–14) on the surface of filamentous phages – probably still the most widely used display technology. In phage display, the bacterial cells are producing the phages, and thus the bacterial cells must first be transformed with the library, limiting the diversity. However, this will not be a comprehensive review on display technologies; such reviews can be found elsewhere (15–20).

In all these technologies that involve cells at any step, including phage display, the genetic information needs to be introduced into cells. This will usually limit the diversity present in the cells to below what has been present in the actual library of DNA molecules. The diversity reached in practice after transformation will depend on the host system (bacteria vs. yeast) and on the efforts made to transform the cells. This large-scale transformation step, while not difficult, can be quite laborious. It is this step that is avoided in technologies that are performed fully in vitro.

This chapter will summarize ribosome display, and mention other technologies that operate without any cells during the library selection. They are not restricted by the effort spent on the transformation step. This advantage is apparent in the primary library, which can be of bigger size, as all DNA (or mRNA) molecules present can in principle give rise to proteins that take part in the selection.

The most important advantage of ribosome display and other "full" in vitro technologies is, however, the easy combination with PCR-based randomization techniques, and thus the creation of a true Darwinian evolution process – in contradistinction to a mere selection from an existing "constant" library. In all technologies that require transformation of cells, after each randomization step the cells have to be transformed with the new library, and the workload is thus potentiated by the number of evolution steps.

In ribosome display, the workload to select binders in the absence or presence of randomization is almost identical. Thus, the more randomization steps are to be carried out in succession, the more attractive ribosome display becomes.

Of course, technologies have been developed to carry out randomization directly in cells as well (21–23). Nonetheless, in vitro methods give the user full control over where mutations should occur in the sequence (by using, e.g., a randomized cassette), which residue types are to be introduced (by using trinucleotide building blocks (24) or suitable mononucleotide mixtures), or how many random mutations should occur on average (25) – a level of control not yet within reach in cellular systems. Furthermore, the "shuffling" of the library (26, 27) can easily be introduced into the procedure if desired, to recombine mutants.

2. Development of Ribosome Display

Key to the development of ribosome display was the observation that rare mRNAs coding for a particular protein can be isolated from a pool of mRNAs by immunoprecipitation of stalled ribosomal complexes containing the nascent polypeptides (28, 29). Apparently, ribosomal stalling is frequent enough to be experimentally exploited.

To create RNA-based "aptamers", Tuerk and Gold (30) had developed a technology called SELEX (Systematic Evolution of Ligands by Exponential Enrichment), where multiple rounds of in vitro transcription of random nucleic acid pools, followed by affinity selection of the RNA aptamers and subsequent RT-PCR, lead to the selection of target-binding RNAs. In SELEX, genotype and phenotype are simultaneously represented by the same RNA molecule, since it exerts its function through its three-dimensional structure, which is in turn determined by its nucleotide sequence.

In their original publication about SELEX, Tuerk and Gold (30) already speculated that a similar approach might be adapted to protein selection, referring to the isolation of stalled translation complexes (28, 29).

The first experimental demonstration of the ribosome display technology was the selection of short peptides from a library using an *Escherichia coli* S30 in vitro translation system (31, 32). Kawasaki (33) had proposed a similar approach to enrich peptides from libraries in a patent application, however, without giving a detailed example, which was only published in 1997 (34).

Meanwhile, Hanes and Plückthun (35) had developed and first reported ribosome display for whole proteins, initially antibody scFv fragments, after significantly improving and modifying the system to increase its efficiency and to allow folding of scFv fragments during in vitro translation (36). The key observation with these longer open reading frames (ORFs) was that mutations were found at great frequency (35) due to the large number of PCR cycles that every gene had undergone after several rounds. Because of the short encoded peptide sequences, Mattheakis et al. (31, 32) did not observe mutations. To play out its strengths, ribosome display needs to be carried out with whole proteins. The appearance of mutations, just by using a normal polymerase without proofreading activity, then made it clear that the unique advantage of ribosome display lies in its potential to do true *Darwinian evolution*, as opposed to mere *selection*, from a given library. This potential was demonstrated in the following year (37) by striving to select antibodies for improved affinity, and by adding additional diversity through random mutagenesis. The first selection from a non-immune synthetic antibody library followed soon thereafter (38).

In the above experiments, the bacterial cell-free translation extracts were home-made, as they needed to be free of reducing agents (because of the intramolecular disulfide bonds within the scFv domains), and they were required in large amounts. The feasibility of carrying out ribosome display in a eukaryotic cell-free translation extracts was subsequently demonstrated with an scFv-kappa fusion (39), and then by selecting an antibody in the same format from immunized transgenic mice (40).

3. The Ribosome Display Methodology

The principle of ribosome display is depicted in Fig. 1. A DNA cassette (typically a PCR fragment) is used that contains a promoter and an ORF, encoding a library of the protein of interest. It is transcribed in vitro, and the resulting mRNA does not contain a stop codon. Cell-free translation can run to the physical end of this mRNA, and complexes consisting of the protein of interest, the ribosome, and mRNA are formed. The ribosome itself serves as the connector. These ternary complexes are exposed to immobilized target molecules, the displayed proteins (library members) binding to the target are enriched on the target and others are washed out. From these bound complexes, the mRNAs are isolated, reverse transcribed and

Fig. 1. (continued) display construct is obtained by PCR amplification of both flanking regions and the library insert from the ligated vector. In vitro transcription of this PCR product yields mRNA that is used for in vitro translation. The ribosome stalls at the end of the mRNA and does not release the encoded and properly folded protein because of the absence of a stop codon. The ternary mRNA–ribosome–protein complexes are used for affinity selection on an immobilized target. The mRNA of bound complexes is recovered after washing from dissociated ribosomes, reverse transcribed and amplified by PCR. Thereby the selected pools of binders can be used directly for the next cycle of ribosome display or analysis of single clones after cloning into expression vectors, which are then used for *Escherichia coli* transformation and small-scale in vivo expression. Adapted from ref. 111; for the most current procedure and details see ref. 97.

1 Ribosome Display: A Perspective

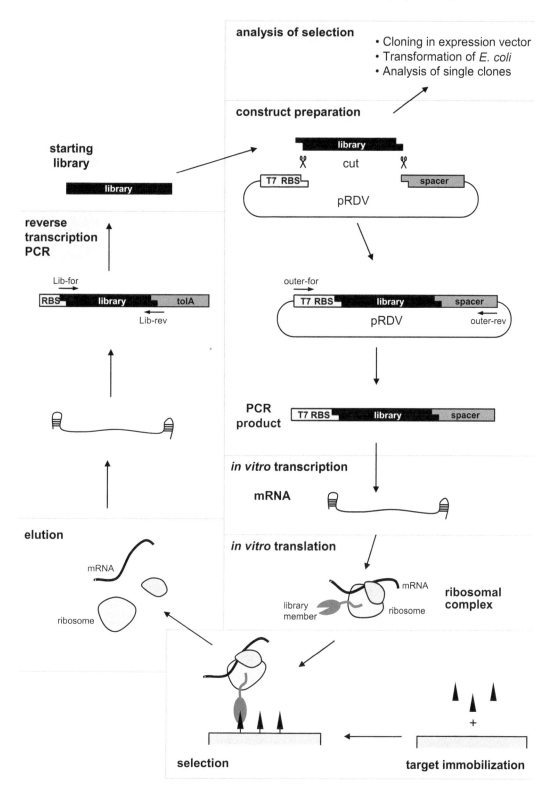

Fig. 1. Overview of the ribosome display selection cycle. A DNA library in the form of a PCR product, coding for binding proteins, is ligated into a ribosome display vector (pRDV), thereby genetically fusing it to a spacer sequence in-frame, and providing a strong promoter (T7) and translation initiation region (ribosome binding site, RBS) at the 5' end. The final ribosome

PCR amplified to serve as the input of another round. After 3–5 rounds, the resulting DNA fragments are ligated into an expression vector and *E. coli* are transformed. The different proteins made by individual *E. coli* clones can then be further evaluated.

The logic of the individual steps will now be discussed. In vitro methods cannot rely on cells to link phenotype and genotype. Instead, a direct physical link between the genetic material and the protein product must be made. (An in vitro alternative is to create compartments in the form of a water-in-oil emulsion, which will be discussed below.) During protein biosynthesis, the encoding mRNA is read by the ribosome and parts of it are almost engulfed by the small subunit, which mediates codon/anti-codon contact to the tRNA. The protein emerges from the ribosomal tunnel within the large ribosomal subunit. During all these steps, the protein chain is covalently connected through an ester bond to the peptidyl-tRNA within the P-site, and thereby tightly maintained within the ribosome. Thus, during protein biosynthesis, neither protein nor mRNA can leave.

Translation normally ends at a stop codon. In *E. coli* ribosomes, the UAG and UAA stop codons are directly recognized by the Release Factor 1 (RF1), the UGA and UAA stop codons by RF2 (41). With the help of RF3, the ester bond between the synthesized protein and the last tRNA is positioned such that it is hydrolyzed, and the last amino acids of the finished protein, still within the exit tunnel, can slide out, leaving the empty tRNA behind. Now the ribosome recycling factor (RRF) and elongation factor G (EF-G) together help separate the large from the small subunit: they remove the tRNA, and after subunit dissociation, the mRNA can then leave as well (42). A similar mechanism also functions in eukaryotic ribosomes (43).

The absence of a stop codon in mRNA for ribosome display thus ensures that this "normal" course of events does not take place. The experimental conditions must furthermore minimize any early unwanted spontaneous hydrolysis of the ester bond between polypeptide and tRNA or any other spontaneous falling apart of the ternary complex. This is usually achieved by rather short translation times, subsequent cooling of the solution and addition of Mg^{2+}. These short translation times are also a compromise between efficient translation and degradation of the mRNA by nucleases present in the extract. It is believed that high Mg^{2+} "condenses" the ribosome by binding to the rRNA, making it difficult for the peptidyl-tRNA to dissociate or be hydrolyzed. While the details differ, the general strategy is the same for eukaryotic ribosomes (44).

The absence of a stop codon then causes "stalled" ribosomes. Once the ribosomes are at the end of the mRNA, they would have a peptidyl-tRNA in the P site and (presumably) an empty A site. As this situation can appear in the bacterial cell as well (if an mRNA molecule is missing its 3′ end containing the stop codon), bacteria have devised a mechanism called *trans*-translation to rescue such stalled ribosomes, which would clog up protein biosynthesis (45, 46). The key molecule to act on the stalled ribosome is the transfer-messenger RNA (tmRNA, also called 10S RNA or SsrA), which has a tRNA-like domain charged with alanine and an mRNA-like domain. Peptide synthesis is resumed, with the additional help of the protein SmpB, by incorporating Ala, and thereby transferring the chain to the tmRNA. Up to now, the tmRNA molecule has acted like tRNA. Now its mRNA domain is being read, and the sequence ANDENYALAA* is appended to the protein, which ends with a stop codon (!), thus leading to regular termination and recycling of the ribosome. Even worse for ribosome display, this C-terminal sequence which has been added serves as a degradation tag in *E. coli*. Thus, the action of tmRNA would be detrimental for ribosome display. Starting from our very first experiments, an antisense oligonucleotide has always been added to titrate out tmRNA (35).

In order for the protein of interest to fold and be able to interact with a target, the whole protein of interest must be outside of the tunnel once the ribosomes have come to the physical end of the mRNA. Thus, to remain connected to the tRNA at the same time, the protein of interest must be fused to an unstructured region at the C-terminus, occupying the ribosomal tunnel. We have called this the "spacer" or "tether". This protein tail, which is the same in all library members, is thus fused in frame to the C-terminus of the randomized protein of interest in the ribosome display construct.

The features of the ribosome display construct are summarized in Fig. 2. On the DNA level, the construct requires a strong promoter for efficient in vitro transcription to mRNA. On the mRNA level, the construct contains, as a regulatory sequence for translation, either a prokaryotic ribosome-binding site (47, 48) if the *E. coli* system is used, or a Kozak consensus and enhancer sequence (49) if the eukaryotic ribosome display system is used. This sequence is followed by the ORF encoding the protein to be displayed, followed by a spacer sequence fused in frame to the protein of interest, as described above.

At both ends of the mRNA, the ribosome display construct should include stemloops. 5'- and 3'-stemloops are known to stabilize mRNA against RNases in vivo as well as in vitro. The presence of stemloops is important, especially in the *E. coli* ribosome display system because the extract used for in vitro translation contains high RNase activities. The efficiency of ribosome display was increased approximately 15-fold (35), when a 5'-stemloop derived from the T7 gene 10 upstream region and a 3'-stemloop derived from the terminator of the *E. coli lpp* lipoprotein were introduced into the ribosome display construct (35). A similar improvement in efficiency was observed when using the same 5'-stemloop and the

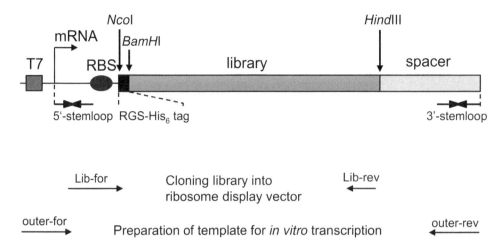

Fig. 2. The construct for ribosome display using *Escherichia coli* ribosomes. A T7 promoter and a ribosome-binding site (RBS) are necessary for in vitro transcription and translation. The coding sequence starts with Met-Arg-Gly-Ser-His$_6$ (the RGS-His$_6$ tag, or any other tag can be used), followed by the DNA library of the binding proteins and a spacer at the C terminus. The stop codon has been removed from the coding sequence. At the mRNA level, the construct is protected against RNases by 5' and 3' stem-loops. Fusion of the promoter and the spacer sequence can best be achieved by cloning into an appropriate vector providing these sequence elements and subsequent amplification by PCR. The oligonucleotides used for the cloning of the library into the ribosome display vector and for the generation of the template for in vitro transcription are indicated at the bottom schematically, the exact oligonucleotides can be found in the corresponding publications. Adapted from ref. 111; for the most current procedure and details see ref. 97.

3'-stemloop derived from the early terminator of phage T3 (35). The stemloop structures may protect the mRNA particularly from degradation by the exonucleases PNPase and RNaseII, which act from the 3'-end of the mRNA, and against RNaseE, which recognizes the 5'-end (50–52).

In the ribosome display cycle, the ribosomal complexes are then exposed to the target of interest. It is either free in solution (to be captured subsequently, e.g. by magnetic beads) or immobilized on plates. A very robust way for both strategies has been to biotinylate the target, as the interaction of biotin with streptavidin, neutravidin or avidin survives stringent washing steps and exposure to detergents. The many variations of this step will be discussed in subsequent chapters. Our laboratory has preferred enzymatic biotinylation at an engineered tail above all other methods (53).

Clearly, the details of how the library is exposed to the target will determine the selection outcome. The use of competitors (to avoid recognition of a similar target) and methods to select for high affinity (discussed below) as well as methods to select for properties other than high affinity (summarized below) are under constant development and refinement. Later chapters will discuss these aspects at length.

One of the great advantages of ribosome display is that the linkage between immobilized target and library member does not have to be broken. After selection on the target, only the mRNA is required from this point onwards. Therefore, it is sufficient to liberate it by the dissociation of small and large ribosomal subunits by the addition of EDTA. This has the advantage that it is not more difficult to isolate complexes of very high affinity (which would be hard to dissociate) than those of lower affinity. Nonetheless, more specific elution procedures can be of interest for selecting binders to particular epitopes or with particular properties.

The mRNA needs to be reverse transcribed and the resulting DNA then amplified and brought back to the initial format containing a promoter for transcription of the next round. In the initial rounds, very few mRNA molecules will be obtained after selection. This step is perhaps one of the few technically demanding steps, as it requires attention to the fragility of RNA (in the presence of nucleases, which can also be introduced by careless laboratory handling). When designing a new ribosome display cassette (with different promoter, ribosome binding site, N-terminal tag on the protein, ORF, C-terminal tether and 5' and 3' stemloops), care must be taken in designing the primers needed for the PCR. Obviously, they must bind with very high specificity. These reverse transcription and PCR steps appear to be the most frequent focus of troubleshooting, when designing a new system from scratch. Even when taking a well-working system, and merely replacing the ORF, it must be considered that new hairpins might form unintentionally, e.g., engaging the start codon or the ribosome binding site, which would compromise translation efficiency. Fortunately, these issues are easily and rapidly evaluated.

4. Protein Folding in Ribosome Display

The most frequently used methods in ribosome display include the use of E. coli S-30 extracts for translation, which do contain ribosome-associated factors important for protein folding such as the trigger factor (54). In addition, molecular chaperones that are not associated to the ribosome are present in the E. coli extract, such as DnaJ/K/GrpE and GroEL/ES, as well as small heat shock proteins and others (55). Additional factors can be added, depending on the requirements of the proteins to be displayed (36). Antibody scFv fragments required the addition of eukaryotic protein disulfide isomerase (36).

Ribosome display has also been used as a tool to define a binding epitope, making use of the somewhat surprising finding that, while still on the ribosome, aggregation of a protein seems to be efficiently prevented. It was found that proteins, such as eukaryotic receptors that could not be expressed in functional form in E. coli nor efficiently refolded from inclusion bodies, nor expressed in functional form by in vitro translation, would fold while still attached to the E. coli ribosome (56). Perhaps the ribosome enhances solubility of the ternary complex and sterically blocks aggregation. To define a binding epitope, a cell surface receptor was subjected to several rounds of random mutagenesis at high error rate, and this ribosome display library of receptor point mutants was selected on the target. The epitope could be recognized as an area devoid of surface mutations – evolutionary pressure in the experiment apparently maintained the residues in the epitope (57).

When using a eukaryotic translation system (39, 44, 58), the corresponding eukaryotic proteins would be expected to be present in the extract. On the other hand, if a system from pure components is used (59–62), it may be necessary to add these proteins relevant for folding, depending on the protein to be displayed (36).

5. Variations in the Ribosome Display System

5.1. Eukaryotic Cell Free Translation System

The ribosome display method can also be carried out with eukaryotic extracts, using a reticulocyte lysate (39, 58). Different methods of sequence recovery have been compared, and it was concluded that a similar procedure as used in the prokaryotic system also performs best in the eukaryotic system (44), even though in situ recovery can also be carried out (39, 58). The wheat germ in vitro translation system has also been used (34). While one might speculate that a eukaryotic translation system should perform better with eukaryotic proteins, there is actually no evidence for this (63). If particular factors are needed, such as, e.g., molecular chaperones and protein disulfide isomerase, they can (and have to) be added to either system (36).

5.2. PURE System

The use of a ribosome display system based on in vitro translation with purified components (PURE system) has also been described (59–62). In this system, no release factor is present. Matsuura et al. examined the efficiency of ribosome display in the absence and presence of a stop codon, as well as when using the *secM* stalling sequence (64) as an alternative means of trapping the ternary protein–ribosome–mRNA complex (60). Interestingly, the efficiency of display was almost identical in all cases. Another encouraging finding from the use of the PURE system is that the *intrinsic* stability of the ternary complexes is actually very high. Even after an incubation of the ternary complexes for 1 h at 50 °C, the display efficiency drops by less than a factor of 10. Presumably, both in the prokaryotic and eukaryotic extracts, RNAses set a practical limit on stability, rather than the intrinsic stability of the ternary complexes themselves.

The system with purified components may thus be of interest where the removal of a stop codon is inconvenient or high temperature is required in the selection procedure. It should be kept in mind, however, that the experiments were carried out with engineered mRNA that had a C-terminal spacer, as in the standard system described above. In a natural non-engineered mRNA, one would expect that the last few amino acids of the protein of interest are still in the ribosomal tunnel, thereby hampering the folding of the protein. It will remain to be seen whether the PURE system is sufficiently cost-efficient to be used for standard selection and evolution experiments, in comparison to the use of translation extracts from *E. coli*.

5.3. Display of mRNA with Stop Codons

To address the display of natural mRNA (which of course all contain a stop codon), as an alternative to the use of the PURE system, engineered suppressor tRNAs have been used (65, 66). This could be another approach useful for future protein–protein interaction studies. Nonetheless, the problem remains that many (if not most) proteins will not fold, if part of their domain structure is still in the ribosomal tunnel. The critical question is therefore whether the translated spacer that results when suppressing the stop codons in natural mRNA will be long enough to allow folding of most proteins.

6. Applications of Ribosome Display to Complex Libraries

6.1. scFv Fragments

Antibody scFv fragments were the first complex library with which ribosome display was tested for selection and affinity maturation (37), initially from a library of immunized mice, later from a synthetic library (38). At that time, the recreated synthetic repertoire of the antibodies was the only general binding protein scaffold available with great diversity.

The folding of antibody fragments in an in vitro translation system must be commensurate with their oxidative folding (many if not most antibody domains need the intradomain V_H and V_L disulfide bond to fold properly), and thus this reaction must be catalyzed (36). In addition, the β-sandwich architecture of antibodies can lead to aggregation, and this may be part of the reason, why more rounds of enrichment appear to be necessary than for some other scaffolds that fold extremely well in an in vitro translation system.

There are more publications on using phage display than ribosome display in the selection from naive antibody libraries, but there simply may be no necessity to break with tradition. Filamentous phage display (12) works very well with secreted one-chain disulfide-containing proteins such as scFv (67), and there is always the option of combining the two methods, as opposed to directly combining selection and affinity maturation in one procedure, as in the ribosome display selection from naive or synthetic libraries (68).

The analysis of ribosome display selection from the fully synthetic antibody library HuCAL leads to the conclusion that the selection is not exhaustive, and the outcome is governed by the occurrence

	Vκ1	Vκ2	Vκ3	Vκ4	Vλ1	Vλ2	Vλ3
VH1A	C(9)					C(2)	
VH1B							
VH2							
VH3		A(9) A(1) B(2)		A(1)	A(1) A(1)	A(1) A(1)	
VH4							
VH5							
VH6					B(6)		B(1)

Fig. 3. Framework usage of the insulin-binding HuCAL scFv fragments in three different ribosome display experiments. The *vertical* and *horizontal axes* denote the HuCAL heavy-chain and light-chain variable domains. ScFvs isolated in experiments (**A**), (**B**), or (**C**) are denoted accordingly. *Numbers in parentheses* represent the number of closely related scFvs with the same CDRs, but different point mutations. It is apparent that in different experiments random mutations lead to the proliferation of particular sequence families, but that this phenomenon occurred in different families in the three experiments. Adapted from ref. 38.

of random mutations; in selections which were repeated against the same target, different frameworks were dominant in the different selections. This suggests that an early beneficial mutation may have given rise to a lot of (further mutated) progeny of a particular clone, while in the next selection experiment on the same target, another framework combination may have acquired such a beneficial mutation (Fig. 3).

6.2. Designed Ankyrin Repeat Proteins

Combinatorial libraries of a new class of small proteins, termed "Designed Ankyrin Repeat Proteins" (DARPins) (69, 70) were developed that can act as an alternative to antibodies, as they are particularly robust to engineering. They are based on a very different structure and are built from consecutive 33-amino acid repeats, each forming a β-turn followed by two antiparallel α-helices. In each repeat, seven residues were randomized, and these internal repeats are flanked by constant capping repeats, to give one contiguous polypeptide chain with a randomized concave, groove-like binding surface, which is randomized in the library. The proteins contain no cysteine, can be expressed in soluble form in the cytoplasm of *E. coli* at very high levels, and are very stable and resistant to aggregation (refs. 71, 72 and references therein).

It may be these favorable biophysical properties, combined with the fact that high affinity binders are obtained at high frequency, that cause the direct selection of binders from the diverse library to work very well with DARPins. Thus, binders against many targets, including difficult ones such as, e.g., detergent-solubilized GPCRs (73) or conformers of DNA (O. Scholz, unpublished), have been selected directly by ribosome display (e.g., see refs. 70, 74–80).

6.3. Other Scaffolds

Binders based on the camelid VHH domains with micromolar affinity have been isolated by ribosome display from a naive library (81), and with nanomolar affinity from an immunized llama (82).

7. Combining Ribosome Display with Other Selection Technologies

Ribosome display has been combined with other selection technologies. It has been used as the affinity maturation step of a phage display library (e.g., see ref. 68), and thus used as the second stage in binder selection.

However, one can also use ribosome display as the first step and follow it up by another technology to simplify the evaluation of individual clones. At the end of the ribosome display procedure, the final selected pool is usually cloned in *E. coli*, and crude extracts of individual *E. coli* expression cultures are then analyzed by ELISA. Instead of going through enough rounds such that most of these clones will be positive, an earlier round can be cloned, and an in vivo selection can be applied to this selected pool. For this purpose, the pools of ribosome display were cloned after the first, second, and third round in a protein fragment complementation assay (PCA) (2), a split enzyme selection system using DHFR. This technology has a low discrimination power for affinity, but essentially serves as a convenient qualitative screen of binding. It can be seen that even after one round of ribosome display, binders can be obtained, albeit with micromolar affinity. These correspond to a random sampling of the library which has been enriched, perhaps 10^3- to 10^4-fold, and high affinity binders are too rare to be expected to be found in this small sampling. However, already after the second round of ribosome display, binders with nanomolar affinity are found (77). This combination of ribosome display with PCA might become of interest in high-throughput applications of ribosome display.

8. Comparison of Ribosome Display with Related In Vitro Methods

8.1. mRNA Display

The first steps of mRNA display are identical to ribosome display. A DNA library, encoding promoter, ribosome-binding site, and the randomized open reading frame of interest without a stop codon is transcribed to yield a library of mRNA molecules. This library is then ligated to a C-terminal linker consisting of DNA which contains at its end a puromycin molecule. In vitro translation is carried out as in ribosome display. The ribosomes are thought to stall at the 3' end of the mRNA where it is linked to DNA. The attached puromycin molecule then enters the P-site and takes on the role of a tRNA, and the growing peptide chain is

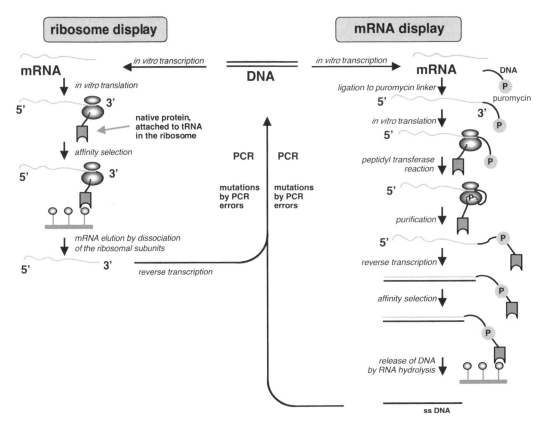

Fig. 4. Comparison of ribosome display (*left*) and mRNA display (*right*). For both, the DNA encoding the library is first transcribed in vitro. In ribosome display, the resulting mRNA lacks a stop codon, giving rise upon translation to linked mRNA–ribosome–protein complexes, which can be directly used for selection against an immobilized target. The resulting mRNA is obtained upon dissociating the ribosomal subunits, reverse transcribed and amplified for the next round. In mRNA display (*right*), the mRNA is first ligated to a DNA linker connected to puromycin. The mRNA is translated in vitro, and the ribosome stalls at the RNA–DNA junction. Puromycin then binds to the ribosomal A-site, and attacks the peptidyl-tRNA at the P-site. The nascent polypeptide is thereby transferred to puromycin, as if it were an aminoacyl-tRNA. The resulting covalently linked mRNA–protein complex has the puromycin-linker-mRNA on one side of the tunnel, the protein on the other side of the tunnel. The mechanism, by which this complex purified from the ribosome is not entirely clear (cf. Fig. 5). It is then reverse transcribed and used for selection experiments. The DNA strand is recovered from target-bound complexes by hydrolyzing the complementary mRNA at high pH, then it is amplified by PCR. Adapted from ref. 112.

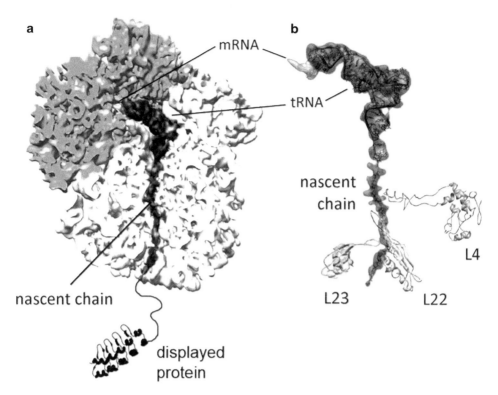

Fig. 5. Schematic illustration of the topology relevant for ribosome display. This figure is adapted from Seidelt et al. (113). (**a**) CryoEM reconstruction of the *Escherichia coli* ribosome 70S complex with the TnaC stalling sequence is shown. For illustration, the folded domain of a DARPin has been added at the N-terminus. The unstructured connector outside the ribosome and the part that is within the tunnel (*nascent chain*) would correspond to the "spacer" or "tether" region in Figs. 1 and 2. After selection, the addition of EDTA disassembles the two subunits, and only the mRNA must be recovered. Note that the mRNA is not shown here – it contacts the tRNA where indicated and is otherwise partially within the small subunit in this representation and thus not highlighted. (**b**) tRNA, attached nascent chain and the contact point of the mRNA are shown. The ribosome has been left out, except for three proteins from the large subunit, *L4, L22,* and *L23* which contact the nascent protein. In ribosome display, the mRNA can be recovered and purified after disassembling the ribosome into small and large subunit. There is no need to elute or recover the protein–tRNA complex. By contrast, in mRNA display, a puromycin molecule, covalently attached to the nascent protein would take the place of the tRNA in this picture. This puromycin is also covalently connected to a DNA spacer, which itself is covalently connected to the mRNA. It is apparent that this will create two large structural units on either side of the ribosomal tunnel: on one side the puromycin, attached to DNA and RNA, on the other side the folded protein, which has emerged from the ribosomal tunnel. Thus, in order to remove the large ribosomal subunit before panning, the ribosome has to be unfolded (potentially unfolding the protein of interest as well), or the protein of interest has to be unfolded (in order to thread backwards through the tunnel), or the whole linker-mRNA has to thread forwards through the tunnel. Alternatively, the large subunit might stay just in place during the selection, such that mRNA display works actually like ribosome display.

transferred to puromycin, and thereby covalently connected, via the linker, to the mRNA (Fig. 4).

The mRNA display procedure leaves us with an interesting topological conundrum. On one side of the ribosomal tunnel is the folded protein, on the other side is puromycin, connected via a linker with the mRNA (Fig. 5). The cartoons of the mRNA display procedure (14) implicitly suggest that the ribosome is removed. There appear some potentially denaturing steps in the procedure after translation (14), even though it is not clear whether this would be sufficient to unfold the large subunit, and especially what effect these conditions would have on the covalently bound protein of interest.

There are thus four possibilities: First, the large subunit of ribosome unfolds, thereby opening the exit tunnel, such that the protein can slide out sideways. It is unclear to what degree the protein of interest would unfold as well. Second, the DNA linker and the whole mRNA thread through the protein tunnel. Third, the protein of interest unfolds before the large subunit and threads backwards through the protein tunnel. Fourth, the large subunit of the ribosome is actually not completely removed, and is still present during panning, similar as in ribosome display. This question is not only of academic interest, as it may have some effect on the protein to be displayed.

A number of protein scaffolds have been used with mRNA display, e.g. some stabilized by metal ions (83–85), which would become unfolded by adding EDTA. However, selections have also been carried out using the fibronectin scaffold (86–88) which probably folds and unfolds reversibly, or scFv fragments (89), as well as diverse other proteins (90) where it is not clear whether selection did require folded domains. The problem of topology in mRNA display has apparently not been solved.

8.2. Water-in-Oil Emulsions

The packaging of the translation extract into small droplets in the form of a water-in-oil emulsion combines the in vitro approach with the compartment concept of coupling genotype and phenotype (91–93). If, on average, each droplet contains only one mRNA molecule, the protein content of each droplet is monoclonal. The most persuasive application of this technology is for enzymatic reactions (94, 95). The basic challenge for evolving enzyme turnover is that the phenotype, namely the production rate of the enzymatic reaction product, cannot be easily linked to the enzyme molecule itself in a mixture of mutants in solution, at least not for reactions with multiple turnovers with high rates. The cellular confinement solves this problem. Nonetheless, the full potential of this approach can be reached only if the reaction can be followed in the compartments directly, requiring optical detection and sorting.

On the other hand, the selection for a binding event in a compartmentalized system is somewhat less compelling. Emulsions have to be broken and selections can be carried out in the bulk phase as in ribosome display and mRNA display. The generic detection of binding events *within* a droplet remains a challenge, and it is less clear how to achieve this in a semiquantitative way for binding *strength*, i.e., affinity.

However, an interesting application of the emulsion technology for the selection of binding proteins was described by Sumida et al. (96), in which they used the emulsion format to express both chains of a Fab fragment from a bicistronic operon within the same droplet. By fusing the heavy chain to streptavidin, and attaching biotin to the DNA fragment via a photocleavable linker, both chains of the Fab fragment and the corresponding DNA stay together after breaking the emulsion. After panning, the DNA has to be recovered by photocleavage and can be amplified. It will be interesting to see how well this system will perform with complex libraries.

9. Comments on Library Size

The absolute functional library size in ribosome display is given by the number of different ternary complexes that are formed from mRNA and ribosomes and give rise to a nascent protein that can fold. This requires that the protein of interest is translated at least to the point that the relevant domain is outside the ribosomal tunnel.

From the amount of PCR fragment that is used as the input for transcription, we can calculate that under standard conditions (97) about $2-3 \times 10^{12}$ molecules input DNA are used. If the library template DNA to make this PCR is of good quality and highly diverse we can assume that these DNA molecules are all different. Also, further mutations will have been introduced while carrying out this very PCR.

The transcription of this linear PCR fragment will usually create multiple copies or mRNA per DNA molecule, and only an aliquot of the resulting mRNA is used for translation. In a standard ribosome display reaction about $1-3 \times 10^{13}$ mRNA molecules are used. It is entirely possible that additional errors are introduced by the RNA polymerase, such that they will contain a greater diversity than the input DNA.

The other critical variable is the number of functional ribosomes. The number of assembled ribosomes in an *E. coli* cell depends on its growth rate and is between 18,000 and 72,000 (98). From the amount of S30 extract used and the number of assembled ribosomes per cell, there should be about $1-4 \times 10^{14}$ assembled ribosomes in the standard ribosome display reaction. Thus, even if not all ribosomes are functional, there should be a sufficient excess to translate most mRNA molecules. Also, while the number of ribosomes per mRNA will follow a binomial distribution, there should be a significant proportion of monosomes (one ribosome per mRNA).

Using multi-ml quantities of S30 extract and more mRNA, almost arbitrarily large numbers for the library size can be stated. For example, in several review articles the library size of ribosome display and mRNA display have been "compared", where the latter has usually referred to an experiment where extreme amounts of S30 extract have been used. This simple, direct relationship between the amount of extract used and the functional diversity of the library has not been recognized by all authors.

More importantly, it remains to be seen whether the best use of the rather valuable S30 extract (and the even more valuable eukaryotic extract, or the truly precious purified components) is to use it all at once in the first round, as opposed to using it in smaller aliquots over a multi-step selection with built-in affinity maturation.

Another important aspect is the functional fraction of the library, and whether the non-functional part is merely inactive, or instead "sticky" and thus becoming enriched during selections. This is a quantity that can almost not be objectively determined. This issue also sets a limit to the number of random mutations that is practically useful before the population becomes extinct.

10. Selections for Higher Affinity

The most attractive exploitation of the built-in possibility of generating mutations is to improve affinity. Affinity is usually quantified by the equilibrium dissociation constant K_D, which is the ratio of the dissociation rate constant k_d (loosely referred to as off-rate) over the association rate constant k_a (loosely referred to as on-rate). The association rate constant for protein–protein complexes falls in a remarkably small window, typically between 1×10^5 and 1×10^6 M^{-1} s^{-1}, as summarized from various experimental studies by Northrup et al. (99) and further computationally analyzed by these authors. The net association rate is often visualized as the collision rate times the fraction of "successful" collisions, in other words, where the two proteins have productive orientations. This means that affinity is largely determined by off-rate, and that measures to improve affinity should normally attempt to decrease the off-rate.

Of course, there are exceptions. A protein pair can be properly oriented upon approach by electrostatic forces, leading to a higher fraction of successful collisions (100), and this can also be engineered (101). However, this electrostatic steering will greatly diminish in importance in physiological buffers with high ionic strength and may thus not be so useful for practical applications. A second class of exceptions will be those interactions which are characterized by an unusually slow observed association rate, much slower than 10^5 M^{-1} s^{-1}. This can be due to two things: either one of the partners is not in a productive conformation, and only a small fraction of molecules are able to interact (conformational selection) (102, 103), or a slow conformational change must occur in one of the partners before or upon binding (induced fit) (104). In summary, if the on-rate of the protein to be improved is not *unusually* slow (say, only 10^3–10^4 M^{-1} s^{-1}), and if high affinity should be achieved also under physiological conditions with considerable ionic strength, then an improvement of off-rate is the most likely route to success.

Nonetheless, we still have at two strategic options. In the first, in round to round, less target is immobilized. The underlying assumption is that the binding molecules to be selected, of different affinities, will equilibrate and the ones of low affinity will be displaced by the ones of high affinity that will eventually occupy all the sites. This approach becomes difficult once the affinities are already quite high, say with K_D in the low or subnanomolar range. Equilibration then becomes slow (see below) and once the amount of immobilized species becomes too low, background binding to the blocked surface or other present molecules such as streptavidin becomes a significant problem. The second, more attractive approach is thus to select for the off-rate directly (105–107).

The typical set-up is to expose the library of binders (the ribosomal complexes in ribosome display, the phages in phage display) to biotinylated target in solution. After some time, an excess of non-biotinylated target is added, with the assumption that a fast dissociating binder (one with fast off-rate) will expose an unoccupied binding site and immediately rebind to soluble competitor target, present in excess. The binders with slow off-rate, on the other hand, remain on the biotinylated target and can thus be isolated by adding capture beads carrying, e.g., streptavidin.

Initially, one might think that for selecting an off-rate as slow as possible one should compete and thus wait as long as possible before collecting the binders with the beads. However, after experimentally finding that this does not lead to the desired outcome, we have recently computationally analyzed this process and found that the optimal selection works quite different (108). Here only a very qualitative summary is given; the interested reader is directed to the original publication.

Let us assume that the initial biotinylated target is sufficient to capture all binders and that the capture beads are sufficient to capture all biotinylated target. If we incubate this library with non-biotinylated target in excess for a very long time, *all* the binders will equilibrate between both forms of target. After sufficiently long time, the distribution of binders on the biotinylated target and on the non-biotinylated target becomes identical – no affinity enrichment at all is achieved!

The enrichment of the binders with the slowest off-rate is thus a transient phenomenon (108). At intermediate times, the immobilized target will indeed carry a population that is enriched for the binders with slow off rate. This enrichment will be highest at the time that is the reciprocal of the best off-rates. When in doubt, it is better to err on the side of shorter times.

The selection pressure is dominated by the ratio of non-biotinylated to biotinylated target, which should be as high as possible. There is usually a practical upper limit, given by the availability of the target. It is not useful to decrease the amount of immobilized target because of the danger of selecting background binders as explained above.

Given a limited amount of target, the practitioner has the choice between few rounds of very stringent selections (using the target in large excess) or many rounds with less stringent selections (using the valuable target over more rounds). The less selective strategy will keep a higher diversity and thus potentially binders with a greater range of biological properties and effects. Yet, when the target is severely limited, one selective round with high target excess is probably the best strategy.

After any highly selective step, the number of binders becomes, by necessity, very small. Thus, background binding by non-functional clones can become significant. In order to rectify this problem, a non-selective round directly following the stringent one has been found to be highly useful (108) (and references therein). In this case, all remaining binders are "collected" and amplified, which thereafter greatly outnumber the non-functional molecules.

11. Selections for Properties Other than Affinity

11.1. Selectivity

Besides high affinity, selective discrimination of a particular target, and non-recognition of a similar molecule, is often desired. This can involve recognition of a particular mutant, a posttranslational modification or a conformation. Like in other display technologies such as phage display, this can best be achieved by immobilizing the desired target, and adding the non-desired target as a competitor, such that all members of the library which recognize both, and thus do not discriminate, will be washed out by binding to the non-immobilized competitor.

11.2. Catalysis

The use of display technologies such as phage display (reviewed in ref. 109) and ribosome display (reviewed in ref. 110) for selecting enzymatic turnover has been attempted. While a number of approaches have been found for carrying out selections to identify active catalysts from among many inactive molecules, it is less apparent how to select for the quantitative improvement of enzymatic turnover with display technologies that ultimately select only for a binding event. Ribosome display might thus play a role in the initial selection of very large libraries to identify active molecules. At the present time, it appears that the use of emulsion techniques might be better suited to select for improvements of in vitro turnover rates (94, 95). Ultimately, a direct sorting of the aqueous compartments as a quantitative measure of turnover will be needed.

11.3. Stability

It is a widespread assumption that ribosome display is unsuitable for evolution of protein stability, and thus its inherent advantages of large library size and facile interfacing with random mutagenesis cannot be exploited for this problem. Fortunately, this is not true. The recent discovery that the intrinsic stability of the ribosomal complexes, e.g., as found in the PURE system (60–62), is rather high is further encouraging for such experiments. Even with the standard *E. coli* system, such experiments have been successfully carried out: Using an antibody scFv fragment of medium stability as a model system, its stability was improved by a succession of random mutagenesis and selection for specific binding in the presence of a suitable buffer favoring unfolding (106). The antibody derives a significant part of its stability from its intradomain disulfide bonds. By increasing the level of reducing agents from round to round, scFv fragments were selected which could fold in the complete absence of disulfides. More importantly, when the disulfides were allowed to form again, the free energy of folding gained by the selected mutations was almost additive. It is likely that similar scenarios can be designed for the selection of high stability variants of other proteins. The main prerequisite will be to select for binding that is strictly coupled to correct folding, and does not allow partially folded "sticky" molecules to become enriched.

12. Conclusions and Future Prospects

Ribosome display has proven to be a robust procedure, used now in academic and industrial laboratories, which comes rather close to experimental protein evolution in the test tube. Undoubtedly, the procedure will be further improved and applied to many new targets and selection goals. Progress in automation, selection on complex targets such as whole cells, as well as applications of deep sequencing are the obvious developments that can be expected to contribute to the further development of this powerful in vitro evolution. Undoubtedly, this evolution technology will itself evolve.

Acknowledgments

My sincere thanks go to the many coworkers mentioned in the references who have developed and continuously improved the ribosome display technology over the years. I am grateful to Drs. Birgit Dreier, Oliver Scholz, Erik Sedlak and to Johannes Schilling for critically reading the manuscript.

References

1. Fields, S. & Song, O. (1989) A novel genetic system to detect protein–protein interactions. *Nature* **340**, 245–246.
2. Pelletier, J. N., Arndt, K. M., Plückthun, A. & Michnick, S. W. (1999) An *in vivo* library-versus-library selection of optimized protein–protein interactions. *Nat. Biotechnol.* **17**, 683–690.
3. Wilson, C. G., Magliery, T. J. & Regan, L. (2004) Detecting protein–protein interactions with GFP-fragment reassembly. *Nat. Methods* **1**, 255–262.
4. Cabantous, S., Pedelacq, J. D., Mark, B. L., Naranjo, C., Terwilliger, T. C. & Waldo, G. S. (2005) Recent advances in GFP folding reporter and split-GFP solubility reporter technologies. Application to improving the folding and solubility of recalcitrant proteins from Mycobacterium tuberculosis. *J. Struct. Funct. Genomics* **6**, 113–119.
5. Rossi, F., Charlton, C. A. & Blau, H. M. (1997) Monitoring protein–protein interactions in intact eukaryotic cells by beta-galactosidase complementation. *Proc. Natl. Acad. Sci. U. S. A.* **94**, 8405–8410.
6. Francisco, J. A., Campbell, R., Iverson, B. L. & Georgiou, G. (1993) Production and fluorescence-activated cell sorting of Escherichia coli expressing a functional antibody fragment on the external surface. *Proc. Natl. Acad. Sci. U. S. A.* **90**, 10444–10448.
7. Boder, E. T. & Wittrup, K. D. (1997) Yeast surface display for screening combinatorial polypeptide libraries. *Nat. Biotechnol.* **15**, 553–557.
8. Smith, G. P. (1985) Filamentous fusion phage: novel expression vectors that display cloned antigens on the virion surface. *Science* **228**, 1315–1317.
9. Cwirla, S. E., Peters, E. A., Barrett, R. W. & Dower, W. J. (1990) Peptides on phage: a vast library of peptides for identifying ligands. *Proc. Natl. Acad. Sci. U. S. A.* **87**, 6378–6382.
10. Devlin, J. J., Panganiban, L. C. & Devlin, P. E. (1990) Random peptide libraries: a source of specific protein binding molecules. *Science* **249**, 404–406.
11. Scott, J. K. & Smith, G. P. (1990) Searching for peptide ligands with an epitope library. *Science* **249**, 386–390.
12. McCafferty, J., Griffiths, A. D., Winter, G. & Chiswell, D. J. (1990) Phage antibodies: filamentous phage displaying antibody variable domains. *Nature* **348**, 552–554.
13. Bass, S., Greene, R. & Wells, J. A. (1990) Hormone phage: an enrichment method for variant proteins with altered binding properties. *Proteins* **8**, 309–314.
14. Takahashi, T. T. & Roberts, R. W. (2009) In vitro selection of protein and peptide libraries using mRNA display. *Methods Mol. Biol.* **535**, 293–314.
15. Levin, A. M. & Weiss, G. A. (2006) Optimizing the affinity and specificity of proteins with molecular display. *Mol. Biosyst.* **2**, 49–57.
16. Leemhuis, H., Stein, V., Griffiths, A. D. & Hollfelder, F. (2005) New genotype-phenotype linkages for directed evolution of functional proteins. *Curr. Opin. Struct. Biol.* **15**, 472–478.
17. Daugherty, P. S. (2007) Protein engineering with bacterial display. *Curr. Opin. Struct. Biol.* **17**, 474–480.
18. Pepper, L. R., Cho, Y. K., Boder, E. T. & Shusta, E. V. (2008) A decade of yeast surface display technology: where are we now? *Comb. Chem. High Throughput Screen.* **11**, 127–134.
19. Mondon, P., Dubreuil, O., Bouayadi, K. & Kharrat, H. (2008) Human antibody libraries: a race to engineer and explore a larger diversity. *Front. Biosci.* **13**, 1117–1129.
20. Bratkovic, T. (2010) Progress in phage display: evolution of the technique and its application. *Cell. Mol. Life Sci.* **67**, 749–767.
21. Labrou, N. E. (2010) Random mutagenesis methods for in vitro directed enzyme evolution. *Curr. Protein Pept. Sci.* **11**, 91–100.
22. Horst, J. P., Wu, T. H. & Marinus, M. G. (1999) Escherichia coli mutator genes. *Trends Microbiol.* **7**, 29–36.
23. Wang, L., Jackson, W. C., Steinbach, P. A. & Tsien, R. Y. (2004) Evolution of new nonantibody proteins via iterative somatic hypermutation. *Proc. Natl. Acad. Sci. U. S. A.* **101**, 16745–16749.
24. Virnekäs, B., Ge, L., Plückthun, A., Schneider, K. C., Wellnhofer, G. & Moroney, S. E. (1994) Trinucleotide phosphoramidites: Ideal reagents for the synthesis of mixed oligonucleotides for random mutagenesis. *Nucleic Acids Res.* **22**, 5600–5607.
25. Zaccolo, M., Williams, D. M., Brown, D. M. & Gherardi, E. (1996) An approach to random mutagenesis of DNA using mixtures of triphosphate derivatives of nucleoside analogues. *J. Mol. Biol.* **255**, 589–603.

26. Zhao, H., Giver, L., Shao, Z., Affholter, J. A. & Arnold, F. H. (1998) Molecular evolution by staggered extension process (StEP) in vitro recombination. *Nat. Biotechnol.* **16**, 258–261.
27. Stemmer, W. P. (1994) Rapid evolution of a protein in vitro by DNA shuffling. *Nature* **370**, 389–391.
28. Kraus, J. P. & Rosenberg, L. E. (1982) Purification of low-abundance messenger RNAs from rat liver by polysome immunoadsorption. *Proc. Natl. Acad. Sci. U. S. A.* **79**, 4015–4019.
29. Korman, A. J., Knudsen, P. J., Kaufman, J. F. & Strominger, J. L. (1982) cDNA clones for the heavy chain of HLA-DR antigens obtained after immunopurification of polysomes by monoclonal antibody. *Proc. Natl. Acad. Sci. U. S. A.* **79**, 1844–1848.
30. Tuerk, C. & Gold, L. (1990) Systematic evolution of ligands by exponential enrichment: RNA ligands to bacteriophage T4 DNA polymerase. *Science* **249**, 505–510.
31. Mattheakis, L. C., Dias, J. M. & Dower, W. J. (1996) Cell-free synthesis of peptide libraries displayed on polysomes. *Methods Enzymol.* **267**, 195–207.
32. Matheakis, L. C., Bhatt, R. R. & Dower, W. J. (1994) An in vitro polysome display system for identifying ligands from very large peptide libraries. *Proc. Natl. Acad. Sci. U. S. A.* **91**, 9022–9026.
33. Kawasaki, G. H. (1991) Cell-free synthesis and isolation of novel genes and polypeptides. *PCT Int. Appl.*, WO 91/05058.
34. Gersuk, G. M., Corey, M. J., Corey, E., Stray, J. E., Kawasaki, G. H. & Vessella, R. L. (1997) High-affinity peptide ligands to prostate-specific antigen identified by polysome selection. *Biochem. Biophys. Res. Commun.* **232**, 578–582.
35. Hanes, J. & Plückthun, A. (1997) In vitro selection and evolution of functional proteins by using ribosome display. *Proc. Natl. Acad. Sci. USA* **94**, 4937–4942.
36. Ryabova, L. A., Desplancq, D., Spirin, A. S. & Plückthun, A. (1997) Functional antibody production using cell-free translation: Effects of protein disulfide isomerase and chaperones. *Nat. Biotechnol.* **15**, 79–84.
37. Hanes, J., Jermutus, L., Weber-Bornhauser, S., Bosshard, H. R. & Plückthun, A. (1998) Ribosome display efficiently selects and evolves high-affinity antibodies *in vitro* from immune libraries. *Proc. Natl. Acad. Sci. USA* **95**, 14130–14135.
38. Hanes, J., Schaffitzel, C., Knappik, A. & Plückthun, A. (2000) Picomolar affinity antibodies from a fully synthetic naive library selected and evolved by ribosome display. *Nat. Biotechnol.* **18**, 1287–1292.
39. He, M. & Taussig, M. J. (1997) Antibody-ribosome-mRNA (ARM) complexes as efficient selection particles for in vitro display and evolution of antibody combining sites. *Nucleic Acids Res.* **25**, 5132–5134.
40. He, M., Menges, M., Groves, M. A., Corps, E., Liu, H., Brüggemann, M. & Taussig, M. J. (1999) Selection of a human anti-progesterone antibody fragment from a transgenic mouse library by ARM ribosome display. *J. Immunol. Methods* **231**, 105–117.
41. Kisselev, L., Ehrenberg, M. & Frolova, L. (2003) Termination of translation: interplay of mRNA, rRNAs and release factors? *EMBO J.* **22**, 175–182.
42. Pavlov, M. Y., Antoun, A., Lovmar, M. & Ehrenberg, M. (2008) Complementary roles of initiation factor 1 and ribosome recycling factor in 70S ribosome splitting. *EMBO J.* **27**, 1706–1717.
43. Hauryliuk, V., Zavialov, A., Kisselev, L. & Ehrenberg, M. (2006) Class-1 release factor eRF1 promotes GTP binding by class-2 release factor eRF3. *Biochimie* **88**, 747–757.
44. Douthwaite, J. A., Groves, M. A., Dufner, P. & Jermutus, L. (2006) An improved method for an efficient and easily accessible eukaryotic ribosome display technology. *Protein Eng. Des. Sel.* **19**, 85–90.
45. Cheng, K., Ivanova, N., Scheres, S. H., Pavlov, M. Y., Carazo, J. M., Hebert, H., Ehrenberg, M. & Lindahl, M. (2010) tmRNA·SmpB complex mimics native aminoacyl-tRNAs in the A site of stalled ribosomes. *J. Struct. Biol.* **169**, 342–348.
46. Moore, S. D. & Sauer, R. T. (2007) The tmRNA system for translational surveillance and ribosome rescue. *Annu. Rev. Biochem.* **76**, 101–124.
47. Shine, J. & Dalgarno, L. (1975) Terminal-sequence analysis of bacterial ribosomal RNA. Correlation between the 3′-terminal-polypyrimidine sequence of 16-S RNA and translational specificity of the ribosome. *Eur. J. Biochem.* **57**, 221–230.
48. Salis, H. M., Mirsky, E. A. & Voigt, C. A. (2009) Automated design of synthetic ribosome binding sites to control protein expression. *Nat. Biotechnol.* **27**, 946–950.
49. Kozak, M. (1999) Initiation of translation in prokaryotes and eukaryotes. *Gene* **234**, 187–208.
50. Carpousis, A. J. (2007) The RNA degradosome of Escherichia coli: an mRNA-degrading

50. machine assembled on RNase E. *Annu. Rev. Microbiol.* **61**, 71–87.
51. Carpousis, A. J., Luisi, B. F. & McDowall, K. J. (2009) Endonucleolytic initiation of mRNA decay in Escherichia coli. *Prog. Mol. Biol. Transl. Sci.* **85**, 91–135.
52. Regnier, P. & Hajnsdorf, E. (2009) Poly(A)-assisted RNA decay and modulators of RNA stability. *Prog. Mol. Biol. Transl. Sci.* **85**, 137–185.
53. Schatz, P. J. (1993) Use of peptide libraries to map the substrate specificity of a peptide-modifying enzyme: a 13 residue consensus peptide specifies biotinylation in Escherichia coli. *Biotechnology (N. Y.)* **11**, 1138–1143.
54. Hoffmann, A., Bukau, B. & Kramer, G. (2010) Structure and function of the molecular chaperone Trigger Factor. *Biochim. Biophys. Acta* **1803**, 650–661.
55. Hoffmann, F. & Rinas, U. (2004) Roles of heat-shock chaperones in the production of recombinant proteins in Escherichia coli. *Adv. Biochem. Eng. Biotechnol.* **89**, 143–161.
56. Schimmele, B., Gräfe, N. & Plückthun, A. (2005) Ribosome display of mammalian receptor domains. *Protein Eng. Des. Sel.* **18**, 285–294.
57. Schimmele, B. & Plückthun, A. (2005) Identification of a functional epitope of the Nogo receptor by a combinatorial approach using ribosome display. *J. Mol. Biol.* **352**, 229–241.
58. He, M. & Taussig, M. J. (2007) Eukaryotic ribosome display with in situ DNA recovery. *Nat. Methods* **4**, 281–288.
59. Villemagne, D., Jackson, R. & Douthwaite, J. A. (2006) Highly efficient ribosome display selection by use of purified components for in vitro translation. *J. Immunol. Methods* **313**, 140–148.
60. Matsuura, T., Yanagida, H., Ushioda, J., Urabe, I. & Yomo, T. (2007) Nascent chain, mRNA, and ribosome complexes generated by a pure translation system. *Biochem. Biophys. Res. Commun.* **352**, 372–377.
61. Ohashi, H., Shimizu, Y., Ying, B. W. & Ueda, T. (2007) Efficient protein selection based on ribosome display system with purified components. *Biochem. Biophys. Res. Commun.* **352**, 270–276.
62. Ueda, T., Kanamori, T. & Ohashi, H. (2010) Ribosome display with the PURE technology. *Methods Mol. Biol.* **607**, 219–225.
63. Hanes, J., Jermutus, L., Schaffitzel, C. & Plückthun, A. (1999) Comparison of Escherichia coli and rabbit reticulocyte ribosome display systems. *FEBS Lett.* **450**, 105–110.
64. Nakatogawa, H. & Ito, K. (2002) The ribosomal exit tunnel functions as a discriminating gate. *Cell* **108**, 629–636.
65. Ogawa, A., Sando, S. & Aoyama, Y. (2005) In vitro read-through polysome/ribosome display of full-length protein ORF and its applications. *Nucleic Acids. Symp. Ser. (Oxf.)*, 267–268.
66. Ogawa, A., Sando, S. & Aoyama, Y. (2006) Termination-free prokaryotic protein translation by using anticodon-adjusted E. coli tRNA^{Ser} as unified suppressors of the UAA/UGA/UAG stop codons. Read-through ribosome display of full-length DHFR with translated UTR as a buried spacer arm. *ChemBioChem* **7**, 249–252.
67. Glockshuber, R., Malia, M., Pfitzinger, I. & Plückthun, A. (1990) A comparison of strategies to stabilize immunoglobulin Fv-fragments. *Biochemistry* **29**, 1362–1367.
68. Groves, M., Lane, S., Douthwaite, J., Lowne, D., Rees, D. G., Edwards, B. & Jackson, R. H. (2006) Affinity maturation of phage display antibody populations using ribosome display. *J. Immunol. Methods* **313**, 129–139.
69. Binz, H. K., Stumpp, M. T., Forrer, P., Amstutz, P. & Plückthun, A. (2003) Designing repeat proteins: well-expressed, soluble and stable proteins from combinatorial libraries of consensus ankyrin repeat proteins. *J. Mol. Biol.* **332**, 489–503.
70. Binz, H. K., Amstutz, P., Kohl, A., Stumpp, M. T., Briand, C., Forrer, P., Grütter, M. G. & Plückthun, A. (2004) High-affinity binders selected from designed ankyrin repeat protein libraries. *Nat. Biotechnol.* **22**, 575–582.
71. Wetzel, S. K., Ewald, C., Settanni, G., Jurt, S., Plückthun, A. & Zerbe, O. (2010) Residue-resolved stability of full-consensus ankyrin repeat proteins probed by NMR. *J. Mol. Biol.* **402**, 241–258.
72. Wetzel, S. K., Settanni, G., Kenig, M., Binz, H. K. & Plückthun, A. (2008) Folding and unfolding mechanism of highly stable full-consensus ankyrin repeat proteins. *J. Mol. Biol.* **376**, 241–257.
73. Milovnik, P., Ferrari, D., Sarkar, C. A. & Plückthun, A. (2009) Selection and characterization of DARPins specific for the neurotensin receptor 1. *Protein Eng. Des. Sel.* **22**, 357–366.
74. Zahnd, C., Wyler, E., Schwenk, J. M., Steiner, D., Lawrence, M. C., McKern, N. M., Pecorari, F., Ward, C. W., Joos, T. O. & Plückthun, A. (2007) A designed ankyrin repeat protein evolved to picomolar affinity to Her2. *J. Mol. Biol.* **369**, 1015–1028.

75. Schweizer, A., Roschitzki-Voser, H., Amstutz, P., Briand, C., Gulotti-Georgieva, M., Prenosil, E., Binz, H. K., Capitani, G., Baici, A., Plückthun, A. & Grütter, M. G. (2007) Inhibition of caspase-2 by a designed ankyrin repeat protein: specificity, structure, and inhibition mechanism. *Structure* **15**, 625–636.

76. Zahnd, C., Pécorari, F., Straumann, N., Wyler, E. & Plückthun, A. (2006) Selection and characterization of Her2 binding-designed ankyrin repeat proteins. *J. Biol. Chem.* **281**, 35167–35175.

77. Amstutz, P., Koch, H., Binz, H. K., Deuber, S. A. & Plückthun, A. (2006) Rapid selection of specific MAP kinase-binders from designed ankyrin repeat protein libraries. *Protein Eng. Des. Sel.* **19**, 219–229.

78. Amstutz, P., Binz, H. K., Parizek, P., Stumpp, M. T., Kohl, A., Grütter, M. G., Forrer, P. & Plückthun, A. (2005) Intracellular kinase inhibitors selected from combinatorial libraries of designed ankyrin repeat proteins. *J. Biol. Chem.* **280**, 24715–24722.

79. Dreier, B., Mikheeva, G., Belousova, N., Parizek, P., Boczek, E., Jelesarov, I., Forrer, P., Plückthun, A. & Krasnykh, V. (2010) Her2-specific multivalent adapters confer designed tropism to adenovirus for gene targeting. *J. Mol. Biol.* **405**, 410–426.

80. Veesler, D., Dreier, B., Blangy, S., Lichière, J., Tremblay, D., Moineau, S., Spinelli, S., Tegoni, M., Plückthun, A., Campanacci, V. & Cambillau, C. (2009) Crystal structure of a DARPin neutralizing inhibitor of lactococcal phage TP901-1: comparison of DARPin and camelid VHH binding mode *J. Biol. Chem.* **384**, 30718–30726.

81. Yau, K. Y., Dubuc, G., Li, S., Hirama, T., Mackenzie, C. R., Jermutus, L., Hall, J. C. & Tanha, J. (2005) Affinity maturation of a V(H)H by mutational hotspot randomization. *J. Immunol. Methods* **297**, 213–224.

82. Perruchini, C., Pecorari, F., Bourgeois, J. P., Duyckaerts, C., Rougeon, F. & Lafaye, P. (2009) Llama VHH antibody fragments against GFAP: better diffusion in fixed tissues than classical monoclonal antibodies. *Acta Neuropathol.* **118**, 685–695.

83. Cho, G. S. & Szostak, J. W. (2006) Directed evolution of ATP binding proteins from a zinc finger domain by using mRNA display. *Chem. Biol.* **13**, 139–147.

84. Keefe, A. D. & Szostak, J. W. (2001) Functional proteins from a random-sequence library. *Nature* **410**, 715–718.

85. Seelig, B. & Szostak, J. W. (2007) Selection and evolution of enzymes from a partially randomized non-catalytic scaffold. *Nature* **448**, 828–831.

86. Parker, M. H., Chen, Y., Danehy, F., Dufu, K., Ekstrom, J., Getmanova, E., Gokemeijer, J., Xu, L. & Lipovsek, D. (2005) Antibody mimics based on human fibronectin type three domain engineered for thermostability and high-affinity binding to vascular endothelial growth factor receptor two. *Protein Eng. Des. Sel.* **18**, 435–444.

87. Xu, L., Aha, P., Gu, K., Kuimelis, R. G., Kurz, M., Lam, T., Lim, A. C., Liu, H., Lohse, P. A., Sun, L., Weng, S., Wagner, R. W. & Lipovsek, D. (2002) Directed evolution of high-affinity antibody mimics using mRNA display. *Chem. Biol.* **9**, 933–942.

88. Olson, C. A. & Roberts, R. W. (2007) Design, expression, and stability of a diverse protein library based on the human fibronectin type III domain. *Protein Sci.* **16**, 476–484.

89. Fukuda, I., Kojoh, K., Tabata, N., Doi, N., Takashima, H., Miyamoto-Sato, E. & Yanagawa, H. (2006) In vitro evolution of single-chain antibodies using mRNA display. *Nucleic Acids Res.* **34**, e127.

90. Shen, X., Valencia, C. A., Szostak, J. W., Dong, B. & Liu, R. (2005) Scanning the human proteome for calmodulin-binding proteins. *Proc. Natl. Acad. Sci. U. S. A.* **102**, 5969–5974.

91. Doi, N. & Yanagawa, H. (1999) STABLE: protein-DNA fusion system for screening of combinatorial protein libraries in vitro. *FEBS Lett.* **457**, 227–230.

92. Yonezawa, M., Doi, N., Higashinakagawa, T. & Yanagawa, H. (2004) DNA display of biologically active proteins for in vitro protein selection. *J. Biochem.* **135**, 285–288.

93. Yonezawa, M., Doi, N., Kawahashi, Y., Higashinakagawa, T. & Yanagawa, H. (2003) DNA display for in vitro selection of diverse peptide libraries. *Nucleic Acids Res.* **31**, e118.

94. Griffiths, A. D. & Tawfik, D. S. (2003) Directed evolution of an extremely fast phosphotriesterase by in vitro compartmentalization. *EMBO J.* **22**, 24–35.

95. Miller, O. J., Bernath, K., Agresti, J. J., Amitai, G., Kelly, B. T., Mastrobattista, E., Taly, V., Magdassi, S., Tawfik, D. S. & Griffiths, A. D. (2006) Directed evolution by in vitro compartmentalization. *Nat. Methods* **3**, 561–570.

96. Sumida, T., Doi, N. & Yanagawa, H. (2009) Bicistronic DNA display for in vitro selection of Fab fragments. *Nucleic Acids Res.* **37**, e147.

97. Dreier, B. & Plückthun, A. (2012) Rapid selection of high affinity binders using ribosome display. *Methods Mol. Biol.*, **805**, 261–286.
98. Bremer, H. & Dennis, P. P. (1996). Modulation of chemical composition and other parameters of the cell by growth rate. In *Escherichia coli and Salmonella typhimurium: Cellular and Molecular Biology* (Neidhard, F. C., Curtiss, R., Ingraham, J. L., Lin, E. C. C., Low, K. B., Magasanik, B., Reznikoff, W. S., Riley, M., Schaechter, M. & Umbarger, H. E., eds.), Vol. 2, pp. 1553–1569. American Society for Microbiology Press, Washington, DC.
99. Northrup, S. H. & Erickson, H. P. (1992) Kinetics of protein-protein association explained by Brownian dynamics computer simulation. *Proc. Natl. Acad. Sci. U. S. A.* **89**, 3338–3342.
100. Schreiber, G. & Fersht, A. R. (1996) Rapid, electrostatically assisted association of proteins. *Nat. Struct. Biol.* **3**, 427–431.
101. Selzer, T., Albeck, S. & Schreiber, G. (2000) Rational design of faster associating and tighter binding protein complexes. *Nat. Struct. Biol.* **7**, 537–541.
102. Berger, C., Weber-Bornhauser, S., Eggenberger, J., Hanes, J., Plückthun, A. & Bosshard, H. R. (1999) Antigen recognition by conformational selection. *FEBS Lett.* **450**, 149–153.
103. Foote, J. & Milstein, C. (1994) Conformational isomerism and the diversity of antibodies. *Proc. Natl. Acad. Sci. U. S. A.* **91**, 10370–10374.
104. Koshland, D. E. (1958) Application of a theory of enzyme specificity to protein synthesis. *Proc. Natl. Acad. Sci. U. S. A.* **44**, 98–104.
105. Hawkins, R. E., Russell, S. J. & Winter, G. (1992) Selection of phage antibodies by binding affinity. Mimicking affinity maturation. *J. Mol. Biol.* **226**, 889–896.
106. Jermutus, L., Honegger, A., Schwesinger, F., Hanes, J. & Plückthun, A. (2001) Tailoring *in vitro* evolution for protein affinity or stability. *Proc. Natl. Acad. Sci. U.S.A.* **98**, 75–80.
107. Zahnd, C., Spinelli, S., Luginbühl, B., Amstutz, P., Cambillau, C. & Plückthun, A. (2004) Directed in vitro evolution and crystallographic analysis of a peptide binding scFv antibody with low picomolar affinity. *J. Biol. Chem.* **279**, 18870–18877.
108. Zahnd, C., Sarkar, C. A. & Plückthun, A. (2010) Computational analysis of off-rate selection experiments to optimize affinity maturation by directed evolution. *Protein Eng Des Sel.* **23**, 175–184.
109. Forrer, P., Jung, S. & Plückthun, A. (1999) Beyond binding: using phage display to select for structure, folding and enzymatic activity in proteins. *Curr. Opin. Struct. Biol.* **9**, 514–520.
110. Amstutz, P., Forrer, P., Zahnd, C. & Plückthun, A. (2001) *In vitro* display technologies: Novel developments and applications. *Curr. Opin. Biotechnol.* **12**, 400–405.
111. Zahnd, C., Amstutz, P. & Plückthun, A. (2007) Ribosome display: selecting and evolving proteins in vitro that specifically bind to a target. *Nat. Methods* **4**, 269–279.
112. Lipovsek, D. & Plückthun, A. (2004) In-vitro protein evolution by ribosome display and mRNA display. *J. Immunol. Methods* **290**, 51–67.
113. Seidelt, B., Innis, C. A., Wilson, D. N., Gartmann, M., Armache, J. P., Villa, E., Trabuco, L. G., Becker, T., Mielke, T., Schulten, K., Steitz, T. A. & Beckmann, R. (2009) Structural insight into nascent polypeptide chain-mediated translational stalling. *Science* **326**, 1412–1415.

Part II

Translation Extract Preparation

# Chapter 2

Preparation and Testing of *E. coli* S30 In Vitro Transcription Translation Extracts

James F. Zawada

Abstract

Crude cell-free extracts are useful tools for investigating biochemical phenomena and exploiting complex enzymatic processes such as protein synthesis. Extracts derived from *E. coli* have been used for over 50 years to study the mechanism of protein synthesis. In addition, these S30 extracts are commonly used as a laboratory tool for protein production. The preparation of S30 extract has been streamlined over the years and now it is a relatively simple process. The procedure described here includes some suggestions for extracts to be used for ribosome display.

Key words: *Escherichia coli*, S30 extract, Cell-free protein synthesis, In vitro transcription translation

1. Introduction

Ribosome display is a powerful and useful technique for protein engineering. One key component of the method is the translation machinery (ribosomes, tRNAs, translation factors, etc.) that must be supplied. These numerous and complex biochemical catalysts can be obtained as a crude mixture extracted from various biomass sources or as a reconstituted mixture of individually purified components. Several vendors sell *E. coli*-derived crude cell-free extracts for translation (Promega, Invitrogen, 5 PRIME, Qiagen, etc.), and a purified system is available from New England Biolabs. In addition to *E. coli* extracts, several other cell-free extracts are commercially available including wheat germ, insect cell, and rabbit reticulocyte-based systems. While buying a kit is the fastest and easiest way to obtain extract, it is difficult or impossible to customize the extract to a specific application. For example, in some cases one may wish to delete a gene in the source organism to reduce the background signal in an assay or to remove a detrimental protease. In addition, the cost of kits may be prohibitive if many experiments are needed. Purifying all the required components for translation provides the greatest control of the system composition; however, it is very labor intensive and time-consuming. Preparation of crude cell-free extracts is relatively quick, simple, and inexpensive. This chapter describes the preparation of *E. coli* crude cell-free extract suitable for in vitro combined transcription and translation reactions.

The goal of extract preparation is to maximize the translational capacity of the final extract, so it is important to start with cells having a high concentration of the translation machinery. This means the cells must be harvested during rapid exponential growth since the intracellular content of ribosomes and other translation components is positively regulated by growth rate (1). After growth, the cells are harvested and washed. The washed cells are processed into extract by lysis, clarification, and pre-incubation.

2. Materials

The amounts here are based on 1 L of culture which can be grown in a single Tunair 2.5-L shake flask (IBI Scientific). Using 2YTPG medium and the A19 strain, each liter of culture should produce about 8 g wet cell weight (harvested at 3 OD) which will yield about 10 mL of extract. The amounts can be scaled up if more extract is needed. In that case, it may be more convenient to grow the culture in a fermenter or bioreactor.

2.1. Cell Growth and Harvest

1. *E. coli* strain A19 (see Note 1).
2. 2.5-L Tunair FUL-BAF flask with a Dri-Gauze filter lining in the cap (see Note 2).
3. 2YT medium for inoculum, 50 mL: 16 g/L tryptone, 10 g/L yeast extract, 5 g/L sodium chloride (see Note 3).
4. 2YTPG medium (2), 1 L: 16 g/L tryptone, 10 g/L yeast extract, 5 g/L sodium chloride, 22 mM sodium phosphate monobasic, 40 mM sodium phosphate dibasic, 100 mM glucose, 100 µL antifoam 204 (Sigma-Aldrich), optional (see Note 4).
5. 1 M potassium, sodium, or ammonium hydroxide for pH control, optional (see Note 5).
6. S30 buffer, approximately 50 mL (about 150 mL total needed for the complete process): 10 mM Tris acetate, 14 mM magnesium acetate, 60 mM potassium acetate (see Note 6).

2.2. Biomass Processing

1. S30 buffer, approximately 100 mL: 10 mM Tris acetate, 14 mM magnesium acetate, 60 mM potassium acetate (see Note 6).
2. Avestin Emulsiflex homogenizer.
3. Pre-incubation mix, about 3 mL: 370 mM Tris acetate pH 8.2, 11.1 mM magnesium acetate, 16.5 mM ATP, 50 µM each of the 20 amino acids, 105 mM phosphoenol pyruvate (PEP), 8.4 U/mL pyruvate kinase (Sigma-Aldrich #P7768).

2.3. Extract Testing

1. 10× salt solution: 1.3 M potassium glutamate, 100 mM ammonium glutamate, 80 mM magnesium glutamate (see Note 7).
2. 10× master mix: 12 mM ATP, 8.5 mM GTP, 8.5 mM CTP, 8.5 mM UTP, 340 µg/mL folinic acid, 1.7 mg/mL *E. coli* total tRNA (Roche).
3. AA mix: 50 mM each of the 20 amino acids (see Note 8).
4. 1 M sodium pyruvate.
5. 1 M sodium oxalate.
6. 200 mM putrescine.
7. 200 mM spermidine.
8. 20 mM coenzyme A (CoA).
9. 100 mM nicotinamide adenine dinucleotide (NAD).
10. L-[^{14}C(U)]-leucine (PerkinElmer).
11. T7 RNA polymerase (see Note 9).
12. Plasmid DNA (see Note 10).
13. Whatman 3MM chromatography paper.
14. Straight pins.
15. 5% (w/v) trichloroacetic acid (TCA).
16. Ethanol (95–100%).
17. Scintillation cocktail.
18. Liquid scintillation counter.

3. Methods

The extract procedure described here is based on the method of Liu et al. (3) which was developed for bulk protein production. Some modifications to this procedure may provide advantages for ribosome display as suggested by Hanes et al. (4). See Notes 13 and 17 for comments on the differences in the procedures.

3.1. Cell Growth and Harvest

1. Inoculate 50 mL of 2YT medium in a 250-mL baffled flask with at least 10 µL from a thawed glycerol stock of *E. coli* A19.

Alternatively, inoculate from a colony on an agar plate or a tube culture. Incubate the culture overnight at 37°C with vigorous shaking (~250–280 rpm).

2. In the morning, transfer the entire 50 mL culture into 1 L of 2YTPG medium in a 2.5-L Tunair FUL-BAF flask with a Dri-Gauze filter lining in the cap (see Note 2). Incubate at 37°C vigorous shaking (~250–280 rpm) and monitor growth by OD at 600 nm.

3. Harvest the culture during exponential phase before the growth rate drops during the transition to stationary phase (see Note 11). The culture should be chilled as quickly as possible either by adding ice directly to the culture or by passing it through a heat exchanger or cooling coil in ice water.

4. Once the culture is chilled, collect the cells by centrifugation at $8,000 \times g$ for 20–30 min. If the entire culture volume cannot be centrifuged at once, pour the supernatant off the cell pellets and add more culture on top the pellet for a second centrifugation. Approximately 2.5 g/L/OD wet cells should be collected (~8 g from 1 L at 3 OD).

5. After all the culture is harvested, resuspend the cell pellet in at least 5 mL of S30 buffer for each gram of wet cell weight.

6. Centrifuge the cell suspension at $8,000 \times g$ for 20–30 min.

7. Discard the supernatant and freeze the washed cell pellet at −80°C (see Note 12).

3.2. Biomass Processing

1. Break the frozen cell paste into small pieces and thaw in 1 mL of room temperature S30 buffer per gram cell paste (see Note 13).

2. Shake and stir the cell suspension periodically until it is well mixed. Once thawed, keep the cell suspension on ice if the lysis step cannot be done immediately. Rinse the homogenizer with S30 buffer before processing the cell suspension.

3. Lyse the cells with a single pass through an Avestin Emulsiflex high pressure homogenizer at 17,500 psi (see Note 14). The lysate should be cooled as quickly as possible after exiting the homogenizer, preferably through a cooling coil or heat exchanger. Keep the lysate on ice until it is all collected.

4. Centrifuge the lysate at $30,000 \times g$ for 30 min at 4°C (see Note 15).

5. Transfer the supernatant to a clean tube and repeat the centrifugation (see Note 16).

6. Add 2 mL of pre-incubation mix for each 10 mL of clarified lysate and place in a closed tube or bottle in a 37°C shaker with gentle shaking (~100 rpm) for 80 min (see Notes 17 and 18).

7. Centrifuge the extract at 4,000×g for 10 min.
8. Aliquot the supernatant into tubes and freeze in liquid nitrogen before storing at −80°C (see Note 19).

3.3. Extract Testing

Testing an extract consists of running a cell-free transcription/translation reaction to produce a model protein. The procedure is described below for the Cytomim system (5) (see Note 20) using chloramphenicol acetyl transferase (CAT) as a model protein. Ideally the model product would be similar to the candidate protein to be used in experiments. Modifications to the reaction conditions may be needed for disulfide bond formation in the product (6–8).

The procedure below uses radiolabeling to quantify the product protein. There is also an enzymatic assay for CAT activity (9, 10). Alternative model proteins can be used to enable other detection methods (i.e. fluorescence for GFP production or luminescence for luciferase production).

The exact optimal expression conditions vary slightly for each batch of extract and potentially for each product, with magnesium concentration being the most sensitive. If the standard conditions do not provide acceptable activity, the magnesium concentration in the cell-free reaction should be varied about ±4 mM in 1 or 2 mM increments to identify the optimum conditions. Optimizing plasmid, T7 RNA polymerase, potassium and extract concentrations can also help improve activity.

1. Thaw an aliquot of extract and all the cell-free reagents in Table 1.
2. Mix reagents in the order listed in Table 1, adding the plasmid and extract last (see Note 21).
3. Incubate at 37°C for 3–5 h (see Note 22).
4. While the reactions are running, cut two small slips of Whatman 3MM filter paper for each reaction and label them with pencil. Pierce each strip with a straight pin and stick the pin in a sheet of Styrofoam covered in aluminum foil (shipping cooler lids work well). Make sure the filter paper is held above the foil by the pin so that the sample will not spread from the filter to the foil.
5. At the desired endpoint, for each reaction, spot 5 µL of cell-free reaction each on two separate filter paper slips.
6. Dry the slips for 15 min under an incandescent lamp or overnight on the bench.
7. Place one set of slips (with the pins still in them) in a small beaker and wash three times with cold 5% (w/v) trichloroacetic acid (TCA) for 10 min. Wash once with ethanol (see Note 23), then return to the Styrofoam sheet. Dry under a lamp for

Table 1
Recipe for a 15 μL cell-free transcription-translation reaction

Reagent	Stock concentration	Reaction concentration	Volume (μL)
Milli-Q water			5.5525
10× salt solution	10×	1×	1.5
10× master mix	10×	1×	1.5
AA mix	50 mM each	2 mM	0.6
Pyruvate	1 M	33 mM	0.5
Oxalate	1 M	4 mM	0.06
Putrescine	200 mM	1 mM	0.075
Spermidine	200 mM	1.5 mM	0.1125
CoA	20 mM	0.27 mM	0.2
NAD	100 mM	0.33 mM	0.05
^{14}C Leu	0.1 mCi/mL	1.7 μCi/mL	0.25
T7 RNA polymerase	0.8 mg/mL	33 μg/mL	0.6
Plasmid	0.5 mg/mL	13.3 μg/mL	0.4
Extract		24% (v/v)	3.6

15 min or overnight on the bench. Ensure that the washed slips can be differentiated from the unwashed slips.

8. Remove the pins from the filters and place each slip into a separate scintillation vial, add cocktail, and count in a liquid scintillation counter.

9. Calculate the protein produced with the following formula:

$$[\text{Product}](\text{mg/L}) = \text{Counts}(\text{washed}) / \text{Counts}(\text{unwashed}) \times [\text{Leucine}] \times Y / X,$$

where [Leucine] is the total concentration of leucine (2 mM in this case), Y is the molecular mass of the product, and X is the number of leucine residues in the product.

4. Notes

1. Other strains can be used (MRE600, BL21, etc.). A strain with reduced RNase activity (i.e. *rna* and/or *rne* mutations) may give better results. Make sure the growth medium contains any required nutritional supplements if an auxotrophic strain is used.

2. Tunair flasks are available from IBI Scientific and several distributors. The flasks have a higher oxygen transfer rate than standard baffled glass flasks, so they support higher growth rates and cell densities. Alternatively, the culture volume can be split between multiple glass flasks (typically no more than 400 mL in a 2-L glass flask).

3. The specific medium used for the inoculum is not critical, but preferably it would be the same as the main culture medium. However, in my experience, 2YT works better for overnight inoculum cultures than 2YTPG. If 2YTPG is used for the inoculum, the lag phase may be longer than with a 2YT inoculum.

4. Other media such as LB, 2YT, or a defined medium can be used, but 2YTPG provides a good balance of cost, simplicity, and reasonable cell density (which affects the yield of extract). Also, 2YTPG will produce better extract than 2YT for longer cell-free reactions (2). The 2YTPG medium should be sterile filtered instead of autoclaved to avoid carmelization of the glucose. Alternatively, a separate 50% (v/v) glucose stock can be made and added to the other medium components after autoclaving. The growth rate of the culture determines the ribosomal content of the extract (11), so a medium which supports rapid growth ($\mu > 0.5$ h^{-1}) should be used for best results. If antifoam 204 is used, samples should be chilled to ≤25°C before measuring the OD to eliminate the cloudiness of the antifoam. Placing a 1 mL sample on ice for a couple minutes works well as long as the sample is mixed before OD measurement.

5. The optimum pH for *E. coli* is typically about 7.2–7.4, but can vary depending on the strain and medium used. Metabolic waste products generated from the glucose in the 2YTPG medium will cause the culture to become acidic as the cell density increases, so pH control can help sustain rapid growth to higher cell densities. If pH control is not possible, high quality extract can still be produced, but the culture may need to be harvested at a lower cell density. In addition to pH control, fed-batch fermentation methods can be used to reach higher cell densities (thus producing more extract per batch), but such methods are more complicated to set up and operate. There are many published strategies for fed-batch fermentations and a detailed description is beyond the scope of this chapter. If a fed-batch method is employed, it is important that it maintains a rapid growth rate to produce the highest ribosome content in the extract (11).

6. S30 buffer is typically prepared as needed from purified water and three separate 100× concentrates (1 M Tris acetate pH 8.2, 1.4 M magnesium acetate, and 6 M potassium acetate) which are sterile filtered for increased stability and stored at room temperature. The pH of the 1 M Tris solution is adjusted to 8.2

with glacial acetic acid. The pH of the final S30 buffer usually is not checked or adjusted. A large container of distilled water can be chilled and used for making cold S30 buffer as needed. If desired, the water can be sterilized prior to chilling.

7. It is convenient to prepare small amounts of the salt solution with varying magnesium concentrations for testing the magnesium optimum of each batch of extract. Typically, 8 mM magnesium in the cell-free reaction (equivalent to 80 mM in the 10× salt solution) gives acceptable performance, but the optimum may vary from about 4 to 12 mM. All reagents for extract testing should be divided into small aliquots, frozen in liquid nitrogen, and stored at −80°C for the greatest stability.

8. Tyrosine is not soluble in the amino acid mixture and other amino acids may precipitate over time. This does not affect the performance as long as the mixture is thoroughly mixed immediately before use.

9. T7 RNA polymerase can be purchased from a number of suppliers. Kigawa recommended Ambion, Takara, and Promega products (12). Alternatively, the polymerase can be prepared in-house as a His-tagged protein and purified with a standard IMAC procedure (13) followed by dialysis or ultrafiltration to remove the imidazole. As another option, Nevin and Pratt (14) as well as Kim et al. (15) have expressed T7 RNA polymerase directly in the source cells used for extract preparation so that the extract already contains the T7 RNA polymerase.

10. Any vector designed for T7-driven expression should work in a cell-free system, but the specific vector design can affect productivity. Invitrogen sells vectors for cell-free expression that should be suitable for this system. Standard plasmid purification kits produce suitable quality plasmid for cell-free reactions.

11. The maximum OD before stationary phase will depend on the strain, medium, and culture vessel used. Virtually any system should support at least 1 OD, so harvesting at this point will likely produce good extract (but less volume of extract). The A19 strain grown in 2YTPG in a Tunair flask with 280 rpm shaking can reach 5–8 OD in exponential phase if the pH is controlled. Without pH control, growth may start to slow around 3 OD as the pH drops. Usually, the culture will have a 1–2 h lag phase followed by 2–3 h of exponential growth. The density and age of the overnight culture will affect the lag time and duration of the growth phase. Checking the OD every 30 min once growth starts will provide good data for growth rate calculation and determination of the maximum OD with rapid growth.

12. It is most convenient to scrape the cell paste into a plastic sample bag (VWR #89004-424) and flatten to a thin sheet

before freezing. This takes up less freezer space than a centrifuge bottle and the frozen sheet can be broken up into smaller pieces to speed thawing. Also, if a large batch of cells is prepared, it can be split into portions for multiple extract preparations, for example, to test variations on the extract procedure. Frozen cells can be stored at −80°C for several months. If the rest of the extract preparation procedure will be done immediately, there is no need to freeze the cells.

13. Hanes et al. (4) used 4 mL of buffer per gram cell paste. Using more buffer will result in a more dilute cell lysate which will improve clarification. However, it also results in a larger volume of cell suspension to be lysed and clarified. Most likely, the buffer volume used does not have a strong impact on the total amount of extract activity produced. So the volume can be chosen to make processing most convenient for the lysis method used and the available equipment. The optimum volume of extract used for transcription/translation reactions will depend on the cell dilution factor. If more buffer is used to resuspend the cells, more extract will be needed in the transcription/translation reaction.

14. Avoid getting any air bubbles or foam in the homogenizer, otherwise the extract activity may be reduced. Other cell lysis methods (bead milling, French press, etc.) have been used for extract preparation. To my knowledge, there is no published study comparing the various lysis techniques, but Kigawa recommended glass bead milling and discouraged sonication (12).

15. The centrifugation can be done at a lower g-force (15). If bottles larger than 50 mL are used, extending the centrifugation time 10–20 min will help improve the clarification.

16. The pellet material is very detrimental to extract performance, and the pellet is usually somewhat loose and fluid. So, the supernatant should be transferred to a clean tube as soon as possible, preferably by using a pipette and leaving the bottom 10% or so on the pellet to avoid transferring any pellet material.

17. This pre-incubation step dissociates the polysomes and ribosomes in the extract into 30S and 50S ribosomal subunits (3). Liu and co-workers showed that omitting the reagent additions to the pre-incubation did not affect bulk protein synthesis by the finished extract. Leaving out the additions simplifies the process and greatly reduces the cost. However, without the reagent additions, the 70S ribosomes do not dissociate into subunits. Since some ribosome display translation reactions last only a few minutes, it may be important to have the ribosomes already dissociated in the extract for the best activity. Also, Hanes et al. (4) suggest better performance is achieved with a pre-incubation of 60 min at 25°C. In addition, their

pre-incubation mix has slightly different concentrations of the components. A few experiments can determine the effects of these variations in any specific application.

18. Most extract preparation procedures include extensive dialysis after the pre-incubation. Typically this is done at 4°C for four 45 min periods versus 20 volumes of S30 buffer (changed every 45 min) using 6–8,000 Da molecular weight cut-off dialysis tubing. The work of Liu et al. (3) showed little if any benefit of the dialysis step for bulk transcription/translation reactions, so it was omitted from the procedure here.

19. Extract is usually stable for at least three freeze-thaw cycles. Frozen extract stored at −80°C is stable for at least a year. If a large batch of extract is made, it is convenient to make a few dozen single-use aliquots and then divide the rest of the extract into larger portions. When needed, a large aliquot can be thawed and split into another set of single-use aliquots. All extract samples should be frozen in liquid nitrogen and stored at −80°C. Slow freezing or storage at higher temperatures can reduce the extract activity.

20. Other energy generation systems have been developed for use in cell-free reactions. For more information, see the review by Calhoun and Swartz (16). The Cytomim system requires inverted membrane vesicles to perform oxidative phosphorylation. These vesicles are formed during the high pressure homogenization lysis step and remain in the final extract as prepared with this procedure. If a different cell lysis method is used, it may be necessary to use a different energy generation system in the cell-free reaction to achieve good performance.

21. If only a few reactions are planned, the recipe should be multiplied to ensure all the required volumes are reasonable to pipette accurately. The standard 15 μL reactions work well in 1.5-mL microcentrifuge tubes. If larger reactions are required, the reaction vessel may need to be changed. Voloshin and Swartz have investigated the effects of volume and reactor geometry (17).

22. Some proteins express better at lower temperatures. If desired, a range of temperatures can be tested for each new product to find the optimum (typically 30°C–37°C). Protein synthesis may continue for 5 h or more (5), so a time course should be run if maximum product titer is important.

23. The ethanol wash speeds up the drying process. It usually does not affect the results if it is omitted.

References

1. Bremer H, Dennis PP (1996) Modulation of Chemical Composition and Other Parameters of the Cell by Growth Rate. In: Neidhardt FC (ed) *Escherichia coli* and *Salmonella*: cellular and molecular biology, 2nd edn. pp 1553–1569, ASM Press, Washington, D.C.
2. Kim RG, Choi CY (2000) Expression-independent consumption of substrates in cell-free expression system from *Escherichia coli*. *J Biotechnol* **84**, 27–32.
3. Liu DV, Zawada JF, Swartz JR (2005) Streamlining *Escherichia coli* S30 extract preparation for economical cell-free protein synthesis. *Biotechnol Prog* **21**, 460–465.
4. Hanes J, Jermutus L, Plückthun A (2000) Selecting and evolving functional proteins *in vitro* by ribosome display. *Methods Enzymol* **328**, 404–430.
5. Jewett MC, Swartz JR (2004) Mimicking the *Escherichia coli* cytoplasmic environment activates long-lived and efficient cell-free protein synthesis. *Biotechnol Bioeng* **86**, 19–26.
6. Kim DM, Swartz JR (2004) Efficient production of a bioactive, multiple disulfide-bonded protein using modified extracts of *Escherichia coli*. *Biotechnol Bioeng* **85**, 122–129.
7. Knapp KG, Goerke AR, Swartz JR (2007) Cell-free synthesis of proteins that require disulfide bonds using glucose as an energy source. *Biotechnol Bioeng* **97**, 901–908.
8. Oh IS, Kim DM, Kim TW, et al (2006) Providing an Oxidizing Environment for the Cell-Free Expression of Disulfide-Containing Proteins by Exhausting the Reducing Activity of *Escherichia coli* S30 Extract. *Biotechnol Prog* **22**, 1225–1228.
9. Shaw WV (1975) Chloramphenicol acetyltransferase from chloramphenicol-resistant bacteria. *Methods Enzymol* **43**, 737–755.
10. Kim D, Swartz J (2001) Regeneration of adenosine triphosphate from glycolytic intermediates for cell-free protein synthesis. *Biotechnol Bioeng* **74**, 309–316.
11. Zawada J, Swartz J (2006) Effects of growth rate on cell extract performance in cell-free protein synthesis. *Biotechnol Bioeng* **94**, 618–624.
12. Kigawa T (2009) Cell-Free Protein Preparation Through Prokaryotic Transcription–Translation Methods. *Methods Mol Biol* **607**, 1–10.
13. He B, Rong M, Lyakhov D, et al (1997) Rapid Mutagenesis and Purification of Phage RNA Polymerases. *Protein Expr Purif* **9**, 142–151.
14. Nevin DE, Pratt JM (1991) A coupled *in vitro* transcription-translation system for the exclusive synthesis of polypeptides expressed from the T7 promoter. *FEBS Lett* **291**, 259–263.
15. Kim TW, Keum JW, Oh IS, et al (2006) Simple procedures for the construction of a robust and cost-effective cell-free protein synthesis system. *J Biotechnol* **126**, 554–561.
16. Calhoun KA, Swartz JR (2007) Energy systems for ATP regeneration in cell-free protein synthesis reactions. *Methods Mol Biol* **375**, 3–17.
17. Voloshin AM, Swartz JR (2005) Efficient and scalable method for scaling up cell free protein synthesis in batch mode. *Biotechnol Bioeng* **91**, 516–521.

Part III

Basic Ribosome Display and Related Selection Methods

Chapter 3

Eukaryotic Ribosome Display Selection Using Rabbit Reticulocyte Lysate

Julie A. Douthwaite

Abstract

Ribosome display is a powerful in vitro technology for the selection and directed evolution of proteins. Cell-free translation is central to the ribosome display process and is performed in such a way that the ribosome provides the link between genotype and phenotype that allows genes encoding proteins with desired properties to be identified by selection. Prokaryotic cell-free translation reagents, based initially on *E. coli* cell extracts and more recently containing purified and recombinant factors, have dominated the ribosome display literature. Eukaryotic cell extracts are also suitable for ribosome display; however, protocols for prokaryotic ribosome display are not directly transferable to the use of eukaryotic cell extracts. This chapter describes an optimised methodology for the use of rabbit reticulocyte lysate for ribosome display selections.

Key words: Eukaryotic ribosome display, Rabbit reticulocyte lysate

1. Introduction

Ribosome display is a powerful tool for the selection of proteins with specific functions, for example the discovery of high-affinity monoclonal antibodies and peptides as well as the optimisation of defined protein characteristics by directed evolution (1). The ribosome display process comprises a series of steps that may be carried out in an iterative manner as required. Firstly, mRNA is transcribed in vitro from DNA encoding a library of molecules. The resulting mRNA library is then translated in vitro in a cell-free translation reaction under conditions that produce an array of stabilised ternary ribosome complexes, where both mRNA and protein remain bound to the ribosome. Stop codons that would otherwise signal release of mRNA and the newly translated protein are omitted from the DNA library. Stalled ribosome complexes are further stabilised by rapid cooling of the cell-free translation reaction and dilution in a high-magnesium-containing buffer. A "tether" sequence encoding a relatively non-structured protein is included downstream within the DNA library to allow the displayed protein to fully exit the ribosome tunnel and fold free from steric hindrance. Stem-loop sequences at the 5′ and 3′ ends of the mRNA increase its stability, and timing of the in vitro translation reaction is optimised and precisely controlled to achieve the most productive balance between translation of protein, degradation of mRNA, and spontaneous ribosome complex dissociation. The ribosome maintains a non-covalent link between genotype and phenotype so that relevant proteins, along with their encoding mRNA, are selected by binding to a target of interest. Recovery of selected mRNA is achieved either by dissociation of ribosome complexes which allows mRNA to be purified prior to reverse transcription (RT) and amplification by polymerase chain reaction (PCR) or by performing RT-PCR directly on the ribosome complexes without a separate ribosome dissociation or mRNA purification step. In either case, the result is an enriched DNA library that can be used for further rounds of selection or can be sub-cloned into suitable vectors for DNA sequencing, protein expression, and subsequent screening and analysis.

The earliest demonstrations of ribosome display used prokaryotic in vitro translation for generation of stable ribosome complexes for selection (2) and described conditions for optimal recovery of mRNA. Experiments performed using rabbit reticulocyte lystae for ribosome display have shown that these prokaryotic elution conditions are not effective for dissociating eukaryotic mRNA complexes (3, 4). A method for performing eukaryotic ribosome display with in situ recovery of mRNA is described in Chapter 6 of this book (5); however, there may be a preference or requirement for ribosome complex disruption and/or mRNA purification in certain situations, for example when selecting on a target immobilised to a surface. This chapter describes a procedure for carrying out ribosome display with mRNA elution that is applicable to all selection scenarios.

2. Materials

2.1. Preparation of the Ribosome Display Construct

1. DNA encoding the protein of interest in ribosome display format obtained by gene synthesis as described in detail in Subheading 3.1, step 1. Store at −20°C.
2. 2× PCR Taq polymerase master mix (or a Taq polymerase with suitable reaction buffer and dNTPs). Store at −20°C.
3. T7KOZ primer: 5′GCAGCTAATACGACTCACTATAGGAA CAGACCA*CCATGG* (see Note 1 and Figs. 1 and 2).

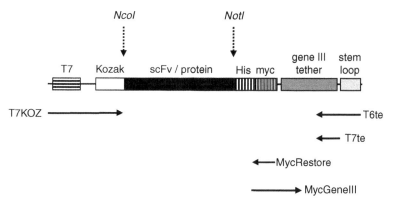

Fig. 1. Schematic illustration of the DNA construct used for eukaryotic ribosome display and the locations of primers used during the process. *T7* T7 RNA polymerase-binding site, *Kozak* eukaryotic ribosome-binding site for translation initiation in eukaryotes, *His* 6× histidine tag, *myc* myc tag. Primers' locations are shown as *horizontal arrows*.

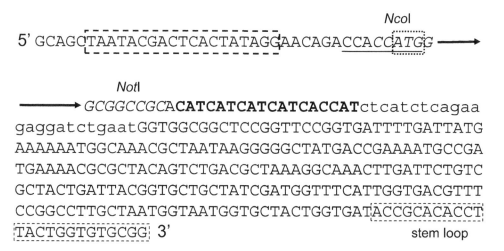

Fig. 2. 5′ and 3′ DNA sequences of the ribosome display construct. The T7 RNA polymerase-binding site is shown by the *dashed box*. The ATG (*dotted box*) in the 5′ region encodes the first methionine of the wild-type molecule (*horizontal arrow*) as well as being part of the NcoI cloning site (CCATGG; shown in *italics*). The 6x His tag is shown in bold, the myc tag sequence is shown in lower case. The gene III tether sequence (see note 18) is located between the myc tag and the 3′ stem-loop.

4. T7te primer: 5′GTAGCACCATTACCATTAGCAAG (see Fig. 1).
5. T6te primer: 5′CCGCACACCAGTAAGGTGTGCGGTATCACCAGTAGCACCATTACCATTAGCAAG (see Fig. 1).
6. Nuclease-free water (see Note 2).
7. 1× TAE gel electrophoresis buffer: 40 mM Tris base, 1.14% (v/v) glacial acetic acid, 1 mM ethylenediaminetetraacetic acid (EDTA), pH 7.6 (harmful/irritant).
8. 1% (w/v) agarose, 1× TAE gels with a DNA staining agent, such as 1× SYBR Safe™.

9. DNA gel loading dye.
10. 1 kb Plus DNA ladder marker (Invitrogen) or equivalent. Store at 4°C.
11. DNA gel purification kit.

2.2. In Vitro Transcription

1. Non-stick, RNAse-free, 1.5-ml microcentrifuge tubes (Sarstedt) (see Note 3).
2. Ribomax Large Scale RNA production system [T7] (Promega).
3. Illustra ProbeQuant G-50 micro column (GE Healthcare).
4. Nuclease-free water (see Note 2).

2.3. In Vitro Translation, Selection, and Recovery of mRNA

1. Non-stick, RNAse-free, 1.5-ml microcentrifuge tubes (Sarstedt) (see Note 3).
2. Nuclease-free water (see Note 2).
3. Flexi Rabbit Reticulocyte Lysate System (Promega) (see Note 4). Store at −70°C.
4. Protein disulphide isomerise (PDI): 5 mg/ml in nuclease-free water. Store at −20°C.
5. 1× phosphate-buffered saline (PBS), nuclease-free, for example sterile for cell culture or prepared with nuclease-free water and chemicals.
6. 10% (w/v) non-fat dried milk in water, autoclaved. Store at 4°C.
7. Relevant biotinylated antigen or binding protein (see Note 5).
8. Dynabeads M-280 streptavidin beads (Invitrogen). Store at 4°C.
9. Wash buffer: 1× PBS, 5 mM magnesium acetate, 1% (v/v) Tween-20. Prepare using nuclease-free reagents; store at 4°C.
10. *Saccharomyces cerevisiae* RNA (Roche) in nuclease-free water at 10 mg/ml. Store at −20°C.
11. High Pure RNA isolation kit (Roche) or equivalent.

2.4. Reverse Transcription to Generate cDNA

1. Non-stick, RNAse-free, 1.5-ml microcentrifuge tubes (Sarstedt) (see Note 3).
2. Nuclease-free water (see Note 2).
3. 5× First Strand Buffer (Invitrogen). Store at −20°C.
4. Dithiothreitol (DTT): 100 mM. Store at −20°C.
5. dNTPmix, 25 mM each. Store at −20°C.
6. T7te primer, see Subheading 2.1, item 4.
7. RNasin: 40 U/μl (Promega). Store at −20°C.
8. Superscript II Reverse Transcriptase: 200 U/μl (Invitrogen). Store at −20°C.

2.5. PCR Amplification of cDNA for Further Selection or for Cloning and Characterisation

1. 2× PCR Taq polymerase master mix, see Subheading 2.1, item 2.
2. Dimethylsulfoxide (DMSO), molecular biology grade.
3. T7KOZ primer, see Subheading 2.1, item 3.
4. T7te primer, see Subheading 2.1, item 4.
5. T6te primer, see Subheading 2.1, item 5.
6. Nuclease-free water (see Note 2).
7. 1× TAE gel electrophoresis buffer, see Subheading 2.1, item 7.
8. 1% (w/v) agarose, 1× TAE gels, see Subheading 2.1, item 8.
9. DNA gel loading dye.
10. 1 kb Plus DNA ladder marker (Invitrogen) or equivalent. Store at 4°C.
11. DNA gel purification kit.

3. Methods

The first step in the process is to prepare a linear DNA construct, encoding the scFv antibody or other protein of interest that is to be optimised, in the appropriate format for ribosome display (Subheading 3.1, step 1). Then, using this "parent" protein construct, perform one cycle of ribosome display selection on the corresponding antigen or binding partner using a concentration that is around the KD (dissociation constant) of the interaction. This step confirms that the parent molecule can be displayed functionally on the ribosome, gives an indication of whether the concentration of target used was appropriate, and also allows the inexperienced user to become familiar with the procedure. In addition, such parent selections are useful for troubleshooting and method optimisation.

The next step is to generate a library of variants by introducing sequence diversity into the parent molecule. A variety of methods for library building are suitable and essentially any PCR-based method can be used (see Note 6). Bear in mind that methods involving cloning may restrict library size due to limits on the transformation efficiency of *E. coli*. Particular regions of the molecule can be targeted for randomisation if there is any structural rationale, as is best exemplified by targeting the complementarily determining regions of an scFv antibody. In the absence of any structural rationale or to find non-predictable improvements, random mutagenesis libraries are recommended and can easily be prepared by the use of commercially available error-prone PCR kits (see Note 7). Additional mutagenesis can also be performed between rounds of selection to maximise the sequence space explored. Once constructed, the library is progressed through rounds of selection with increasing stringency, for example using lower and lower concentrations of target protein or adding competitors. During the selection cascade, selection outputs should be sub-cloned to allow DNA sequencing and analysis of the variant proteins using appropriate assays to identify improved variants.

3.1. Preparation of Parent Protein in Ribosome Display Format

1. Obtain DNA encoding the protein to be optimised in the correct format for ribosome display by gene synthesis. Figure 1 shows the elements required in the ribosome display construct, as well as the locations of primers used throughout the process, and Fig. 2 shows the nucleotide sequences that need to be added upstream and downstream of the scFv antibody of protein coding sequence of interest. Once one construct has been prepared and inserted into a suitable vector for storage and propagation, this can be used to easily convert other molecules to ribosome display format by sub-cloning alternative inserts between the *Nco*I and *Not*I restriction sites. *Nco*I and *Not*I are also used for sub-cloning selection outputs into other vectors for analysis; therefore, ensure that the DNA sequence contains no internal *Nco*I or *Not*I restriction sites. The DNA sequence should also be free from any stop codons. The restriction sites can be modified to suit other required cloning strategies if required.

2. Amplify the full-length linear construct with the T7KOZ and T6te primers. Set up the following PCR reaction, including a control reaction with no template present: 50 μl of 2× PCR master mix, 2 μl of 10 μM T7KOZ primer, 2 μl of 10 μM T6te primer, 50–200 ng of DNA template, and nuclease-free water to final volume of 100 μl.

3. PCR amplify DNA using the following conditions: 94°C for 3 min, 25 cycles of 94°C for 30 s, 55°C for 30 s, and 72°C for 105 s, 72°C for 5 min.

4. Check a 5 μl sample of the PCR product on a 1% (w/v) agarose, 1× TAE gel, for size and purity. Proceed to in vitro transcription (Subheading 3.2) or store the PCR product at −20°C.

3.2. In Vitro Transcription to Generate mRNA

1. Assemble the transcription reaction as follows in a non-stick, RNAse-free microcentrifuge tube in the order listed at room temperature: 10 μl of 5× transcription buffer, 15 μl of 25 mM each rNTP mix (prepared by pooling equal volumes of 100 mM rUTP, rATP, rGTP, and rCTP solutions provided in the kit), 20 μl of linear DNA template (non-purified PCR product), and 5 μl of T7 polymerase enzyme mix.

2. Mix well by pipetting and incubate at 37°C for 2–3 h.

3. Purify the mRNA using a ProbeQuant G50 micro column as follows: Vortex the column to re-suspend the matrix, break off

the tip, and place into a microcentrifuge tube. Loosen the cap ¼ of a turn and centrifuge for 1 min at $735 \times g$ to remove the storage buffer. Discard the flow-through and place the column into a fresh RNAse-free microcentrifuge tube. Carefully pipette the 50 µl transcription reaction onto the centre of the surface of the resin. Replace the cap, loosen it ¼ turn, and centrifuge for 2 min at $735 \times g$ to elute the purified mRNA.

4. Place the tube containing the mRNA immediately on ice.

5. Quantify the mRNA by measuring A260 of a sample (for example, a 1 in 150 dilution) in a spectrophotometer. The concentration should be higher than 2,000 ng/µl. If the concentration is lower, the transcription should be repeated. If the concentration is persistently low, repeat the PCR amplification of the template using more PCR cycles.

6. Use the mRNA immediately for in vitro translation (Subheading 3.3) or freeze the mRNA and store at −70°C until required.

3.3. In Vitro Translation, Selection, and Recovery of mRNA

1. Pre-chill the PBS that is required to stop the translation reactions and stabilise the ribosome complexes prior to selection. For each selection, prepare 100 µl of 1× PBS in a non-stick, RNAse-free microcentrifuge on ice.

2. Prepare de-biotinylated milk for selection (see Note 8). Add 100 µl of streptavidin beads to 1 ml of 10% (w/v) autoclaved, non-fat, dried milk in water in an RNAse-free microcentrifuge tube. Incubate with end-over-end mixing for at least 10 min. Collect the beads using a magnetic particle concentrator and transfer the milk to a fresh tube and keep at 4°C or on ice until required.

3. A 100 µl in vitro translation reaction is required for each individual selection and control. Typically, include two duplicate selections of the library and one or more negative control selections (a selection without the addition of antigen or binding partner, or with the addition of an irrelevant protein, to test for specificity). Prepare a translation master mix for the required number of translations being performed. Combine the following in the order listed below in an RNAse-free microcentrifuge tube on ice per 100 µl translation: 16.7 µl of nuclease-free water, 2 µl of complete amino acid mixture (prepared by pooling equal volumes of all three amino acid mixtures provided in the kit), 1.6 µl of 2.5 M KCl, 2 µl of 5 mg/ml PDI (see Note 9), and 66 µl of rabbit reticulocyte lysate. Mix gently by pipetting (do not vortex).

4. If mRNA has been produced on a previous day and stored at −70°C, thaw it quickly by holding between the fingertips and place immediately on ice when thawed. For each translation

reaction, add 10 μl of mRNA at 1 μg/μl (approximately 2×10^{13} molecules) to the bottom of a pre-chilled, 1.5-ml, RNAse-free microcentrifuge tube on ice. Immediately refreeze any unused mRNA.

5. Add 100 μl of the translation master mix to each 10 μl aliquot of mRNA and mix gently by pipetting. Incubate at 30°C (preferably on a heat block to ensure rapid equilibration to temperature) for 20 min (see Note 10).

6. After translation, immediately pipette the reactions into the pre-prepared RNAse-free microcentrifuge tubes containing ice-cold PBS on ice to stop the reaction and stall the ribosome complexes.

7. Add biotinylated antigen or binding partner at the required concentration to the positive selections. For blocking non-specific interactions, add 50 μl of de-biotinylated autoclaved milk from step 2 to each selection if desired (see Note 11).

8. Incubate the selections at 4°C for 2 h or overnight (see Note 12) with gentle end-over-end rotation.

9. Prepare streptavidin-coated magnetic beads for capture of target-bound ribosome complexes. Use 1 μl of beads per "nM" of biotinylated cognate binding partner used, down to a minimum volume of 50 μl. Wash the beads four times in 1× PBS in an RNAse-free microcentrifuge tube and re-suspend in 1× PBS to the original volume. Store on ice until required.

10. Capture the bound complexes by addition of washed streptavidin-coated beads to each selection. Incubate for 1–2 min at 4°C (see Note 13).

11. Wash the beads five times with 500 μl of wash buffer at 4°C to remove non-specifically bound complexes (see Note 14). Re-suspend the beads in 200 μl of PBS containing *S. cerevisiae* mRNA (at 10 μg/ml final concentration). Incubate for 2–5 min at 50°C and mix by vortexing occasionally.

12. Transfer the 200 μl of PBS which now contains the eluted mRNA into a pre-prepared 400-μl aliquot of lysis buffer from the High Pure RNA Isolation kit in a fresh RNAse-free microcentrifuge tube. Vortex immediately to mix well.

13. Isolate mRNA using the High Pure RNA Isolation kit according to manufacturer's instructions (see Note 15). Include the DNase I digestion step and elute the mRNA in 40 μl of elution buffer (or nuclease-free water). Proceed immediately to reverse transcription (Subheading 3.4).

3.4. Reverse Transcription to Generate cDNA

Perform reverse transcription of mRNA using the primer T7te (see Note 16) to generate full-length cDNA that is ready for PCR amplification with T7KOZ and T6te for a subsequent round of selection without any additional processing.

1. Prepare a master mix for the required number of RT reactions to be performed (one per selection plus a negative "no-template" control). Mix on ice in a fresh RNAse-free microcentrifuge tube per reaction: 4 μl of 5× first-strand buffer, 2 μl of 100 mM DTT, 0.25 μl of 100 μM T7te primer, 0.5 μl of 25 mM dNTP mix, 0.5 μl of 40 U/μl Rnasin, and 0.5 μl of 200 U/μl Superscript II Reverse Transcriptase.

2. Mix well and aliquot 7.75 μl of the RT master mix per reaction into the bottom of an RNAse-free, thin-walled, 0.2-ml PCR tube or a microcentrifuge tube.

3. Mix the eluted mRNA from Subheading 3.3, step 13, by pipetting up and down and add 12.25 μl of mRNA to the tube. Mix well but gently by pipetting up and down. Freeze any unused mRNA as soon as possible and store at −70°C.

4. Incubate RT reactions at 50°C for 30 min in a thermal cycler or on a heat block.

5. Proceed immediately to PCR amplification or store cDNA at −20°C if PCR is to be performed at a later time.

3.5. PCR Amplification of cDNA for Further Selection or for Cloning and Characterisation

1. Prepare a PCR master mix for the required number of reactions to be performed (one per RT reaction plus a negative "no-template" control). Mix on ice in a fresh RNAse-free microcentrifuge tube per reaction: 34.5 μl of nuclease-free water, 50 μl of 2× PCR master mix, 0.25 μl of 100 μM T7KOZ primer, 0.25 μl of 100 μM RT primer (e.g. T7te), and 5 μl of DMSO.

2. Mix well by pipetting and aliquot 90 μl of master mix into the bottom of a thin-walled, 0.2-ml PCR tube on ice.

3. Add 10 μl of cDNA from Subheading 3.4, step 4 to the PCR tubes and mix well. Store any unused cDNA at −20°C.

4. Amplify using the following conditions: 94°C for 3 min, 25–40 cycles (see Note 17) of 94°C for 30 s, 55°C for 30 s, and 72°C for 105 s, 72°C for 5 min.

5. Compare 5 μl of the PCR sample on a 1% (w/v) agarose, 1× TAE electrophoresis gel. There should be one band per lane corresponding to the appropriate size for the construct and primers used. There should be a visible "selection window", i.e. the positive selections should show more PCR product than the negative selection and there should be no amplification in the RT and PCR no-template controls. Low-molecular-weight products may suggest degradation of the specific product and this should be avoided since these smaller products may be amplified preferentially at the next PCR and eliminate the specific product. A high-quality PCR product is critical to the success of the selections, and in some cases the PCR step may require optimisation.

6. To process selection outputs for further rounds of selection or for vector sub-cloning, run the remaining PCR reaction on a large 1% (w/v) agarose, 1× TAE gel. Cut the appropriate band from the gel and purify. Store purified DNA at −20°C. This DNA can be used for sub-cloning to analyse the sequence and function of individual clones in this selection output as desired. To carry out a further selection on this output, proceed to the next step.

7. Re-amplify the output DNA by PCR as follows: 50 μl of 2× PCR master mix, 2 μl of 10 μM T7KOZ primer, 2 μl of 10 μM T6te, 5 μl of gel-purified product from step 6, and 41 μl nuclease-free water.

8. Amplify DNA using the following conditions: 94°C for 3 min, 20 cycles of 94°C for 30 s, 55°C for 30 s, and 72°C for 105 s, 72°C for 5 min.

9. Check a 5 μl PCR sample on a 1% (w/v) agarose, 1× TAE electrophoresis gel, for size and purity using the same criteria as in step 5. Store DNA at −20°C.

10. Begin the next round of selection as described in Subheading 3.2.

4. Notes

1. This primer contains the upstream elements required for in vitro transcription (T7 promoter) and eukaryotic in vitro translation (Kozak sequence). The ATG encoding the first methionine of the protein of interest is shown in italics. The underlined region shows the *Nco*I restriction site used for cloning (other restriction sites can be used if preferred).

2. mRNA is extremely susceptible to degradation by nuclease enzymes; therefore, all solutions should be as nuclease-free as possible. For water coming into direct contact with purified mRNA, we highly recommend the use of commercially available nuclease-free water. We store nuclease-free water at 4°C for convenience since many steps require the use of chilled water.

3. We prefer the use of non-stick, RNAse-free microcentrifuge tubes for all steps involving pure mRNA, i.e. transcription, cell-free translation and selection steps, and for elution of purified mRNA, although other nuclease-free microcentrifuge tubes can be used.

4. Other rabbit reticulocyte lysate systems are available and may be used. However, the Flexi system allows reaction conditions to be tailored if required.

5. The quality of this molecule is critical for the success of ribosome display selections. The molecule should be as pure as possible and ideally tagged in some way to allow capture. Biotinylation for capture using streptavidin-coated magnetic beads is highly recommended. Activity of the tagged target molecule (at least retained binding to relevant molecules) should be confirmed to ensure that the target is still appropriate for selection. Alternatively, surface or panning selections using non-tagged target are possible, although it is still recommended to check for activity of the protein once it has been coated onto the desired surface.

6. Targeted libraries are easily prepared by overlap PCR using the ribosome display construct as a template. Use the oligonucleotides T7KOZ and T7te with internal primers containing an overlapping region, and one containing degenerate codons, to introduce diversity as required to prepare two PCR fragments for assembly. Amplify the full-length library with T7KOZ and T6te. Successful library construction should be verified by cloning a sample of the library (e.g. using *Nco*I and *Not*I) into a suitable vector and DNA sequencing of a representative number of clones (e.g. 88).

7. When performing error-prone mutagenesis, take care to create mutations only in the parent protein sequence and not in the upstream or downstream regions since mutations here may disrupt display efficiency and downstream processing. To ensure this, use the T7KOZ primer and a reverse primer annealing at the end of the sequence to be randomised (for example, the MycRestore primer: 5′ATTCAGATCCTCTTCTGAGATGAG) (see Fig. 2) for the error-prone PCR amplification. Create the full-length ribosome display construct (i.e. add the tether sequence) by assembly PCR of this product with the geneIII tether DNA prepared (both gel purified) by PCR using the MycGeneIII primer 5′ATCTCAGAAGAGGATCTGAATGGTGGCGGCTCCGGTTCCGGTGAT (see Fig. 2) and the T7te primer. Finally, amplify the gel-purified T7KOZ-T7te product with T7KOZ and T6te. Do not attempt to use the T7KOZ and T7te or T6te primers directly for error-prone PCR. Successful library construction should be verified by cloning a sample of the library (e.g. using *Nco*I and *Not*I) into a suitable vector and DNA sequencing of a representative number of clones (e.g. 88).

8. It is possible that biotin found naturally in milk could interfere with the selection process when using streptavidin-biotinylated antigen capture. For this reason, biotin is removed from the autoclaved milk using streptavidin beads prior to using the milk in selections.

9. PDI is included for the correct folding of disulphide bonds, for example in the display of scFv antibody fragments, but may be omitted if this is not required.

10. Incubation for 20 min at 30°C is suitable for most applications. If desired, the precise optimal translation time and temperature can be determined empirically using test selections of the parent protein.

11. In some cases, this blocking step is not necessary and may even be detrimental to selection. In the event of poor results, for example high background, the effect of blocking should be investigated by performing selections with and without the addition of milk or if desired alternative blocking agents.

12. The time required for functional selection is dependent upon the binding affinity between the wild-type molecule and the cognate binding partner. For interactions in the nM range, this should be achieved within 2 h. The selection can also be successfully run overnight, and this can make the work load more manageable for performing complicated or large experiments.

13. For use of other tags and capture reagents, this incubation may need to be longer. A short incubation is sufficient for biotin–streptavidin capture because of the very high affinity of this interaction. Low-affinity interactions should be allowed more time, for example capture of Fc-tagged proteins with protein G-coated beads is ideally performed for 20 min.

14. To wash the beads, gently re-suspend them in wash buffer (do not vortex) and then collect on a magnetic particle concentrator prior to removing the supernatant. Automated washing and elution (e.g. using the Thermo Kingfisher ML) is recommended to improve the efficiency and reproducibility of the process, as well as being more user friendly for bead washing at 4°C.

15. The use of a vacuum aspirator is recommended to remove column flow-through waste to increase speed and consistency and reduce sample cross-contamination.

16. T7te is a shortened form of T6te that does not contain the stem-loop sequence. It is used for RT because it generally results in better yields. The 3′ stem-loop sequence is added in the final PCR amplification prior to performing transcription for the next round of selection.

17. The optimal number of PCR cycles varies for different selections depending on the yield and quality of cDNA. For a first round, some guidance can be obtained from the model selection of the parent clone, although bear in mind that the selection output from a library should be expected to be lower then the cDNA recovered from "model selection" of a single clone. Alternatively, small-scale PCRs of the selection output

can be performed to determine a suitable number of cycles, where a good positive-to-negative ratio is seen without over-amplification or excessive amplification of background. In general, 25 or 30 PCR cycles is a good starting point.

18. The gene III tether sequence is derived from the gene III sequence of filamentous phage M13, spanning amino acids 269–336 (SwissProt P69168). The tether is essentially used to provide an unstructured portion at the C terminus of the protein such that the ribosomal tunnel can cover at least 20–30 amino acid residues of the emerging polypeptide without interfering with the folding of the protein of interest. Other similarly non-structured sequences can be employed.

References

1. Douthwaite J, Jermutus L, Jackson R (2007) Accelerated protein evolution using ribosome display. In: Kudlicki WA, Katzen F, Bennett RB (ed) Cell-free protein expression. Landes Bioscience
2. Mattheakis LC, Bhatt RR and Dower WJ (1994) An in vitro polysome display system for identifying ligands from very large peptide libraries. Proc Natl Acad Sci USA 91: 9022–9026
3. He M and Taussig MJ (2005) Ribosome display of antibodies: expression, specificity and recovery in a eukaryotic system. J Immunol Methods 297: 73–82
4. Douthwaite JA, Groves MA, Dufner P et al (2006) An improved method for an efficient and easily accessible eukaryotic ribosome display technology. PEDS 19: 85–90
5. He M, Edwards BM, Kastelic D et al (2011) Eukaryotic ribosome display with in situ DNA recovery. In: Jackson R, Douthwaite JA (ed) Ribosome display and related technologies: methods and protocols. Humana Press

Chapter 4

Stabilized Ribosome Display for In Vitro Selection

Shuta Hara, Mingzhe Liu, Wei Wang, Muye Xu, Zha Li, and Yoshihiro Ito

Abstract

Ribosome display is a very effective and powerful technology for screening functional peptides or polypeptides in vitro. In ribosome display, each peptide or polypeptide (phenotype) links with its corresponding mRNA (genotype) through a ribosome. This link can be achieved by the absence of a stop codon in the mRNA, therefore stalling the ribosome at the end of translation with the nascent random sequence peptide extended by a spacer outside of the ribosome tunnel. In this chapter, we describe a method for the use of a further stabilized peptide–ribosome–mRNA complex for ribosome display.

Key words: In vitro selection, Stabilized ribosome display, Peptide library, Peptide aptamer

1. Introduction

Over the last two decades, in vitro selection has become a general technology for screening and isolating functional peptides (or proteins) and isolating their encoding nucleic acids. Ribosome display is a cell-free, translation-based method that has been used to select antibodies and other protein-binding partners for factors including other proteins and metals (1–8). The principle is to display a protein library on the surface of a ribosome by translation and link the protein (phenotype) to its corresponding mRNA (genotype) via the ribosome in forming a complex. For screening and isolating functional peptides or proteins successfully using ribosome display, several issues are important. One critical aspect is to ensure the stability of the peptide–ribosome–mRNA (PRM) complexes. Sawata and Taira (9) have reported a ribosome display method that introduces a specific protein–RNA interaction to enhance the stability of the PRM complexes. This method uses the *Escherichia coli* bacteriophage MS2 coat protein (MSp) and the C-variant (Cv)

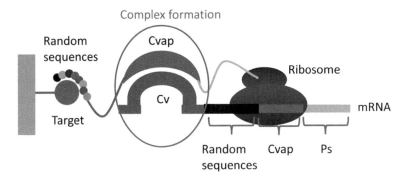

Fig. 1. Complex formation for stabilizing the ribosome display linkage.

RNA motif to which it binds. The MSp coding sequence and the Cv RNA motif are placed at the N-terminal end of the nascent peptide and at the 5′ end of the mRNA, respectively. The resulting complex is illustrated in Fig. 1. Binding of MSp to Cv increases the stability of the PRM complex as shown by (1) selection of the peptide of interest even in the presence of a stop codon and (2) improved yield of selection-positive mRNA by the presence of the MSp–Cv interaction. We have succeeded in in vitro selection to isolate a metal-binding motif from a random sequence peptide by employing MSp–Cv interaction (8). Here, we describe a step-by-step procedure to perform this stabilized method of ribosome display selection.

2. Materials

2.1. Construction of the Random Peptide Library DNA Construct

1. Single-strand (ss) DNA oligonucleotide encoding the random peptide library: 5′-ATATGGCCATGCAGGCC(NNN)nGGC CAGCTAGGCCAGTT, where N=G, C, T, or A (see Note 1).
2. Primer rp-1: 5′-AAACAGCTATGACCATGATTA.
3. Primer rp-2: 5′-AACTGGCCTAGCTGG.
4. Primer fp-1: 5′-TAATACGACTCACTATA-GAACATGAGG-ATCACCCATGTAAAAGTCGACAATAATTTTGTT TAACTT.
5. SD-Cvap-Ps plasmid (see Note 2) (8, 9).
6. PrimeSTAR® GXL DNA polymerase kit (Takara Otsu, Shiga, Japan) or equivalent DNA polymerase with reaction buffer (various suppliers). The kit contains 1.25 U/μL PrimeSTAR® GXL DNA polymerase and dNTP mix (2.5 mM each of dATP, dCTP, dGTP, and dTTP). Store at −20°C.
7. QIAquick PCR purification kit (Qiagen, Hilden, Germany) or equivalent.

8. *Sfi*I (20 U/μL) (New England Biolabs, Ipswich, MA, USA) or equivalent with reaction buffer (10× NE buffer: 500 mM NaCl, 100 mM Tris–HCl, 100 mM $MgCl_2$, 10 mM 1,4-dithiothreitol (DTT), pH 7.9) and 100× BSA (10 mg/mL). Store at −20°C.

9. Mighty Mix DNA Ligation kit (Takara) or equivalent T4 DNA ligase with reaction buffer (various suppliers).

10. Nuclease-free water (see Note 3).

11. 1× TAE buffer: 40 mM Tris base, 20 mM glacial acetic acid, and 1 mM ethylenediaminetetraacetic acid (EDTA) (pH 8.0). Store at room temperature.

12. 6% PAGE gel: 6% acrylamide/bis (19:1) gel in 1× TAE buffer. Unpolymerized acrylamide is a neurotoxin; wear gloves and avoid exposure.

13. 10% (w/v) ammonium persulfate (APS) for preparation of PAGE gels. Prepare using nuclease-free water. Store at 4°C for up to 1 month.

14. N,N,N′,N′-Tetramethylethylenediamine (TEMED) for preparation of PAGE gels. Store at 4°C.

15. 0.5 μg/mL ethidium bromide. Store in a dark bottle. Ethidium bromide is a mutagen; wear gloves, avoid exposure, and dispose of waste accordingly.

2.2. In Vitro Transcription

1. T7 RiboMAX™ Large Scale RNA Production System-T7 (Promega, San Luis Obispo, CA, USA) or equivalent. The kit contains enzyme mix (T7 RNA polymerase, recombinant inorganic pyrophosphatase, and Recombinant RNasin® Ribonuclease inhibitor), DNase (1 U/μL), rNTPs (100 mM rATP, rCTP, rUTP, and rGTP), and 5× T7 transcription buffer. Store at −20°C.

2. Recombinant RNasin® Ribonuclease inhibitor (40 U/μL) (Promega).

3. RNeasy® Mini kit (Qiagen) or equivalent.

4. 2% denaturing mRNA agarose gel: 2% (w/v) agarose, 26 mM MOPS, 6.5 mM sodium acetate (pH 7.0), 0.6 mM EDTA, 64% (v/v) formamide in nuclease-free water.

5. Denaturing mRNA running buffer: 26 mM MOPS, 6.5 mM sodium acetate (pH 7.0), 0.6 mM EDTA in nuclease-free water.

6. 10× mRNA loading buffer: 50% (v/v) glycerol, 1% (v/v) SDS, and 0.05% (w/v) bromophenol blue in nuclease-free water.

7. 0.5 μg/mL ethidium bromide, see Subheading 3.1 item 18.

8. Nuclease-free water.

2.3. In Vitro Translation

1. PURESYSTEM Classic II kit (Wako, Osaka, Japan) or equivalent for cell-free translation. Store at −80°C.
2. RNasin® Ribonuclease inhibitor (40 U/μL) (Promega, San Luis Obispo, CA, USA). Store at −20°C.
3. Nuclease-free water.

2.4. Affinity Selection

1. Target of interest: The method is exemplified here by use of the small-molecular-weight target 6-[hydroxy(4-nitrobenzyl) phosphonyl] hexanoic acid (PHA) (10) (see Notes 4 and 5).
2. Solid support for immobilization of target, for example Dynabeads M-270 Amine (Invitrogen Carlsbad, CA, USA) or other beads suitable for the target of interest (see Note 6).
3. MES buffer: 0.1 M 2-[N-morpholino]ethane sulfonic acid (MES), 0.5 M NaCl, pH 6.0.
4. Coupling agents for immobilization of the target molecule as appropriate, for example 1.8 mM 1-Ethyl-3-(3-dimethyla-minopropyl) carbodiimide hydrochloride (EDC) and 3.6 mM N-hydroxy succinimide (NHS) for amine coupling of PHA (see Note 6).
5. Tris–acetate buffer: 50 mM Tris–acetate, 150 mM NaCl, 50 mM magnesium acetate, pH 7.5.
6. For amine coupling, ninhydrin solution: 0.3 g ninhydrin and 3 mL acetic acid in 100 mL n-butanol.
7. Tris–acetate EDTA: 660 mM EDTA in Tris–acetate buffer.
8. Selection buffer (see Note 7).
9. RNeasy® Mini kit (Qiagen) or equivalent.

2.5. Reverse Transcription and PCR

1. RNasin® Ribonuclease inhibitor, see Subheading 2.3, item 8.
2. Primer rp-1, see Subheading 2.1, item 2.
3. 5× PrimeScript buffer: 250 mM Tris–HCl, 375 mM KCl, and 15 mM $MgCl_2$.
4. PrimeScript™ Reverse Transcriptase (200 U/μL) (Takara) or equivalent. Store at −20°C.
5. Nuclease-free water.
6. Primer fp-1, see Subheading 2.1, item 4.
7. PrimeSTAR® GXL DNA polymerase kit or equivalent, see Subheading 2.1, item 6.
8. 1× TAE buffer, see Subheading 2.1, item 11.
9. 5% PAGE gel: 5% acrylamide/bis (19:1) gel in 1× TAE buffer. Unpolymerized acrylamide is a neurotoxin; wear gloves and avoid exposure.
10. 10% (w/v) APS, see Subheading 2.1, item 13.

11. TEMED, see Subheading 2.1, item 14.
12. 0.5 μg/mL ethidium bromide, see Subheading 2.1, item 15.
13. QIAquick PCR purification kit.

2.6. Cloning and Sequencing

1. Primer fp-2: 5′-ATATGGCCATGCAGGCC.
2. Primer rp-2, see Subheading 3.1, item 3.
3. PrimeSTAR® GXL DNA polymerase kit or equivalent, see Subheading 2.1, item 6.
4. Nuclease-free water.
5. QIAquick PCR purification kit.
6. 1× TAE buffer, see Subheading 2.1, item 11.
7. 5% PAGE, see Subheading 2.5, item 9.
8. 10% (w/v) APS, see Subheading 2.1, item 13.
9. TEMED, see Subheading 2.1, item 14.
10. 0.5 μg/mL ethidium bromide, see Subheading 2.1, item 15.
11. *Sfi*I (20 U/μL), see Subheading 2.1, item 8.
12. SD-Cvap-Ps plasmid (see Note 2) (8, 9).
13. Mighty Mix DNA Ligation kit, see Subheading 2.1, item 9.
14. Competent *E. coli* cells for transformation and appropriate culture medium (various suppliers).

2.7. SDS-PAGE Gel for Western Blotting

1. Primer fp-3: 5′-TAATACGACTCACTATAGGGGTCGACAATAATTTTGTTTAACTT.
2. Primer, rp-flag: 5′-CTACTTGTCGTCATCGTCCTTGTAGTCCGCAATCAGACTGATCATACC.
3. PrimeSTAR® GXL DNA polymerase kit or equivalent, see Subheading 2.1, item 6.
4. Nuclease-free water.
5. QIAquick PCR purification kit.
6. 2× SDS loading buffer: 0.125 M Tris–HCl (pH 6.8), 20% (w/v) glycerol, 4% (w/v) SDS, 10% (v/v) 2-mercaptoethanol, 0.004% (w/v) bromophenol blue.
7. 10× SDS-PAGE buffer: 2.5 M Tris, 19.2 M glycine, 10% (w/v) SDS.
8. Resolving gel: 15% (w/v) acrylamide/bis solution (19:1), 0.4 M Tris–HCl (pH 8.0), 0.1% (w/v) SDS, 0.1% (w/v) APS, 0.04% (v/v) TEMED in nuclease-free water. The solution can be stored for more than 12 months at room temperature.
9. Stacking gel: 5% (w/v) acrylamide/bis solution (19:1), 1.2 M Tris–HCl (pH 6.8), 0.1% (w/v) SDS, 0.1% (w/v) APS, 0.1% (v/v) TEMED in nuclease-free water. The solution can be stored for more than 12 months at room temperature.

10. PVDF membrane (Milipore, Billerica, MA, USA) or equivalent.
11. 1× blotting buffer: 0.25 M Tris, 1.92 M glycine, 1% (w/v) SDS, and 20% (v/v) methanol.
12. TBS-T: 50 mM Tris–HCl, 150 mM NaCl, 0.05% (v/v) Tween 20, pH 7.5.
13. Anti-FLAG M2 monoclonal antibody–peroxidase conjugate (Sigma–Aldrich, St. Louis, MO, USA) or equivalent.
14. ECL Plus Western Blotting Detection System (GE/Amersham Biosciences, Little Chalfont, Buckinghamshire, UK) or equivalent.
15. Semi-dry blotting system model BE-310 (BIO CRAFT, Tokyo, Japan) or equivalent.
16. Target of interest immobilized in an enzyme-linked immunosorbent assay (ELISA) plate, for example a 96-well NH microtiter plates (CovaLink™ NH Module) (Thermo Fisher Scientific/Nalge Nunc International, Rochester, New York, USA).
17. Tris–acetate buffer, see Subheading 2.4, item 5.
18. Tris–acetate buffer containing 0.01% (v/v) Tween 20.
19. Blocking buffer: 4% (w/v) Block ACE® powder (DS Pharma Biomedical, Osaka, Japan) or equivalent blocking agent (see Note 8) in selection buffer (see Subheading 2.4, item 8).

3. Methods

The principle of ribosome display we have developed is shown in Fig. 2.

3.1. Generation of the DNA Template

Figure 3 shows the DNA template for in vitro selection. First, ssDNA containing the sequence encoding RPL is synthesized and converted to double-strand (ds) DNA using DNA polymerase in the presence of the reverse primer rp-2. The resulting dsDNA products are digested using *Sfi*I and inserted between the *Sfi*I sites located between T7-Cv and Cvap-Ps of the plasmid vector SD-Cvap-Ps (see Note 9). Finally, the DNA templates for in vitro selection are prepared by PCR using the ligated product as template and the primers, fp-1 and rp-1.

1. Prepare dsDNA from the ssDNA oligonucleotide using the reverse primer rp-2. Prepare a reaction mixture containing 10 pmol of ssDNA, 100–200 pmol of rp-2, 5× PrimeSTAR® GXL buffer, 4 μL of dNTP mix, 0.5–1.0 μL of PrimeSTAR GLX DNA Polymerase, and nuclease-free water up to 50 μL. Incubate the reaction in a thermal cycler at 98°C for 10 s, 55°C for 15 s, and 68°C for 30 s.

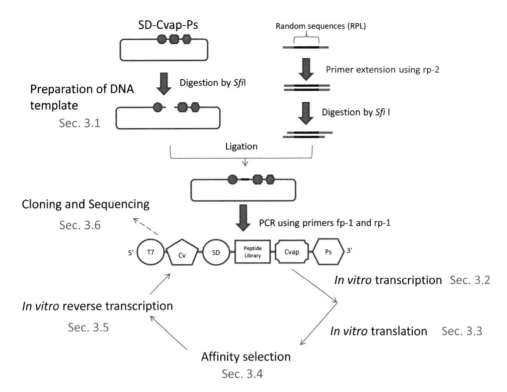

Fig. 2. The ribosome display cycle illustrating the steps of ribosome complex generation from the PCR library by in vitro translation, selection of complexes by ligand-binding, and in vitro reverse transcription for regenerating full-length cDNA.

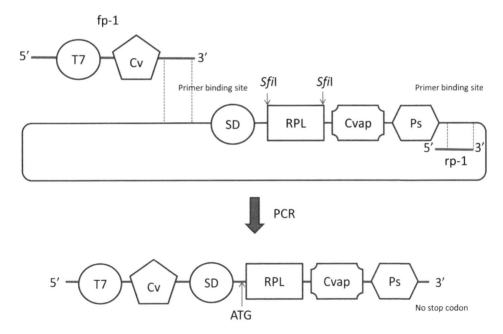

Fig. 3. The prepared plasmid from plasmid vector SD-Cvap-Ps and primers (**a**). DNA template containing a T7 promoter 5′-TAATACGACTCACTATA-3′ and a Shine–Dalgarno (SD) sequence, necessary for in vitro transcription and translation, respectively (**b**). The RPL consisted of several NNN codons inserted between *Sfi*I restriction sites, followed by the encoding sequences for C-variant (Cv) RNA-associating protein (Cvap) and a protein spacer (Ps). Cvap is the gene for the dimer of mutant MS2 coat proteins of *Escherichia coli* bacteriophage encoding a Cv-associating protein, and Ps is the gene for the protein spacer derived from dehydrofolate reductase. The start codon ATG follows the 5′ constant region and is necessary for the initiation of translation.

2. Purify the dsDNA using a QIAquick PCR purification kit.

3. Digest the PCR product with *Sfi*I as follows. Prepare a reaction mixture containing 2 μg of dsDNA from step 2, 1 μL of 100 × BSA, and 10 μL of 10× NE buffer, 7 μL of *Sfi*I and make up to 100 μL with nuclease-free water. Incubate at 50°C overnight.

4. Purify the *Sfi*I-digested dsDNA using a QIAquick PCR purification kit.

5. Confirm the digestion on a 6% PAGE, 1× TAE gel, and stain the gel by incubation in ethidium bromide solution for 10 min. If the digestion is not complete, perform a further 6-h digestion with *Sfi*I and purify the product (see Note 10).

6. Determine the DNA concentration by UV absorbance at 260 nm.

7. Digest the SD-Cvap-Ps plasmid with *Sfi*I as described in steps 3–6 above.

8. Ligate more than 300 fmol of the dsDNA with 25 fmol of vector fragment (see Note 11) by combining these DNA samples and adding an equal volume of Mighty Mix DNA Ligation mix (or ligate using standard T4 DNA ligase conditions).

9. Incubate at 16°C overnight.

10. Amplify the random peptide-encoding DNA library construct from the ligated product by PCR (ten cycles of 10 s at 98°C, 15 s at 55°C and 30 s at 68°C) using a reaction mixture containing 5 μL of ligated template (5 fmol), 1 μL of fp-1 primer (10 pmol), 1 μL of rp-1 primer (10 pmol), 4 μL of dNTP mix, 5 μL of PrimeSTAR® GXL DNA Polymerase, and nuclease-free water up to 50 μL (see Note 12).

11. Purify the DNA using a QIAquick PCR product purification kit and determine the DNA concentration by UV absorbance at 260 nm (see Note 13).

3.2. In Vitro Transcription

1. Perform in vitro transcription with the dsDNA (prepared in Subheading 3.1) as a template by using T7 RiboMAX™ Large Scale RNA Production System. Use a nuclease-free tube to prepare a reaction mixture containing 2 μg of DNA template, 10 μL of enzyme mix, 20 μL of rNTP mix, 1 μL of RNasin® Ribonuclease inhibitor, and 20 μL of 5× T7 transcription buffer and make up to 100 μL with nuclease-free water.

2. Mix the reaction well using a pipette and incubate at 37°C for 3 h.

3. Add 5 μL of DNase (1 U/μL) to the reaction and incubate for another 30 min at 37°C.

4. Purify the mRNA using an RNeasy® Mini kit according to the manufacturer's manual.

5. Check the mRNA by electrophoresis on a 2% (w/v) denaturing agarose gel. Denature the mRNA samples in gel loading buffer prior to electrophoresis by incubating at 65°C for 15 min. Visualize the mRNA on the gel by staining with ethidium bromide.

6. Determine the concentration of mRNA by UV absorbance at 260 nm (see Note 14).

3.3. In Vitro Translation

Prepare the immobilized target for selection (Subheading 2.4 steps 1–4) prior to beginning in vitro translation.

1. Perform in vitro translation using the PURESYSTEM classic II kit (see Note 15). Thaw solutions A and B on ice and prepare a reaction mixture in a nuclease-free tube containing 25 μL of solution A, 10 μL of solution B, 4 pmol of mRNA template, and 1 μL of RNasin® Ribonuclease inhibitor and make up to 50 μL with nuclease-free water.

2. Mix gently and incubate the reaction at 37°C for 15 min.

3. Stop the translation reaction by transferring the tube to ice for 10 min.

3.4. Affinity Selection

1. Chill the selection buffer to 4°C.

2. Prepare the target of interest appropriately immobilized for selection. As an example, selection on PHA covalently immobilized on amine beads (Dynabeads M-270 Amine) is described. Add 100 μL of amine beads to a coupling reaction mix containing 150 μL of MES buffer, 50 μL of 1.8 mM PHA, 50 μL of 1.8 mM EDC, and 50 μL of 3.6 mM NHS.

3. Incubate the suspension at 25°C overnight.

4. Wash the beads twice with Tris–acetate buffer.

5. For amine coupling, determine the coupling ratio by a decrease in the amino groups of microbeads using the ninhydrin reaction. Mix 10 μL of beads and 90 μL of ninhydrin solution and incubate the mixture at 25°C for 1 h. Estimate the end of coupling reaction from the decrease in absorbance at 570 nm.

6. Mix the 50 μL translation reaction with prechilled selection buffer up to 300 μL and add this mixture to 5 μL of target-immobilized microbeads. Incubate the selection for 1 h at 4°C with gentle rotation.

7. Wash the beads with 200 μL of prechilled selection buffer at 4°C. Repeat the wash ten times. After the final wash, allow a small amount of buffer to remain on the beads to prevent them from drying out.

8. Add 100 μL of selection buffer containing 2.5 mM PHA (or target of interest) to elute the bound RPM. Incubate at

4°C for 30 min with gentle rotation. Separate the beads from the solution using a magnet and transfer the 100 μL elution solution to a fresh tube.

9. Add 100 μL of Tris–acetate EDTA to the recovered solution and incubate the mixture at room temperature for 15 min.

10. Purify mRNA using an RNeasy® Mini kit.

3.5. Reverse Transcription and PCR

1. Prepare a mixture containing 100 μL of purified mRNA template (2 μg), 40 μL of rp-1 primer (1 μM), and 2 μL of RNasin® Ribonuclease inhibitor in a nuclease-free tube.

2. Incubate the mixture at 70°C for 5 min and then immediately cool it on ice for 5 min.

3. Complete the RT reaction mixture by combining the above mixture and a solution containing 10 μL of dNTP mix, 40 μL of 5× PrimeScript buffer, and 8 μL of PrimeScript Reverse Transcriptase and make up to 200 μL with nuclease-free water. Gently pipette to mix and incubate at 50°C for 1 h.

4. In order to minimize by-products of PCR, optimize the number of PCR cycles using a reaction containing 10 μL of reverse transcription reaction, 1 μL of fp-1 primer (10 μM), 1 μL of rp-1 primer (10 μM), 4 μL of dNTP mix, 10 μL of 5× PrimeSTAR GXL buffer, 1 μL of PrimeSTAR GXL DNA Polymerase, and nuclease-free water up to 50 μL.

5. Perform PCR using the thermal cycling conditions described in Subheading 3.1, step 10, and remove 5 μL of the reaction mixture every two cycles from the tenth to the last cycle. Analyze the samples on a 5% PAGE, 1× TAE gel, to visualize the optimum number of cycles required for the production of a good-quality PCR product.

6. Amplify the remainder of the reverse transcription product using the optimized PCR conditions determined in steps 4 and 5.

7. Purify the DNA product using a QIAquick PCR purification kit.

8. Verify the quality of the PCR product on a 5% PAGE gel, 1× TAE, and measure its concentration by absorbance at 260 nm. The DNA is now ready for transcription (Subheading 3.2) for the next round of selection. Alternatively, the DNA population can be subcloned for DNA sequencing (Subheading 3.6) or can be subject to PCR to incorporate tags and a stop codon to allow analysis for binding (Subheading 3.7). This is illustrated schematically in Fig. 4.

3.6. Cloning and Sequencing

1. In order to analyze the selected DNA sequences, the selection output must be subcloned to produce individual clones for analysis. For this purpose, PCR amplify the DNA recovered from the selection (Subheading 3.5, step 3) using a reaction

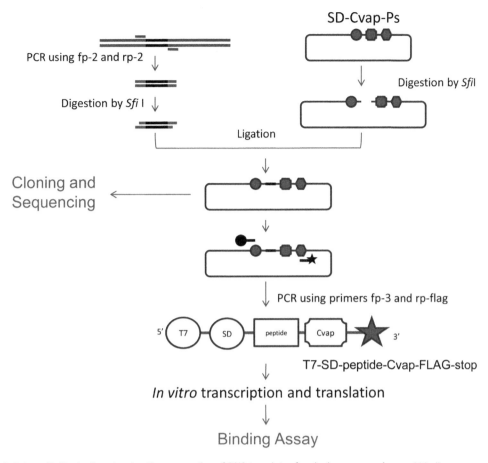

Fig. 4. Schematic illustration showing the preparation of DNA templates for cloning, sequencing, and binding assays.

containing 66 μL of recovered DNA (8 ng), 2 μL of fp-2 primer (10 μM), 2 μL of rp-2 primer (10 μM), 8 μL of dNTP mix, 20 μL of 5× PrimeSTAR® GXL buffer, and 2 μL of PrimeSTAR® GXL DNA Polymerase in nuclease-free water up to 100 μL. Use the thermal cycling conditions determined in Subheading 3.5, step 5.

2. Purify the DNA product using a QIAquick PCR purification kit. Verify the quality of the PCR product on a 5% PAGE, 1× TAE gel, and measure its concentration by absorbance at 260 nm.

3. Prepare two restriction digestion reactions containing DNA from step 2 and SD-Cvap-Ps plasmid. Prepare the former solution by mixing 82 μL of recovered DNA (2 μg), 10 μL of 10× NE buffer, 1 μL of 100× BSA buffer, and 7 μL of *Sfi*I and make up to 100 μL with nuclease-free water. The latter solution has the same composition as the former solution but containing 82 μL of plasmid (4 μg) instead of recovered DNA.

4. Incubate the reactions at 50°C overnight and purify the products using a QIAquick PCR purification kit. Verify the quality

of the DNA products on a 5% PAGE, 1× TAE gel, and measure its concentration by absorbance at 260 nm.

5. Insert the digested DNA fragments into the vector fragment by DNA ligation. Prepare a ligation reaction mixture containing 2.5 μL of vector (25 fmol), 1 μL of digested DNA fragments (250 fmol), and 3.5 μL of Ligase Mix from the Mighty Mix DNA Ligase kit.

6. Incubate at 16°C overnight and transform into competent *E. coli* cells according to manufacturer's instructions.

7. DNA sequence a representative number of clones for analysis.

3.7. Confirmation of Binding by Western Blot and ELISA

To confirm binding of the selected peptides to their target, a Western blot or an ELISA can be performed (or alternatively any suitable assay can be used). To avoid Cv–Cvap complex formation, DNA template lacking the Cv sequence must first be prepared. The DNA fragment encoding the selected peptide sequences is inserted between *Sfi*I sites in the original plasmid to form a Cv-selected peptide-Cvap-Ps construct. Subsequently, a DNA fragment encoding peptide-FLAG without the Cv sequence is obtained by PCR using the plasmid and the primers fp-3 and rp-flag, which introduce a FLAG tag and a stop codon for FLAG-tagged peptide expression (Fig. 4).

1. Prepare a reaction mixture containing 5 μL of ligated templates (5 fmol) from Subheading 3.6, 1 μL of fp-3 primer (10 pmol), 1 μL of rp-flag primer (10 pmol), 4 μL of dNTP mix, 10 μL of 5× PrimeSTAR® GXL buffer, 1 μL of PrimeSTAR® GXL DNA Polymerase, and nuclease-free water up to 50 μL.

2. Amplify the DNA templates from the ligated product by PCR (ten cycles of 10 s at 98°C, 15 s at 55°C, and 30 s at 68°C).

3. Purify the DNA using a QIAquick PCR product purification kit.

4. Transcribe and translate the DNA product in vitro as described in Subheadings 3.2 and 3.3, respectively.

5. For Western blotting, mix the translated product 1:1 with 2× SDS-PAGE loading buffer and incubate at 95°C for 5 min.

6. Fractionate on a suitable SDS-PAGE gel and transfer the products to a PVDF membrane in 1× blotting buffer according to equipment manufacturer's instructions.

7. Block the membrane with blocking buffer for 30 min.

8. Detect proteins with a 1/1,000 dilution of anti-FLAG M2 monoclonal antibody–peroxidase conjugate for 30 min.

9. Wash the membrane three times with TBS-T.

10. Visualize the protein band (37 kDa) using an ECL Plus Western Blotting Detection System with a suitable light capture instrument.

11. For ELISA, immobilize the target on a microtiter plate as described in Subheading 3.5.
12. Dilute 25 μL of the translation reaction up to a final volume of 100 μL with Tris–acetate buffer.
13. Add the solution to the target-immobilized plate and incubate for 1 h at 4°C.
14. Remove unbound polypeptide by washing with Tris–acetate buffer five times.
15. Add 100 μL of blocking buffer per well and incubate for 30 min at 4°C.
16. Remove the blocking buffer and add 100 μL of anti-Flag M2 monoclonal antibody–peroxidase diluted 1:4,000 in blocking buffer per well and incubate for 30 min at 4°C.
17. Wash five times with 0.01% (v/v) Tween in Tris–acetate buffer and then perform a chemiluminescence reaction using the ECL Plus Western Blotting Detection System and analyze by chemiluminescent imaging.

4. Notes

1. The peptide library is generated using random codons (11). NNN (where $N=A/T/G$ or C) produces random codons for unbiased access to all 20 amino acids for a translated library. However, NNN codons contain a higher frequency of stop codons than other codons, such as NNS and NNK. NNK (where $K=G/T$) are less frequently used than NNS (where $S=G/C$) in this system. The choice of codon usage for the in vitro translation system being used can also be important, for example it has been reported that NNS codons (where $S=G/C$) are the best random codons for unbiased access to all 20 amino acids for a library translated in rabbit reticulocyte lysate (from *Oryctolagus cuniculus*) (12). A biased random sequence can also be utilized, for example VVN (where $V=A/C/G$) for hydrophilic amino acids, and if more specific amino acids are desired, randomized units of three nucleotides are also commercially available (13).

2. The SD-Cvap-Ps plasmid vector is described in the literature (9) and is available from Nano Medical Engineering Laboratory, RIKEN Advanced Science Institute (http://www.riken.jp/engn/r-world/research/lab/wako/medical/index.html).

3. We use a commercially available water purification system capable of producing nuclease-free water for every experiment. Working with RNA requires several basic precautions, including

wearing gloves and using disposable materials as much as possible. Tips and tubes can be purchased that are RNase/DNase free and used without autoclaving.

4. Various selection targets can be employed. In addition to PHA, hemin as a low-molecular-weight target, calmodulin as a protein, and carbon nanotube and polystyrene as materials have been utilized as targets by us.

5. The method used for immobilization of target depends on the particular requirements of the target and the chosen matrix. Various immobilization methods have been reported (14). Generally, if the target molecule has amino groups, carboxylic acid, and thiol, it can be immobilized on activated ester-, aldehyde-, or epoxide-, amino-, and thiol-containing solid supports, respectively. If the target is solid, the target can be utilized without further modification. Here, since PHA has a caroboxic acid, it was coupled to amine-containing beads by coupling agents EDC and NHS.

6. Various types of solid support are available for target immobilization. Other covalent immobilization methods include the use of sepharose or cyanogen bromide-activated sepharose. The target can also be immobilized by noncovalent affinity methods, for example use of a biotinylated target and capture on neutravidin acrylamide, streptavidin agarose, or another biotin-binding matrix.

7. The selection buffer depends on the particular requirements of the target and expected functions of selected sequence. We have successfully used Tris–acetate buffer (50 mM Tris–acetate, 150 mM NaCl, 50 mM magnesium acetate, pH 7.5) as selection binding buffer, wash buffer, and elution buffer (selection buffer plus target).

8. Other blocking agents, for example nonfat milk powder or BSA, can be used instead of Block Ace.

9. By designing the insertion point of the original *Sfi*I restriction site at both ends of a library, it is possible to clone the library using a single restriction enzyme (8).

10. Digestion with *Sfi*I can be difficult and additional treatment is usually required to complete the digestion reaction. Digestion should be confirmed by PAGE showing a 24 bp size difference.

11. Assuming that the insertion efficiencies into the plasmids are 100%, the diversity is estimated to be about 10^{10} from the calculated amount of plasmid.

12. The optimal condition is the sample that yields the largest amount of PCR product as a single band with very little smearing present. We usually take a series of samples from the PCR products with different cycle numbers (e.g., 10, 12, and 14).

By performing PAGE on PCR products, the number of cycles is determined by the lane having a single obvious amplified band.

13. A yield of more than 1 μg of DNA is required for further experiments.
14. A yield of more than 2 μg of RNA is required for further experiments.
15. Translation in vitro is produced by combining recombinant *E. coli* protein factors and purified 70S ribosomes. Thus, the reconstituted systems enable a highly efficient, robust, and accessible prokaryotic ribosome display technology (15, 16).

References

1. Hanes J, Pluckthun (1997) In vitro selection and evolution of functional proteins by using ribosome display. Proc Natl Acad Sci USA 94:4937–4942.
2. He M, Taussig MJ (1997) Antibody-ribosome-mRNA (ARM) complexes as efficient selection particles for in vitro display and evolution of antibody combining sites. Nucleic Acids Res. 25:5132–5134.
3. Amstutz P, Forrer P, Zahnd C et al (2001) In vitro display technologies: novel developments and applications. Curr Opin Biotechnol 12:400–405.
4. Schaffitzel C, Berger I, Postberg J et al (2001) In vitro generated antibodies specific for telomeric guanine-quadruplex DNA react with Stylonychia lemnae macronuclei. Proc Natl Acad Sci USA 98:8572–8577.
5. Amstutz P, Pelletier JN, Guggisberg A et al (2002) In vitro selection for catalytic activity with ribosome display. J Am Chem Soc 124: 9396–9403.
6. Binz HK, Amstutz P, Kohl A et al (2004) High-affinity binders selected from designed ankyrin repeat protein libraries. Nat Biotechnol 22:575–582.
7. Ohashi H, Shimizu Y, Ying BW et al (2007) Efficient protein selection based on ribosome display system with purified components. Biochem Biophys Res Commun 352:270–276.
8. Wada A, Sawata SY, Ito Y. (2008) Ribosome display selection of a metal-binding motif from an artificial peptide library. Biotechnol Bioeng 101:1102–1107.
9. Sawata SY, Taira K (2003) Modified peptide selection in vitro by introduction of a protein–RNA interaction. Protein Eng 16:1115–1124.
10. Kurihara S, Tsumuraya T, Suzuki K (2000) Antibody-catalyzed removal of the p-nitrobenzyl ester protecting group: the molecular basis of broad substrate specificity. Chemistry 6:1656–1662.
11. Tanaka J, Doi N, Takahashima H (2010) Comparative characterization of random-sequence proteins consisting of 5, 12, 20 kinds of amino acids. Protein Sci 19:786–795.
12. Takahashi TT, Roberts RW (2009) In vitro selection of protein and peptide libraries using mRNA display. In: Meyer G (Ed) Nucleic Acid and Peptide Aptamers: Methods and Protocols, **535**:293–314.
13. Kayushin AL, Korosteleva MD, Miroshnikov AI (1997) A convenient approach to the synthesis of trinucleotide phosphoramidites-synthons for the generation of oligonucleotide/peptide libraries. Nucleic Acids Res 24:3748–3755.
14. Rusmini F, Zhong Z, Feijen J (2007) Protein immobilization strategies for protein biochips. Biomacromolecules 8:1775–1789.
15. Villemagne D, Jackson R, Douthwaite JA (2006) Highly efficient ribosome display selection by use of purified components for in vitro selection. J Immunol Meth 313:140–148.
16. Osada E, Shimizu Y, Akbar BK et al (2009) Epitope mapping using ribosome display in a reconstituted cell-free protein synthesis system. J Biochem 145:693–700.

Chapter 5

Eukaryotic Ribosome Display with In Situ DNA Recovery

Mingyue He, Bryan M. Edwards, Damjana Kastelic, and Michael J. Taussig

Abstract

Ribosome display is a cell-free display technology for in vitro selection and optimisation of proteins from large diversified libraries. It operates through the formation of stable *p*rotein-*r*ibosome-*m*RNA (PRM) complexes and selection of ligand-binding proteins, followed by DNA recovery from the selected genetic information. Both prokaryotic and eukaryotic ribosome display systems have been developed. In this chapter, we describe the eukaryotic rabbit reticulocyte method in which a distinct in situ single-primer RT-PCR procedure is used to recover DNA from the selected PRM complexes without the need for prior disruption of the ribosome.

Key words: Ribosome display, In situ RT-PCR, Single-chain antibody, Cell-free expression

1. Introduction

Ribosome display produces stable *p*rotein-*r*ibosome-*m*RNA (PRM) complexes in a cell-free system for selection of required proteins expressed from large DNA libraries (1–3). The PRM complex is formed by deleting the stop codon, which causes stalling of the translating ribosome at the end of mRNA together with the non-released nascent protein (1–3). Through the interaction of ribosome-associated polypeptides with an immobilised ligand, specific PRM complexes are captured, from which the ribosome-attached mRNA is recovered as DNA by RT-PCR for subsequent manipulations. This process can be repeated to enrich ligand-specific binding molecules from a very large library. A major advantage of ribosome display over cell-based display methods is that it directly screens larger PCR-generated libraries without the need for DNA cloning. In combination with PCR-based mutagenesis, this feature allows continuous introduction of new diversity into the selected DNA pool between display cycles, providing a particularly efficient tool for in vitro molecular evolution of proteins.

Ribosome display has been widely applied to select different proteins, including antibody fragments, peptides, scaffolds, novel tags, enzymes, DNA-binding proteins, receptors, membrane proteins, and vaccine candidates (2). Both prokaryotic and eukaryotic cell-free systems have been developed for ribosome display, each with its own DNA recovery protocol and modifications (1, 4). A "pure" cell-free system composed of purified components and enzymes has also been adapted for display of proteins (5). In this chapter, we focus on the eukaryotic rabbit reticulocyte lysate system using single-chain antibody fragment (scFv) as the example. This method has a distinctive feature of using an in situ single primer RT-PCR procedure to recover DNA from PRM complexes without the need for dissociation of ribosome complexes (1–3) (Fig. 1).

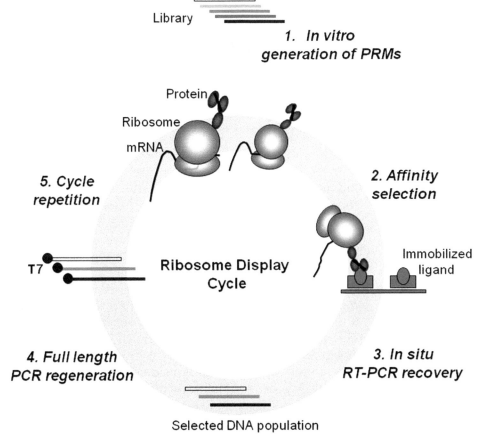

Fig. 1. The ribosome display cycle. The cycle is comprised of (1) cell-free expression and generation of PRM complexes from a library, (2) affinity selection of binders on immobilised ligand, (3) in situ RT-PCR recovery of the selected genetic information and (4) regeneration of the full-length PCR construct for (5) the subsequent selection cycle.

2. Materials

All solutions, tubes and tips used must be sterilised. Reagents should be nuclease-free. Precautions should be taken to avoid any contamination. It is recommended that primers, RT-PCR reagents and dNTP solutions are stored at −20°C in aliquots.

2.1. Molecular Biology Reagents and Kits

1. The primers used for PCR and RT-PCR are synthesised by Sigma (Table 1).
2. GenElute™ Gel Extraction Kit (Sigma).
3. Rabbit Reticulocyte TNT T7 Quick for PCR DNA (Promega).
4. Taq DNA polymerase (Qiagen).
5. 25 mM dNTPs: Mix equal volumes of each 100 mM dNTP stock solution (Sigma).
6. SuperScript™ III Reverse Transcriptase (100 mM DTT and first-strand buffer included) (Invitrogen).
7. RNase-free DNase I (Boehringer Mannheim).
8. RNase inhibitor (Ambion).
9. Agarose.
10. Glutathione (Sigma, reduced form and oxidised form).
11. 5× gel loading buffer (40% w/v sucrose, 0.25% bromophenol blue).

Table 1
Primers for PCR and DNA recovery

Primer	Sequence (from 5′ to 3′)
1. N-Ab/B	GGAACAG*ACCACC*ATGSARGTNSARCTBGWRSAGTCYGG
2. scFv-link/F[a]	GCTACCGCCACCCTCGAGAGATGGTGCAGCCACAG
3. Link-Cκ/B	CTCGAGGGTGGCGGTAGCACTGTGGCTGCACCATCTGTC
4. Cκ/F	GCACTCTCCCCTGTTGAAGCT
5. T7A1/B	GCAGCTAATACGACTCACTATAGGGAACAGACCACCATG
6. RTKz1	GAACAGACCACCATGACTTCGCAGGCGTAGAC
7. Kz1	GAACAGACCACCATG
8. Ck-f/F	GCACTCTCCCCTGTTGAAGCTCTTTGTGACGGGCGAGCTCAGGCCCTGATGGGTGACTTCGCAGGCGTAGACTTTG′

Underlined are the overlapping sequences for PCR assembly. B = G + T, Y = C + T, N = A + C + T + G, R = A + G, S = G + C, W = A + T

[a]This primer is designed to anneal at 5′ end of human Cκ region when antibody format scFv-Cκ is used as the template (3)

12. TopYield Strips (NUNC).
13. Nuclease-free water.
14. BSA.
15. Magnesium acetate (MgAc).

2.2. Solutions

1. Single-primer RT-PCR *Solution 1* (per 12 μl).

Primer RTKz1 (8 μM)	1 μl
dNTPs (10 mM)	2 μl
dH$_2$O	9 μl

2. Single-primer RT-PCR *Solution 2* (per 8 μl)

5× First-strand buffer	4 μl
100 mM DTT	1 μl
RNase inhibitor (20 U)	1 μl
SuperScript III (200 U)	1 μl
dH$_2$O	1 μl

3. Blocking buffer: 1% BSA and 0.05% Tween 20 in PBS (Phosphate-buffered saline pH 7.4).
4. Antigen solution (1–100 μg/ml) in PBS.
5. 100 mM magnesium acetate (MgAc).
6. Washing buffer: PBS containing 0.05% Tween 20 and 5 mM MgAc, stored at 4°C.
7. 2× dilution buffer: 4 mM Glutathione (GSSG: GSH = 1:1) and 10 mM MgAc, stored at 4°C.
8. 10× DNase I digestion buffer: 400 mM Tris–HCl, pH 7.5, 60 mM MgCl$_2$, 100 mM NaCl. Autoclaved and stored at 4°C.

3. Methods

Figure 1 presents the ribosome display cycle comprising the key steps. The following sections describe the use of ribosome display for selection of single-chain antibodies (scFvs).

3.1. Construction of the PCR DNA Template

The PCR DNA construct contains a T7 promoter and a Kozak sequence upstream of the gene of interest for synthesis of proteins in a rabbit reticulocyte lysate. A spacer domain is placed at the C-terminus of the gene for both displaying the protein on the surface of ribosome and providing a known priming site for RT-PCR recovery after selection (1–3). The stop codon at the 3′

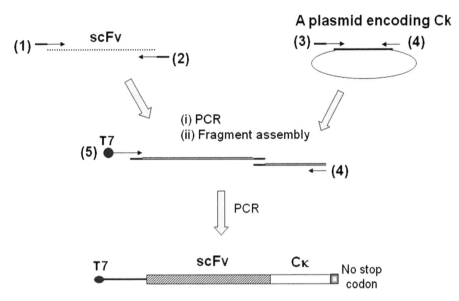

Fig. 2. Construction of scFv-Cκ fragment(s) by PCR for ribosome display. The numbers in the brackets indicate the primers used for PCR (see Table 1). *T7* T7 promoter, *scFv* single-chain antibody fragment, *Cκ* the constant region of κ chain.

end of mRNA is removed by using a primer lacking a stop codon in the final PCR assembly. Figure 2 shows a generalised strategy to make PCR construct(s) for eukaryotic ribosome display. For display of scFv fragments, we used a three-domain format of scFv-Cκ, where Cκ is used as the spacer (3) (see Note 1).

1. Generate scFv fragment(s) by PCR (see Note 2).

   Set up PCR mixture as follows:

10× PCR buffer (supplied with Taq)	2.5 μl
5× Q solution (supplied with Taq)	5 μl
dNTPs (2.5 mM)	2 μl
Primer N-Ab/B (16 μM)	0.8 μl
Primer scFv-link/F (16 μM)	0.8 μl
scFv DNA template(s) (10 ng/μl) (see Note 2)	1 μl
Taq polymerase	1 U
dH$_2$O	to 25 μl

   Carry out 30 thermal cycles: 94°C 30 s; 54°C 1 min; 72°C 1 min. Then 72°C for 7 min. Finally, hold at 10°C.

2. Generate the Cκ domain by PCR, using primers Link-Cκ/B and Cκ/F (Table 1) on a plasmid template encoding the κ light chain (1, 5), applying the PCR protocol as above.

3. Analyse the products by agarose (1%) gel electrophoresis. Extract the amplified scFv fragment(s) (~700 bp) and the Cκ domain (~300 bp) using the gel extraction kit.

4. Assemble equal amounts of scFv and Cκ DNA fragments, 10–50 ng DNA in total, without the presence of primers:

scFv PCR fragment	x μl
Cκ fragment	y μl
10× PCR buffer	2.5 μl
5× Q solution	5 μl
dNTPs (2.5 mM)	1 μl
Taq polymerase	1 U
dH$_2$O	to 25 μl

Carry out eight thermal cycles: 94°C 30 s; 54°C 1 min; 72°C 1 min. Finally, hold at 10°C.

5. Amplify the assembled template from step 4:

The above assembled PCR product	1–2 μl
10× PCR buffer	5 μl
5× Q solution	10 μl
dNTPs (2.5 mM)	4 μl
Primer T7A1/B (16 μM)	1.5 μl
Primer Cκ/F (16 μM)	1.5 μl
Taq polymerase	1–2.5 U
dH$_2$O to	50 μl

Carry out 30 thermal cycles: 94°C 30 s; 54°C 1 min; 72°C 1 min. Then 72°C for 7 min. Finally, hold at 10°C.

6. Analyse the PCR product (scFv-Cκ, size 1 kb) by agarose gel electrophoresis (see Note 3).

3.2. Preparation of Immobilised Antigens

Immobilised antigen for selection of specific PRM complexes is prepared by coating the antigen onto a TopYield well as follows (see Note 4):

1. Add 20 μl protein (at 1–100 μg/ml in PBS, pH 7–8.5) to each well of TopYield Strips and incubate at 4°C overnight.

2. Remove the solution and block the well with 200 μl 1% BSA in PBS for 1 h at RT.

3. Wash three times with PBS and the well can be used directly or stored at 4°C for 2 weeks. Wash wells briefly with ice-cold washing buffer before use.

3.3. Ribosome Display and Selection

To generate PRM complexes for selection, PCR DNA construct (in the format of scFv-Cκ, see Subheading 3.1) is directly added into a coupled rabbit reticulocyte lysate (TNT) system. Typically, for display of an antibody library, 0.5–1 μg of PCR DNA (~0.5–1 × 10^{12} molecules) is used in a standard 50 μl reaction.

1. Set up the TNT rabbit reticulocyte cell-free system to generate PRM complexes:

TNT T7 Quick for PCR	40 μl (see Note 5)
PCR DNA fragment(s)	0.5–1 μg
Methionine (1 mM) (from TNT kit)	1 μl
MgAc (100 mM)	1 μl (see Note 6)
Distilled H_2O	to 50 μl

   Incubate at 30°C for 60 min.

2. Remove the input PCR DNA fragment by adding 120 U RNase-free DNase I together with 7 μl 10× DNase I digestion buffer and H_2O to final 70 μl. Incubate at 30°C for a further 20 min (see Note 7).

3. Dilute with 70 μl of ice-cold 2× dilution buffer.

4. Add 50–140 μl of the TNT mixture containing the generated PRM complexes to an antigen-coated well and incubate at 4°C for 2 h with a gentle shaking.

5. Wash the wells five times with 200 μl ice-cold washing buffer, followed by two quick washes with 100 μl ice-cold sterilised H_2O (see Note 8). The wells carrying selected PRM complexes are either used directly for RT-PCR or stored at –20°C.

3.4. In Situ RT-PCR Recovery

After selection, in situ RT-PCR recovery is performed using a single-primer procedure (Fig. 3). In this method, a novel internal primer RTKz1 is designed to contain both a sequence for hybridising to the upstream region of 3′ mRNA (to avoid the stalling ribosome) and a sequence identical to the 5′ region of mRNA (Table 1). cDNA synthesis using RTKz1 leads to the generation of single-stranded cDNAs with a complementary flanking sequence at both 5′ and 3′ ends, which can be effectively amplified using a single primer Kz1 (Fig. 3).

1. Set up reverse transcription reaction by adding 12 μl *Solution 1* to each PRM-bound well. Incubate at 65°C for 5 min; then quickly place on ice for at least 30 s.

2. Add 8 μl of *Solution 2* containing SuperScript III and incubate the mixture at 50°C for 45 min followed by 5 min at 85°C. Transfer the RT mixture to a fresh tube for subsequent single-primer PCR (see Note 9).

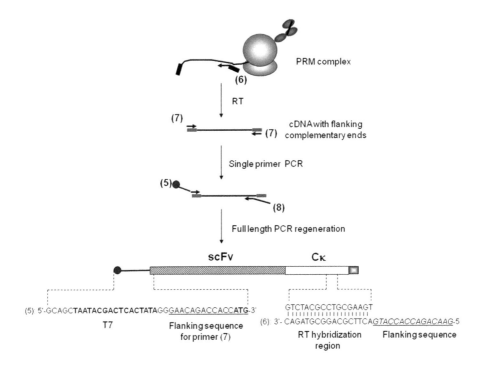

Fig. 3. In situ single primer RT-PCR recovery and the full length DNA regeneration. The *numbers in the brackets* indicate the primers used (see Table 1). The RT hybridising region and the flanking sequence are indicated. *RT* reverse transcription, *T7* T7 promoter, *scFv* single-chain antibody fragment, *Cκ* the constant region of κ chain.

3. Set up single-primer PCR mixture as follows (see Note 10):

10× PCR buffer	2.5 µl
5× Q solution	5 µl
dNTPs (2.5 mM)	2 µl
Primer Kz1 (16 µM)	1.5 µl
Taq DNA polymerase	1 U
cDNA from step 2	0.1–0.5 µl
dH$_2$O	to 25 µl

Carry out 30–35 cycles of thermal cycling as follows: 94°C 30 s, 48°C 1 min, 72°C 1 min; then 72°C for 7 min. Finally hold at 10°C.

4. Analyse the PCR by loading 5 µl of the sample onto a 1% agarose gel containing 0.5 µg/ml ethidium bromide. The PCR fragment can be used either for generation of the full length construct (see below) or for cloning and expression in *E. coli*.

3.5. Regeneration of the Full-Length Construct

The use of an internal primer RTKz1 in the in situ RT-PCR recovery leads to shortening of the DNA fragment compared to the original fragment; therefore, a further PCR step is required to regenerate the full-length construct using a long 3′ primer Ck-f/F in combination with the 5′ primer T7A1/B (Fig. 3).

1. Set up PCR mixture as follows:

10× PCR buffer	5 μl
5× Q solution	10 μl
dNTPs (2.5 mM)	4 μl
Primer T7A1/B (16 μM)	1.5 μl
Primer Ck-f/F (16 μM)	1.5 μl
Taq DNA polymerase	2 U
PCR template from Subheading 3.4	1–10 ng
dH$_2$O	to 50 μl

Carry out 30 thermal cycles: 94°C 30 s, 54°C 1 min, 72°C 1 min; then, 72°C for 7 min. Finally hold at 10°C.

2. Analyse the PCR by loading 5 μl of the sample onto a 1% agarose gel containing 0.5 μg/ml ethidium bromide. The full-length PCR product (~1 kb) can be used directly for subsequent ribosome display cycles (see Note 11).

4. Notes

1. In addition to the Cκ domain described here, a number of different spacers have been employed, including gene III of filamentous phage M13, the CH$_3$ domain of human IgM, streptavidin and GST (2).

2. The choice of a Taq polymerase depends on the downstream application. Any PCR Kits from different companies can be used in this step according to the manufacturer's instructions. scFv DNA template(s) for the PCR can originate from either a single construct or various libraries.

3. A clean PCR fragment of the expected size indicates the successful construction. To confirm the construct, direct DNA sequencing of the PCR product is performed.

4. Alternatively, magnetic beads (Dynabeads M-280 streptavidin (Dynal UK; 6.5×10^8/ml or 10 mg/ml)) can be used for antigen immobilisation (1).

5. This system can be carried out in the range between 10 and 250 μl of TNT without detectable reduction in recovery efficiency. Besides using the coupled system, the uncoupled rabbit reticulocyte lysate (Promega) can also be employed using mRNA template(s) generated by a separate in vitro transcription step.

6. MgAc concentration added in the TNT mixture during translation has an influence on the generation and recovery of PRM complexes. We have shown that antibodies can be more efficiently displayed by adding MgAc in the concentration ranging between 0.5 and 2 mM (6).

7. It is important to avoid any input DNA contamination during the RT-PCR step, although the single primer system described has been designed not to amplify the input DNA construct (1).

8. We have noticed that quick washes with water produced cleaner RT-PCR without affecting the efficiency of DNA recovery.

9. SuperScript II reverse transcriptase (Invitrogen) can be used as an alternative to SuperScript III. In this case, the following procedure should be used:

    (a) Add 12 µl *Solution 1* to each PRM-bound well. Incubate at 65°C for 5 min; then chill on ice for at least 30 s.

    (b) Add 8 µl of *Solution 3* (containing 4 µl 5× first-strand buffer, 1 µl 100 mM DTT, 1 µl RNase Inhibitor (20 U), 1 µl SuperScript II (200 U) and 1 µl dH$_2$O) and incubate the mixture at 42°C for 45 min followed by 5 min at 85°C. Transfer the RT mixture to a fresh tube for subsequent single-primer PCR.

10. The PCR volume can be scaled down to 10 µl. Negative controls lacking a template should always be included in every PCR experiment to assess possible contamination. A real-time PCR can be performed at this step to analyse the amount of cDNA recovered from ligand-selected PRM complexes. By comparing the DNA recovery between ligand and control wells, it is possible to validate the selected population and thus decide whether or not to repeat the cycles prior to DNA cloning and expression in *E. coli*.

11. The number of ribosome display cycles required to enrich the ligand-binding molecules depends on the nature of the ligand, as well as the quality and diversity of the library used. Generally, three to five display cycles are sufficient to enrich the required protein from a library.

Acknowledgements

The Babraham Institute is an institute of the Biotechnology and Biological Sciences Research Council (BBSRC), UK. Damjana Kastelic was supported by an exchange grant from the ESF Frontiers of Functional Genomics Research Networking Programme and Slovenian Research Agency (ARRS); Research Programme P1-0104. We thank Ms Hong Liu for technical assistance.

References

1. He, M and Taussig, M (2007) Eukaryotic ribosome display with *in situ* DNA recovery *Nature Methods* **4**, 281–288.
2. He, M and Khan, F (2005) (Review) Ribosome display: next-generation display technologies for production of antibodies *in vitro*. *Expert Rev. Proteomics* **2**, 421–430.
3. He, M and Taussig, M (1997) Antibody-ribosome-mRNA (ARM) complexes as efficient selection particles for *in vitro* display and evolution of antibody combining sites. *Nucleic Acids Res.* **25**, 5132–5134.
4. Zahnd, C, Amstutz, P, Plückthun, A (2007) Ribosome display: selecting and evolving proteins *in vitro* that specifically binds to a target. *Nature Methods* **4**, 269–279.
5. Ohashi, H, Shimizu, Y, Ying, BW, Ueda, T (2007) Efficient protein selection based on ribosome display system with purified components. *Biochem. Biophys. Res. Commun.* **352**, 270–276.
6. He, M and Taussig, M.J (2005) Ribosome display of antibodies: expression, specificity and recovery in a eukaryotic system. *J. Immunol. Methods* **297**, 73–82.

Chapter 6

mRNA Display Using Covalent Coupling of mRNA to Translated Proteins

Rong Wang, Steve W. Cotten, and Rihe Liu

Abstract

mRNA display is a powerful technique that allows for covalent coupling of a translated protein with its coding mRNA. The resulting conjugation between genotype and phenotype can be used for the efficient selection and identification of peptides or proteins with desired properties from an mRNA-displayed peptide or protein library with high diversity. This protocol outlines the principle of mRNA display and the detailed procedures for the synthesis of mRNA–protein fusions. Some special considerations for library construction, generation, and purification are discussed.

Key words: mRNA display, Covalent coupling of mRNA and protein, Genotype–phenotype conjugation, In vitro protein selection, High diversity

1. Introduction

The ability to rapidly identify and select proteins with desired properties from synthetic polypeptide or natural proteome libraries has become increasingly useful in biological and biomedical studies. A number of approaches, such as phage display, ribosome display, and yeast two-hybrid, have been developed to address the challenge (1–4). Phage display is a widely used method to isolate peptide sequences with desired functions, often from a short combinatorial peptide library (3). Yeast two-hybrid is often used to isolate interacting protein sequences of a target protein from natural cDNA libraries (4). Ribosomal display is another powerful genotype–phenotype conjugation method that allows for the selection of polypeptide sequences with desired properties from a highly diversified polypeptide library displayed on the ribosome (2).

mRNA display is an in vitro selection technique that allows for the identification of polypeptide sequences with desired properties from both a natural proteome library and a synthetic combinatorial peptide library (5–9). The central feature of this method is that the polypeptide chain is covalently linked to the 3′ end of its own mRNA. This is accomplished by synthesis and in vitro translation of an mRNA template with puromycin attached to its 3′ end via a short oligo linker. During in vitro translation, when the ribosome reaches the RNA–oligo junction and translation pauses, puromycin, an antibiotic that mimics the aminoacyl moiety of tRNA, enters the ribosome 'A' site and accepts the nascent polypeptide chain by forming a peptide bond. This results in tethering the nascent polypeptide to its own mRNA (Fig. 1). When the initial mRNAs are composed of many different sequences, the corresponding protein or proteome library is generated. Since the genotype coding sequence and the phenotype polypeptide sequence are covalently combined within the same molecule; the selected protein can be revealed by DNA sequencing after reverse transcription and PCR amplification. Therefore, mRNA display provides a powerful means for reading and amplifying a peptide or protein sequence after it has been functionally isolated from a library with high diversity. Multiple rounds of selection and amplification can be performed, enabling enrichment of rare sequences with desired properties. Compared to prior peptide or protein selection methods, mRNA display has several major advantages. First, the genotype is covalently linked to and is always present with the phenotype. This stable linkage makes it possible to use any arbitrary and stringent conditions in the functional selection. Second, unlike cell-based systems, such

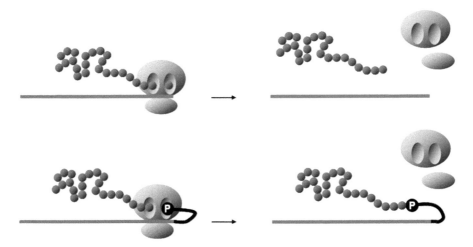

Fig. 1. The formation of mRNA–protein fusion. Without puromycin-containing oligo linker (*the black line*), mRNA (*gray line*) and newly synthesized polypeptide (*dotted chain*) are separated from each other (*top*). Puromycin (P), which mimics the aminoacyl-tRNA, can enter the ribosome 'A' site and be incorporated into the nascent polypeptide chain as the last amino acid, mediating the covalent conjugation between mRNA and its coded polypeptide (*bottom*).

as yeast two-hybrid or phage display that are limited by the transformation efficiency, the complexity of the peptide or protein library that is allowed by using cell-free system can be close to that of the mRNA or cDNA pools. The reaction scale is tunable, typically from microliters to milliliters. Peptide or protein libraries containing as many as 10^{12}–10^{14} unique sequences can be readily generated and selected, a few orders of magnitude higher than that can be achieved using phage display or other peptide/protein selection platforms. Therefore, both the likelihood of isolating rare sequences and the diversity of the sequences isolated in a given selection are significantly increased.

The generation of mRNA–protein fusion molecules using mRNA display consists of the following steps: library construction and amplification, in vitro transcription, DNase digestion, conjugation with puromycin oligo linker, in vitro translation/fusion formation, oligo(dT) mRNA purification, reverse transcription, and protein affinity purification. Specifically, a cDNA library is first in vitro transcribed to generate mRNAs using T7, T3, or SP6 RNA polymerase. The resulting mRNA templates are modified by covalently linking to a short oligo linker containing a puromycin at the 3′ ends. Such a linkage can be achieved by photo-cross-linking, splint ligation, or Y ligation (8, 10–12). Creation of the mRNA–protein fusion is accomplished by in vitro translation in a cell-free system using rabbit reticulocyte lysate that has low nuclease activity. Efficient mRNA–protein fusion formation can be accomplished through a post-translational incubation with high concentrations of Mg^{2+} and K^+ (8). mRNA templates and mRNA–protein fusion molecules can be readily purified from the lysate by using an oligo(dT) column, taking advantage of the oligo(dA) residues in the puromycin-containing oligo linker. To remove the secondary structures of mRNAs that may interfere with the selection step, the fusion molecules are often converted to DNA/RNA hybrids by reverse transcription. The resulting mRNA-displayed protein library is then purified on the basis of the affinity tags engineered at the N- and C-termini and used for subsequent selection. We describe here the general procedures for the generation of mRNA–protein fusion molecules that can be used for the selection of peptide or protein sequences with desired properties. The procedures for the functional selection using mRNA-displayed peptide or protein libraries are detailed in Chapter 16.

2. Materials

2.1. Reagents

1. Expand long template PCR system (Roche).
2. T7 RNA polymerase (NEB).

3. 10× RNA polymerase buffer: 400 mM Tris-HCl, pH 7.9, 60 mM MgCl$_2$, 100 mM DTT, 20 mM spermidine.
4. RNase-free DNase (Promega).
5. 10× DNase buffer: 400 mM Tris–HCl, pH 8.0, 100 mM MgSO$_4$, 10 mM CaCl$_2$.
6. Acidic phenol chloroform: Isoamyl alcohol (Ambion).
7. 7.5 M LiCl RNA precipitation solution (Ambion).
8. Puromycin oligo linker: 5′-Psoralen-(TAGCCGGTG)$_2$′$_{OMe}$-dA$_{15}$ C9C9dAdCdC-Puromycin-3′.
9. Retic lysate IVT™ kit (Ambion).
10. [^{35}S]-L-methionine (Perkin Elmer).
11. Oligo(dT) cellulose (Ambion).
12. 1× Oligo(dT) binding buffer: 100 mM Tris–HCl, pH 8.0, 1 M NaCl, 10 mM EDTA, 0.2% Triton X-100.
13. Oligo(dT) wash buffer: 20 mM Tris–HCl, pH 8.0, 300 mM KCl, 0.1% Tween-20.
14. RNase-free 10 ml poly-prep chromatography column (Biorad).
15. SuperScript II RNase H$^-$ reverse transcriptase (Invitrogen).
16. Reverse transcription primer: TTTTTTTTTTNNCCAGATCCAGACATTCCCAT.
17. Anti-FLAG M2 affinity gel (Sigma).
18. FLAG peptide (Sigma).
19. 100 mM glycine buffer, pH 3.5.
20. 1× TBST buffer: 20 mM Tris–HCl, pH 8.0, 150 mM NaCl, 0.2% Tween-20.
21. 1× TE buffer: 10 mM Tris–HCl, pH 8.0, 1 mM EDTA.
22. NAP-5 column (GE Healthcare).
23. NAP-10 column (GE Healthcare).
24. QIAquick PCR purification kit (Qiagen).
25. QIAquick gel extraction kit (Qiagen).

2.2. Equipment

1. Thermal cycler.
2. Nanodrop spectrophotometer.
3. UV lamp (Black Ray Lamp 365 nm, 0.16 Amps).
4. Barnstead labquake shaker/rotator.
5. Scintillation counter.

3. Methods

3.1. General Design of the cDNA Library

Both natural cDNA and synthetic cDNA libraries can be used for mRNA display-based selections (13, 14). In general, the variable region is flanked with two consensus regions at the 5′ and 3′ ends, respectively. The 5′ consensus region contains a transcription promoter and a short UTR that facilitate efficient in vitro transcription and translation, respectively. Depending on the choice of RNA polymerase used for in vitro transcription, T7, T3, or SP6 promoter can be used. A short 5′-UTR originated from tobacco mosaic virus (TMV) results in efficient in vitro translation when rabbit reticulocyte lysate is used. If T7 RNA polymerase is chosen, the 5′UTR should begin with GGG to facilitate transcription initiation. A general 5′ consensus sequence upstream of the start codon has the following sequence: TTC <u>TAA TAC GAC TCA CTA TAG</u> GGA **CAA TTA CTA TTT ACA ATT ACA**, in which the T7 promoter is underlined and the TMV 5′-UTR is bolded. If necessary, a sequence encoding a 5-amino-acid recognition site (RRASV) by protein kinase A (PKA) can be incorporated for universal radiolabeling with ^{32}P. The right consensus region contains a short sequence for hybridizing and cross-linking with the puromycin-containing oligo linker (8, 10). In addition, various affinity tags, such as His×6, E, HA, and FLAG tags, can be incorporated at the N- and/or C-termini to facilitate affinity purification of mRNA–protein fusion molecules from in vitro translation reaction mixture.

Because mRNA is covalently linked with the C-terminus of the translated polypeptide, several flexible and hydrophilic amino acid residues, such as Gly or Ser, are often engineered at the very C-terminal region of the protein sequence to minimize the possible rigid structure and steric hindrance. Some special sequences that facilitate subsequent selections are optional, as discussed in Chapter 16 that focuses on functional selections from mRNA-displayed peptide or protein libraries.

3.2. Amplification of the Initial Library

With the availability of a high-quality cDNA library that is compatible for mRNA display, the initial library should be amplified by PCR. One critical issue is to avoid overamplification that could result in preferential enrichment of some sequences before the selection. The PCR conditions should be optimized prior to large-scale amplification, particularly the annealing temperature, number of cycles, concentrations of cDNA template, primers, Mg^{2+}, and dNTPs.

1. Set up reactions of less than 100 µl for optimization.
2. Run a hot-start PCR program with desired parameters. Save 5–10 µl samples every 2–3 cycles.
3. Run 1–2% agarose gel to compare the quality and quantity of each PCR product.

The best PCR conditions should allow for efficient amplification of the initial library with an amplification factor close to 2 each cycle. The number of PCR cycles should be chosen so that the amount of desired full-length products is maximal, whereas that of undesired shorter by-products is minimal. The optimized PCR conditions are applied to the large-scale amplification of the whole library. The amplified cDNA library should be cleaned up by using a PCR purification kit to remove PCR primers and salts. Further ethanol precipitation is recommended. The final cDNA library is dissolved in a buffer that contains 10 mM Tris–HCl, pH 8.0, and 75 mM NaCl and stored at −20°C before use.

3.3. Generation of mRNAs by In Vitro Transcription

The next step for the generation of mRNA–protein fusion molecules is the synthesis of large quantities of mRNAs by in vitro transcription. The following procedures describe the synthesis and purification of mRNAs corresponding to the initial cDNA library using T7 RNA polymerase. Typically, the concentration of cDNA library used for in vitro transcription is around 200 nM. The scale of in vitro transcription is tunable from 0.25 to 10 ml, depending on the diversity of the initial library and the average copy number of each unique sequence. Typically, 0.5–1 ml of in vitro transcription is performed for the first round of selection from natural proteome libraries, whereas 3–10 ml reaction is required for the first round of selection from synthetic peptide libraries with very high diversities (14, 15).

1. Assemble an in vitro transcription reaction mixture in a nuclease-free Eppendorf tube on ice that contains 1× T7 RNA polymerase buffer, 25 mM $MgCl_2$, 5 mM each rNTP, 200 nM initial cDNA library, and 3 U/μl T7 RNA polymerase. Nuclease-free DEPC water should be used to obtain the desired final volume (see Note 1).

2. Incubate the reaction mixture at 37°C for 6–12 h. The reaction mixture becomes opaque if in vitro transcription is successful, presumably due to the generation of magnesium pyrophosphate as by-product.

3. Add EDTA to a final concentration of 30 mM (1.2 equivalent of (Mg^{2+})) and incubate at room temperature for 5 min to dissolve the white precipitate. The solution should become clear after such treatment.

4. Extract the reaction mixture with 1 volume of acidic phenol/chloroform followed by 1 volume of chloroform.

5. Desalt the product using NAP-5 column and collect the product in 1 ml nuclease-free DEPC water. Concentrate the sample to 400 μl in a Speedvac.

6. Load 1–5 μl of product onto a denaturing PAGE gel to check the quality of purified mRNAs (see Note 2).

3.4. Removal of cDNAs from mRNAs by DNase Digestion

Due to the amplification nature of mRNA display-based selection, the presence of even trace amount of cDNAs in the selected pool could result in enrichment of sequences that do not possess desired properties. Therefore, the cDNA templates in the transcription reaction mixture should be completely removed by DNase digestion using the following procedures.

1. Assemble a 500 µl reaction in a nuclease-free Eppendorf tube that contains 1× DNase buffer, 400 µl desalted in vitro transcription product from Subheading 3.3, and 0.05 U/µl RNase-free DNase.
2. Incubate at 37°C for 30 min to perform DNase digestion.
3. Add EDTA to a final concentration of 13.2 mM (1.2 equivalents of (Mg^{2+})) to stop the reaction.
4. Extract the reaction mixture with 1 volume of acid phenol/chloroform followed by 1 volume of chloroform.
5. Desalt using NAP-5 column and collect the product in 1 ml nuclease-free DEPC water.
6. Add 0.5 volume of 7.5 M LiCl RNA precipitate solution and incubate at −80°C for 2 h.
7. Spin at $16,800 \times g$ for 15 min at 4°C, and wash the pellet with 500 µl 75% ethanol twice.
8. Allow the pellet to air dry until no ethanol is detectable. Resuspend the purified mRNAs in 150 µl nuclease-free DEPC water.
9. Measure OD using Nanodrop, and calculate the molar concentration of mRNAs based on the (average) length of cDNA library. Load 1–5 µl of product onto a denaturing PAGE gel to check the quality of purified mRNAs (see Note 2).

3.5. Conjugation with Puromycin-Containing Oligo Linker

The mRNA templates with puromycin at 3′ ends can be generated by conjugating with an oligonucleotide containing a puromycin residue at the 3′ end. Various conjugation methods can be used, including UV cross-linking and enzymatic ligation (8, 10, 16). In general, psoralen-mediated UV cross-linking is much simpler to perform (Fig. 2), without the need of purifying the products from the reaction mixture. The conjugation efficiency is typically in the range of 40–60%, and further purification is not necessary. The following procedures describe the introduction of a puromycin-containing oligo linker to the 3′ end of mRNAs by irradiation under ~365-nm UV light.

1. Assemble a reaction mixture that contains 20 mM HEPES, pH 7.4, 100 mM KCl, 5 µM purified mRNAs from Subheading 3.4, and 12.5 µM (2.5 equivalent of mRNAs) puromycin-containing oligo linker.

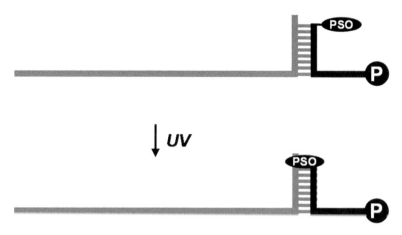

Fig. 2. Psoralen (PSO)-mediated photo-cross-linking between mRNA (*gray line*) and puromycin (P)-containing oligo linker (*black line*). The base pairing between the 3′ constant region of mRNA and the 5′ portion of oligo linker is necessary for site-specific cross-linking.

2. Mix well and aliquot 50 μl sample into an RNase-free 8-strip PCR tube.

3. Anneal mRNA and puromycin-containing oligo linker on a thermal cycler under the following conditions: 85°C, 8 min; cool from 80°C to 25°C at a rate of 1°C/20 s; 25°C, 25 min.

4. Transfer strip tubes to an ice bath and irradiate under 365-nm UV light in darkness for 20 min at 4°C.

5. Pool the samples into an RNase-free 1.5 ml Eppendorf tube.

6. Add 0.5 volume of LiCl precipitate solution and incubate at −20°C for 4 h to precipitate mRNA.

7. Wash pellet with 75% ethanol twice, air dry the pellet, and resuspend mRNA product in 100 μl DEPC water.

8. Measure OD using Nanodrop and calculate the molar concentration of mRNAs. If necessary, run a denaturing PAGE to check the conjugation efficiency and the quality of the resulting mRNAs (see Note 2).

3.6. In Vitro Translation and Fusion Formation

For the synthesis of protein sequences using in vitro translation, we recommend rabbit reticulocyte lysate from Ambion or Novagen. Wheat germ lysate can also be used, but bacterial lysate is not suitable due to the high levels of nucleases (9). It is critical to first optimize the translation conditions in a small scale, particularly the concentrations of Mg^{2+}, K^+, and mRNA, to get the highest yield. Radiolabeling of proteins can be achieved by adding [^{35}S]-methionine into the translation mixture in the absence of endogenous methionine. We strongly recommend radiolabeling that greatly facilitates the quantification of each step by liquid scintillation counting and/or autoradiography (see Note 3).

After translation, addition of Mg^{2+} and K$^+$ to the optimized concentrations followed by incubation at –20°C overnight are critical to promote the formation of mRNA–protein fusions (8, 9). Depending on the lengths and sequences of the mRNA templates, the fusion efficiency varies, typically in the range of 5–40%, if high-quality mRNAs and reticulocyte lysate are used for in vitro translation. The volume of in vitro translation is dependent on the diversity of the initial library. Typically, 1.0 ml (for natural proteome libraries) or 10 ml (for synthetic peptide libraries) of in vitro translation is performed for the first round of selection.

1. Assemble a reaction mixture that contains 1× translation mix without methionine, 50–120 mM KOAc or KCl, 0.3–1.0 mM Mg(OAc)$_2$, 200 nM puromycin-containing mRNAs, 0.5 µCi/µl [^{35}S]-L-methionine, and 40% (volume) reticulocyte lysate. It is critical that nuclease-free DEPC water should be used, and lysate should be as fresh as possible. Radiolabeling is optional but greatly facilitates signal monitoring at every step (see Note 3).

2. Incubate the reaction mixture at 30°C for 60–90 min.

3. Add MgCl$_2$ and KCl to a final concentration of 50 mM and 580 mM, respectively. Mix gently and incubate at room temperature for 30 min.

4. Incubate at –20°C overnight. The fusion mixture can be stored at –20°C for several days prior to purification. The translation and fusion efficiencies can be estimated by SDS-PAGE and autoradiography if a radioisotope is used to label the proteins or/and mRNAs (see Note 4).

3.7. Oligo(dT) Purification to Remove Free Proteins

After translation and fusion formation, the free mRNA templates and mRNA–protein fusions should be purified from lysate by using an oligo(dT) column, taking advantage of oligo(dA) residues in the puromycin-containing oligo linker. It is critical to dilute the translation reaction mixture at least 20 times using oligo(dT)-binding buffer to reduce the possible degradation of small amount of mRNA–protein fusion molecules by proteases and nucleases that are present in the lysate. The manipulation should be performed as fast as possible in cold room. The base pairing between oligo(dT) on cellulose and poly(dA) linker at the 3′ end of mRNAs are optimal under high-salt conditions, but can be readily disrupted by using a buffer that contains low concentration of salt.

1. For a 500 µl translation reaction, weigh 60 mg of oligo(dT) cellulose into a 1.5-ml nuclease-free Eppendorf tube.

2. Wash cellulose beads three times with 1 ml nuclease-free DEPC water and two times with 1 ml Oligo(dT) binding buffer.

3. Resuspend oligo(dT) cellulose in 1 ml binding buffer and transfer to a 15-ml tube containing 9 ml of oligo(dT) binding buffer. Add DTT to a final concentration of 1 mM.

4. Add the post-translation reaction mixture from Subheading 3.6 to the slurry, wrap the tube in aluminum foil, and rotate for 2 h at 4°C.

5. Load the slurry mixture into a 10 ml nuclease-free poly-prep chromatography column.

6. Collect flow-through and reload to resin bed. Retain the final flow-through for analysis.

7. Wash oligo(dT) cellulose on column twice each with 1 ml of oligo(dT)-binding buffer, followed by three times each with 1 ml of Oligo(dT) wash buffer. Retain each wash fraction for analysis.

8. Elute mRNA and mRNA–protein fusion molecules four times each with 600 µl of DEPC plus 1 mM DTT.

9. If [^{35}S]-methionine is used to radiolabel the protein, take 1/100 of each fraction and 1/10 of the beads to quantify the radioactivity using a liquid scintillation counter. The fusion efficiency can be estimated by SDS-PAGE and autoradiography (see Note 5).

3.8. Reverse Transcription to Remove Secondary Structure of mRNAs

The single-stranded mRNAs on mRNA–protein fusions could adopt complicated structures and therefore likely interfere with the subsequent functional selection. One effective way to address the problem is by converting the fusion molecules into rigid DNA/RNA hybrids through reverse transcription. Additional advantages of this step include the protection of mRNAs from degradation and direct amplification of the selected library by PCR for iterative rounds of selection. However, the reverse transcriptase should not have RNase H activity to keep the mRNA/DNA hybrids from degradation.

Since reverse transcriptase is sensitive to the salt concentration, the best time to perform this reaction is after oligo(dT) purification when the eluted fusion molecules are in a buffer with relatively low-salt concentrations. In general, the reverse transcription is efficient and can be performed at 37°C if necessary.

1. Add reverse transcription primer (final concentration 2 µM, see Note 6) to the mRNA–protein fusions purified from Subheading 3.7. Mix and incubate at room temperature for 15 min before addition of the first-strand synthesis buffer (1×), dNTPs (0.5 mM each), and DTT (10 mM). The mixture is incubated at 37°C or 42°C for 2 min.

2. Initiate the reverse transcription by adding an RNaseH$^-$ reverse transcriptase (e.g., Superscript II, Invitrogen) to a final

concentration of 2 U/μl and the reaction mixture is incubated at 37°C or 42°C for 50 min.

3. Terminate the reaction by adding EDTA to a final concentration of 3.6 mM (1.2× (Mg^{2+})).

4. Change to TBST buffer by applying to an NAP-10 column equilibrated with 1× TBST buffer.

5. Collect the reverse-transcribed product in 1.35 ml TBST buffer for anti-FLAG purification.

3.9. Anti-FLAG Purification to Remove Free mRNAs

While purification using an oligo(dT) column effectively removes free proteins that are not fused with their own mRNAs, the free mRNAs that are not fused with their encoded proteins are isolated together with mRNA–protein fusions. In some cases, such mixture can be directly used for selection, assuming that the free mRNA/DNA hybrids do not possess secondary structures and are less likely to interfere with the selection. However, the presence of a trace amount of free mRNA/DNA hybrids in the selected pool increases the background when the selected molecules are PCR amplified for next round of selection. To remove free mRNA/DNA hybrids, the reverse-transcribed fusion molecules can be further purified if an affinity tag is engineered at the N- or C-terminus of the protein sequences. We found that FLAG and E tags work well for this purpose. The following procedure describes the purification of mRNA–protein fusion molecules from free mRNAs using an anti-FLAG column.

1. Quickly wash 600 μl of anti-FLAG M2 affinity resin in a 10 ml poly-prep chromatography column five times each with 1 ml of 100 mM glycine buffer at pH 3.5.

2. Wash the resin three times each with 5 ml of 1× TBST buffer.

3. Cap the bottom of the column and load the reverse-transcribed product from Subheading 3.8.

4. Cap the top of the column with lid and rotate at 4°C for 2 h at an angle so that the resin is mixed well but attachment of solution to the top and side of the chromatography column is minimized.

5. Stand column upright and allow the resin to settle for 2 min. Remove top and bottom caps and collect and retain flow-through.

6. Wash the resin five times each with 1 ml of 1× TBST, and collect and retain each wash fraction.

7. Cap the bottom of the column and add 600 μl of 1× TBST plus 30 μl of 5 mg/ml FLAG peptide to slurry and rotate at 4°C for 30 min to elute the captured fusion molecules.

8. Collect the elution fraction and repeat step 7 three times.

9. If [^{35}S]-methionine is used to label the protein, take 1/100 of flow-through, wash, and elution fractions and 1/10 of the beads to quantify the amount of fusion molecules recovered using a liquid scintillation counter (see Note 7).

10. Change to a desired buffer using an NAP-10 column for the selection step.

4. Notes

1. Nuclease inhibitors can be added but are usually not necessary if all the procedures are strictly performed under RNase-free conditions.

2. The quality of the RNAs can be judged by running a denaturing PAGE gel. The bands on the gel should have the molecular weights as shown by the corresponding cDNA library. Short RNA bands usually suggest the presence of abortive transcription by-products or RNA degradation. Because the mRNAs conjugated with the puromycin-containing oligo linker are larger in size than the unconjugated ones, they can be easily separated by using a 5–10% denaturing PAGE. There should be two bands for the cross-linking products. The lower band corresponds to the unconjugated mRNAs while the higher band corresponds to the conjugation products. The ratio of these two bands can be used to estimate the conjugation efficiency, which is typically around 40–60%.

3. Radiolabeling of the translation products is optional, but very helpful for monitoring the signals at every step. It is not necessary to radiolabel all the proteins in the reaction. Typically, two translation reactions, one hot and one cold, are performed in parallel. The hot reaction (10% of the reaction volume) includes [^{35}S]-methionine while the cold reaction (90% of the reaction volume) uses normal methionine. After translation, the two reactions are mixed as one pot for all the subsequent procedures.

4. The puromycin-mediated fusion between the mRNA and its coding protein sequence can be monitored based on comparing migration of radiolabeled fusion molecules on an SDS-PAGE gel. The resulting fusion molecules usually migrate near the top of the gel while the free proteins migrate as expected based on the MWs.

5. Purification of mRNA–protein fusion molecules using oligo(dT) cellulose can be readily monitored by quantification of the radioactive counts present in 1% of the sample volume. Graphing the CPM data for the flow-through, washes, elution,

and bead samples should illustrate an elution profile showing radioactive counts returning to background for the wash samples followed by a large increase in counts for the elution samples. The presence of a large excess of unincorporated [^{35}S]-methionine always imparts high counts to the flow-through sample.

6. To make the reverse transcription more efficient, a unique primer with the following sequence can be used as an example: TTTTTTTTTTNN<u>CCAGATCCAGACATTCCCAT</u>, in which oligo(dT) is used to hybridize with the oligo(dA)-containing oligo linker at the 3′ end of mRNAs, NN used to stride over the mRNA–oligo junction, and underlined sequence used to hybridize with the very 3′ end of the mRNAs.

7. Anti-FLAG affinity purification of fusion molecules shows a similar elution profile to that of the oligo(dT) purification step. Radioactive counts should return to baseline during the wash steps and there should be a significant increase in counts in the elution fractions.

References

1. Lin, H., Cornish, V.W. (2002) Screening and selection methods for large-scale analysis of protein function. *Angew Chem Int Ed Engl.* **41**, 4402–4425
2. Amstutz, P., Forrer, P., Zahnd, C., Pluckthun, A. (2001) In vitro display technologies: novel developments and applications. *Curr Opin Biotechnol.* **12**, 400–405
3. Kay, B.K., Kasanov, J., Yamabhai, M. (2001) Screening phage-displayed combinatorial peptide libraries. *Methods* **24**, 240–246
4. Fields, S., Sternglanz, R. (1994) The two-hybrid system: an assay for protein-protein interactions. *Trends Genet.* **10**, 286–292
5. Roberts, R.W., Szostak, J.W. (1997) RNA-peptide fusions for the in vitro selection of peptides and proteins. *Proc Natl Acad Sci USA* **94**, 12297–12302
6. Nemoto, N., Miyamoto-Sato, E., Husimi, Y., Yanagawa, H. (1997) In vitro virus: bonding of mRNA bearing puromycin at the 3′-terminal end to the C-terminal end of its encoded protein on the ribosome in vitro. *FEBS Lett.* **414**, 405–408
7. Roberts, R.W. (1999) Totally in vitro protein selection using mRNA-protein fusions and ribosome display. *Curr Opin Chem Biol.* **3**, 268–273
8. Liu, R., Barrick, J.E., Szostak, J.W., Roberts, R.W. (2000) Optimized synthesis of RNA-protein fusions for in vitro protein selection. *Methods Enzymol* **318**, 268–293
9. Szostak, J., Roberts, R., Liu, R. in WO/1998/031700 (1998); WO/2000/047775 (2000); U.S. Patent 6,207,446 (2001); U.S. Patent 6,214,553 (2001); U.S. Patent 6,258,558 (2001); U.S. Patent 6,261,804 (2001); U.S. Patent 6,281,344 (2001).
10. Kurz, M., Gu, K., Lohse, P.A. (2000) Psoralen photo-crosslinked mRNA-puromycin conjugates: a novel template for the rapid and facile preparation of mRNA-protein fusions. *Nucleic Acids Res* **28**, E83
11. Leemhuis, H., Stein, V., Griffiths, A.D., Hollfelder, F. (2005) New genotype–phenotype linkages for directed evolution of functional proteins. *Curr Opin Struct Biol.* **15**, 472–478
12. Tabuchi, I., Soramoto S., Suzuki, M., Nishigaki, K., Nemoto, N., Husimi, Y. (2002) An efficient ligation method in the making of an in vitro virus for in vitro protein evolution. *Biol Proced Online.* **4**, 49–54
13. Cho, G., Keefe, A.D., Liu, R., Wilson, D.S., Szostak, J.W. Constructing high complexity synthetic libraries of long ORFs using in vitro selection. *J Mol Biol* **297**, 309–319 (2000).
14. Shen, X., Valencia, C.A., Szostak, J.W., Dong, B., Liu, R. Scanning the human proteome for calmodulin-binding proteins. *Proc Natl Acad Sci U S A.* **102**, 5969–5974; 5919 (2005).

15. Huang, B.C., Liu, R. Comparison of mRNA-display-based selections using synthetic peptide and natural protein libraries. *Biochemistry.* **46**, 10102–10112; 10109 (2007).

16. Tabata, N., Sakuma, Y., Honda, Y., Doi, N., Takashima, H., Miyamoto-Sato, E., Yanagawa, H. Rapid antibody selection by mRNA display on a microfluidic chip. *Nucleic Acids Res.* **37**, e64; 2030 (2009).

Chapter 7

SNAP Display: In Vitro Protein Evolution in Microdroplets

Miriam Kaltenbach and Florian Hollfelder

Abstract

SNAP display is based on the covalent reaction of the DNA repair protein AGT (O^6-alkylguanine DNA alkyltransferase, the "SNAP-tag") with its substrate benzylguanine (BG). Linear, BG-labelled template DNA is encapsulated in water-in-oil emulsion droplets with a diameter of a few micrometres (i.e. 1 mL of emulsion contains ~10^{10} compartments). Each droplet contains only a single DNA copy, which is transcribed and translated in vitro. The expressed AGT fusion proteins attach to their coding DNA via the BG label inside the droplet, which ensures that a specific genotype–phenotype linkage is established. Subsequently, the emulsion is broken and protein-DNA conjugates, which constitute a DNA-tagged protein library, selected via affinity panning. This method will prove a useful addition to the array of in vitro display systems, distinguished by the stability of DNA as the coding nucleic acid and the covalent link between gene and protein.

Key words: In vitro compartmentalisation, Emulsion, DNA display, SNAP-tag, Genotype–phenotype linkage, Panning, Protein engineering, Directed evolution

1. Introduction

SNAP display is a system for the directed evolution of protein binders in vitro. The link between the genotype (the coding nucleic acid) and the phenotype (the observable functional trait) is achieved by compartmentalizing genes in water droplets immersed in oil that serve as simple, artificial cells as introduced by Griffiths and Tawfik (1–4). The protein of interest is expressed as a SNAP-tag fusion from the compartmentalized genes by coupled in vitro transcription/translation (IVTT). "SNAP-tag" is the commercial name for the DNA repair protein AGT (O^6-alkylguanine DNA alkyltransferase),[1] which has been developed as a general protein tag to react with an artificial substrate, benzylguanine (BG) (5–7).

---

[1] The SNAP-tag technology was commercialised by Covalys AG in 2007. Since 2009, it is distributed by New England Biolabs.

When the coding DNA contains a BG-label, the fusion protein covalently attaches to it after expression. By working at DNA concentrations sufficiently low so that each droplet contains no more than one copy of the gene, one ensures this link is unambiguous. Subsequently, the emulsion can be broken and binders recovered via affinity panning. The attachment of the coding DNA to the protein of interest allows identification and amplification of the selected clones by PCR.

Like other in vitro display systems, SNAP display overcomes restrictions affecting traditional in vivo systems such as *E. coli* or yeast as well as phage display. Most notably, library size is not limited by the generally low transformation efficiencies (e.g. usually $<10^9$ in *E. coli*). Also, in vitro methods may give access to proteins that are otherwise difficult to express, e.g. due to their toxic or growth-limiting properties. Several alternative expression systems are available (bacterial, wheat germ, rabbit reticulocyte) for which codon usage can be optimized. Furthermore, in vitro expression systems can be supplemented with unnatural amino acids.

From other in vitro systems, such as the more familiar ribosome display or mRNA-display methods, SNAP display is distinguished by a potentially more stable and versatile set-up. First, the coding nucleic acid is DNA rather than the chemically much less stable RNA (about 3×10^5-fold in water at pH 7 (8, 9)), which can simplify handling enormously. Second, the covalent nature of the genotype–phenotype linkage will allow selections under conditions that may be incompatible with other in vitro display systems, e.g. at extremes of pH and temperature, in the presence of organic solvents or drug delivery formulations. By contrast, the most widely used in vitro method, ribosome display, is limited in its scope as the ternary complex consisting of mRNA, ribosome and protein is non-covalent and requires stabilisation by low temperature and low magnesium ion concentrations. Furthermore, post-display processing such as chemical modification of displayed peptides or proteins is conceivable in SNAP display. Finally, it is possible to switch the SNAP-tag system to a multivalent format in which multiple copies of protein are displayed per DNA, taking advantage of avidity effects during affinity panning with the ligand-coated surface to increase the recovery of low-affinity binders (10).

Thus far, SNAP display has only been demonstrated in model selections (11). However, due to the intrinsic advantages discussed above, we envision that it can be used in real selections; especially those that have proven impossible or difficult using other more established methods. The procedures elaborated below should equip the reader with detailed instructions to attempt such challenging experiments.

It should be noted that two systems conceptually similar to SNAP display have also been reported. In STABLE, the nucleic acid-protein linkage is formed non-covalently between biotin and streptavidin, and in M.*Hae* III display, the link is established covalently via a suicide substrate sequence on the DNA. For more information on these systems, the reader is referred to refs. 12–16.

2. Materials

2.1. Reagents

1. pIVEX vector for cell-free protein expression containing SNAP-tag fused to the gene of interest. A pIVEX-SNAP plasmid can be obtained from us to generate flanking regions for linear template assembly or cloning. Compared to wild type human AGT, our SNAP mutant contains the mutations K32I, L33F, C62A, Q115S, Q116H, K125A, A127T, R128A, G131K, G132T, M134L, R135S, C150N, S151I, S152N, A154D, N157G, S159E and has been shortened by 25 residues at its C-terminus (6, 17–20).
2. LMB2-6(long) (5′-ATGTGCTGCAAGGCGATTAAG, forward) and pIV-B1(long) (5′-GCGTTGATGCAATTTCTATGC, reverse) primers for linear template generation, additional real time PCR primers as appropriate. E.g. for reco-very of DHFR (dihydrofolate reductase), use DHFR_forward (5′-CTGACG CATATCGACGCAGAA) and DHFR_reverse (5′-CCGCTCCA GAATCTCAAAGCA). For additional detection of GST (glutathione S-transferase) to determine the enrichment, use GST_forward (5′-TCCAAAAGAGCGTGCAGAGAT) and GST_reverse (5′-TGCAATTCTCGAAACACCGTAT).
3. 100 mM Tris–HCl, pH 8.5.
4. 1 M DTT.
5. 100 mM phosphate buffer, pH 6.9.
6. BG-maleimide (NEB).
7. DMF.
8. Gel filtration columns (illustra NAP-5, Sephadex G-25 DNA grade, GE Healthcare).
9. Liquid nitrogen.
10. Mineral oil mix (Mineral oil, Span 80, Tween 80, all Sigma).
11. Linear, non-coding DNA.
12. RTS 100 *E. coli* HY kit for cell-free protein expression (5 Prime or RiNA).

Buffers and reagents for breaking the emulsions

13. 1-mM stock solution of benzylguanine (BG, Sigma) in DMSO.
14. Diethyl ether.

15. Recovery & Binding buffer (RBB), Binding buffer (BB) and Elution buffer (EB) suitable for the protein of interest. e.g. for DHFR binding and elution, use PBS (10 mM NaH_2PO_4, 1.8 mM KH_2PO_4, 140 mM NaCl, 2.7 mM KCl, pH 7.4) supplemented with 5 mM EDTA and 10 μM BG (RBB), PBS supplemented with 5 mM EDTA and 0.05 wt.% Tween 20 (BB) and 3 mM folic acid, 20 mM NaH_2PO_4, pH 7.4 (EB).

16. Ligand-coated superparamagnetic microbeads as appropriate for the protein of interest. For the functionalization of Dynabeads (Invitrogen), see the corresponding Dynabead manuals or ref. 17. E.g. use Dynabeads MyOne Carboxylic Acid (Invitrogen) and functionalize with methotrexate as described in ref. 17 for DHFR binding.

17. Proofreading polymerase and/or Taq (e.g. Biotaq, Bioline) for PCR amplifcation, SensiMix SYBR No-ROX kit, Bioline for real time PCR.

18. Reagents for agarose gel electrophoresis.

2.2. Equipment

1. Thermomixer.
2. Lyophilizer.
3. Spectrophotometer for DNA concentration determination.
4. Thermal cycler.
5. 1.8-mL cryotubes (Nalgene).
6. Omni International tissue homogeniser and Omni Tips Clear Plastic Homogenising Probes (7×110 mm).
7. Benchtop centrifuge with cooling function (e.g. Eppendorf Centrifuge 5415R).
8. Vortexer.
9. Vacuum concentrator (e.g. Eppendorf Concentrator 5301).
10. Magnet (e.g. Dynamag-2, Invitrogen).
11. Roller mixer (e.g. Stuart RollerMixer SRT6, Progen Scientific).
12. Real time thermal cycler (e.g. Rotor-Gene Q, Qiagen).
13. Apparatus for agarose gel electrophoresis.

3. Methods

3.1. Preparation of Linear Templates Containing BG

To begin, clone the gene corresponding to the protein of interest into the pIVEX-SNAP plasmid (see Notes 1–4). From this, you can generate a stock of a DNA fragment containing the 5′-regulatory region and the SNAP gene (1) and a 3′-regulatory fragment (2) via PCR. You can also generate the diversified region (3) from this, e.g. by error-prone PCR. The three fragments can then be reassembled

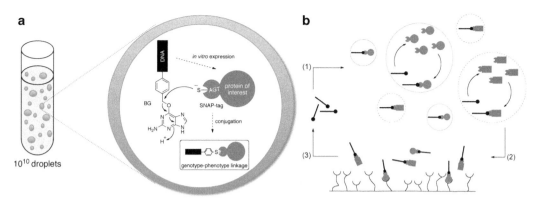

Fig. 1. SNAP display. (**a**) One millilitre of emulsion accommodates around 10^{10} expression-competent droplets, in each of which the protein SNAP-tag fusion is linked to its coding DNA. (**b**) Outline of the selection system. (1) BG-labelled DNA templates are emulsified with IVTT mix. The protein is expressed and the genotype–phenotype linkage established as shown in (**a**). (2) The emulsion is broken and binders recovered via affinity panning on a magnetic bead surface. (3) Recovered DNA is amplified by PCR and reassembled into the full-length templates. Note: BG is shown here as *small dot* at the end of the linear DNA. The protein of interest is symbolised by a *circle*, non-binding proteins by a *square*.

into the full-length template library between rounds. A fast (~90 min) assembly method is described in detail in refs. 21 and 22. Here, we will describe a model selection with only two types of templates (a binder, AGT-DHFR and a non-binding control, AGT-GST), which can directly be amplified from the respective plasmids (11). Importantly, BG, which subsequently serves as connector between genotype (DNA) and phenotype (SNAP-fusion) (Fig. 1) is introduced during this PCR by using BG-labelled primers. The labelling of primers is described in steps 1–6.

1. Use LMB2-6(long), pIV-B1(long) or both primers on a 1 μmole synthesis scale with a 5′-thiol modification.

2. If the thiol is purchased as a disulfide, deprotect as follows: Resuspend the oligonucleotide in 240 μL 100 mM Tris–HCl, pH 8.5, add 60 μL 1 M DTT and incubate at 25°C for 1 h. If the oligonucleotide is purchased as a free thiol, we also recommend performing the deprotection reaction to reduce oligonucleotide dimers that might have formed.

3. Remove excess DTT by gel filtration over a NAP-5 column equilibrated in 100 mM phosphate buffer, pH 6.9 to prevent quenching of BG-maleimide in the subsequent coupling reaction. Elute in 700 μL.

4. Dissolve 1–2 mg BG-maleimide in 300 μL fresh DMF, add to the oligonucleotide directly after elution (to prevent re-oxidation of thiols) and incubate at 40°C for 2 h (see Notes 5 and 6).

5. Remove excess BG-maleimide and DMF over two NAP-5 columns (500 μL maximum capacity) equilibrated in Milli-Q water. Elute in 750 μL each, snap-freeze in liquid nitrogen and lyophilize the combined fractions.

6. Resuspend the white powder in ca. 100 μL Milli-Q water, determine the concentration in a spectrophotometer and adjust as appropriate for your PCR reactions (e.g. to 100 μM).

7. Amplify linear templates with BG-labelled primers LMB2-6(long) and pIV-B1(long) from pIVEX plasmid and purify with a PCR purification kit (see Note 7).

3.2. Emulsification for Compartmentalized Expression and Conjugation

Before beginning the experiment, prepare the following: Mineral oil mix (95% mineral oil, 4.5% Span 80 and 0.5% Tween 80, all wt./wt.), aliquots of the RTS 100 *E. coli* HY kit to avoid repeated freeze–thaw cycles (see Notes 8 and 9) and clean homogeniser tips (soak in diluted HCl, wash with water and ethanol three times, dry thoroughly).

1. Add 950 μL of mineral oil mix to a 1.8 mL cryotube, cool down on ice.

2. Prepare a 50 μL IVTT reaction on ice (see Note 10). For model selections, we use a concentration of 100 pM for the non-binding control template and 100 pM or lower for the binder. Also add 60 ng of linear, noncoding DNA (see Note 11). Mix by pipetting up and down; avoid foaming.

3. Apply to the top of the oil phase, insert homogenizer tip as low as possible without it scratching the bottom of the tube and homogenise at 5,000 rpm for 3 min. Keep cryotube surrounded by ice to avoid heating up of the sample during homogenisation.

4. Transfer the emulsion to a 1.5-mL tube and express for 3–4 h at 25°C without shaking (see Note 12).

3.3. Breaking the Emulsions

1. Add 1 μL of a 1 mM BG solution to the bottom of a 1.5-mL tube; keep.

2. Spin down the emulsion at $13,000 \times g$ for 5 min at 4°C.

3. Remove the supernatant oil phase by pipetting.

4. Add 100 μL RBB and subsequently 1 mL water-saturated diethyl ether to the white pellet. Vortex at full speed just until the emulsion is broken.

5. Remove diethyl ether and repeat step 4.

6. Directly pipette out the aqueous phase from underneath the ether into the BG-containing tube (see Note 13). The free BG serves to quench unreacted SNAP tag to prevent cross-reaction between DNA and protein from different droplets (leading to an incorrect genotype–phenotype linkage).

7. Dry for 5 min in vacuo to remove traces of diethyl ether.

3.4. Affinity Panning

1. Before breaking the emulsion Subheading 3.3, prepare magnetic beads by washing four times in BB. Use 5 μL of beads (10^6–10^7 beads/μL) for each 50 μL IVTT reaction.

2. Add the beads to the recovered aqueous phase and let protein-DNA conjugates bind to the beads, e.g. by incubating on a roller mixer at a desired temperature.

3. Place the tube in a magnet for 1 min, take off the supernatant S and keep it for analysis by competitive PCR if applicable.

4. Resuspend the beads in 100 μL BB supplemented with 0.05% Tween 20 to reduce non-specific binding and transfer to a fresh tube. In this way, contamination with drops containing high concentrations of unbound conjugates that still stick to the top of the tube is avoided.

5. Remove supernatant; wash two more times in 100 μL BB.

6. Resuspend in 30 μL EB and elute, e.g. with EB containing a competitive eluting agent.

7. Remove the eluted fraction E and keep it.

3.5. Quantifying Recovery and Amplification of the Diversified Region by Real Time PCR

The recovery is an absolute measure of the performance of the selection system. It can be used to monitor the progress of adaptation and adjust the selective pressure accordingly. In theory, the recovery should first increase over successive selection cycles and eventually become saturated at a constant level when adaptation is completed. Better mutants can then only be enriched if the selective pressure is increased. If the recovery is low or decreases over several rounds, then the selection pressure is too high for the library (i.e. binders are not sufficiently strong or occur too rarely in the starting library (23)).

1. Make a standard ladder from 1:10 dilutions of template DNA in a way that the amount of recovered DNA resides within the range of the ladder, e.g. eight dilutions starting from 2 nM template.

2. If the elution buffer inhibits PCR performance, dilute sample in Milli-Q water prior to PCR amplification. It is advisable to dilute the standard ladder into an equivalent mix of EB/water for comparison.

3. Amplify each sample in triplicate using the SensiMix SYBR No-ROX kit (see Note 14). In model selections, use primers specific for your binder, e.g. use DHFR_forward and DHFR_reverse and cycle as follows: Initial denaturation at 95°C for 10 min followed by 40 cycles of denaturation at 95°C for 15 s, annealing at 54°C for 1 min and extension at 72°C for 20 s. In real selections, use primers flanking the diversified region and use the amplified fragment for template reassembly to prepare full-length templates for the next selection round.

4. Calculate the recovery in percent as the number of recovered DNA molecules relative to the input DNA.

3.6. Quantifying Enrichment by Real Time and Competitive PCR in Model Selections

When conducting model selections, an important measure for the success of the selection system, i.e. the specificity of the selected interaction and the fidelity of the genotype–phenotype linkage is the enrichment of the binder relative to the non-binding control.

1. Perform real time PCR as described in Subheading 3.5 but using primers specific for the non-binding control, i.e. GST_forward and GST_reverse. Use the cycling parameters described in step 3 in Subheading 3.5, but adjust the annealing temperature to 60°C.

2. Calculate the enrichment as follows:

    Number of recovered binders times dilution relative to the control/number of recovered non binders.

3. In addition, the enrichment can be visualized and estimated on agarose gel after performing competitive PCR. Use primers that hybridize to both target and control and amplify DNA from the eluted fraction E and the supernatant S. You will have performed the selection for different dilutions of target relative to the non-binding control. The dilution at which the bands of target and control in the eluted fraction E appear at equal intensities on gel corresponds to the enrichment.

4. Notes

1. pIVEX vectors have been especially designed for cell-free protein expression. Like the pET vector series, they contain all regulatory elements necessary for transcription by T7 RNA polymerase. The main difference is that there is no LacI binding site in pIVEX vectors, so that expression is always active (no induction by IPTG necessary).

2. If positioned at the N-terminus, SNAP is expressed and active in emulsions for all fusions we have tested, whereas for C-terminal SNAP, the expression is dependent on the fusion partner. We therefore suggest positioning the gene of interest downstream of the SNAP-tag.

3. To test protein expression, perform the reaction with unlabelled DNA in the presence of an excess of a BG derivative such as SNAP-Biotin, SNAP-Vista Blue or SNAP-Vista Green (NEB). These compounds all label SNAP tag at its active site and can therefore be used to measure the amount of functionally expressed SNAP tag. For comparison between samples, always recover the same amount of aqueous phase by pipetting. Precipitate proteins in the aqueous phase by acetone, separate by SDS-PAGE and visualize as appropriate for the label. For SNAP-biotin, conduct a Western Blot and detect with Streptavidin

Horseradish peroxidase (Streptavidin-Peroxidase Polymer, Sigma). For BG-fluorophores, detect directly on a gel scanner. One possibility to monitor total protein expression is to stain all proteins with SYPRO Orange (Sigma) and analyze on the gel scanner.

4. In addition to the above-mentioned SNAP substrates, a range of SNAP-fluorophores, SNAP-encoding vectors (for expression in *E. coli* or mammalian cells) and related reagents are available from NEB. As an alternative to obtaining the pIVEX-SNAP plasmid from us, the SNAP-tag gene can also be extracted from the vector pSNAP-tag(T7) available from NEB. This gene differs from pIVEX-SNAP in that it lacks mutations decreasing the interaction of SNAP with DNA and increasing the rate of reaction with BG derivatives (6, 17–20).

5. For the generation of protein-DNA conjugates, labelling template DNA on one end suffices. However, the recovery can be improved by labelling both ends. If both primers are labelled in parallel, just one batch of 2 mg BG-maleimide can be used to reduce the cost (dissolve in 600 μL DMF).

6. For labelling of oligonucleotides with BG, it is important that the DMF used is of high purity and fresh to avoid chemical degradation, which may increase the pH of the conjugation reaction and quench maleimides at high pH > 8.

7. For a model selection, directly proceed to Subheading 3.2. For a selection from a library, treat with *Dpn*I to avoid contamination with plasmid DNA and remove primer-dimers by PEG-$MgCl_2$ precipitation as described in refs. 21 and 22.

8. As stated in the RTS manual, do not combine aliquots/buffers from different RTS lots.

9. In principle, it is possible to use IVTT kits other than the RTS, but the expression level in emulsions and the integrity of the selection system should be checked and optimized as necessary. For example, we have tested the PURExpress In Vitro Protein Synthesis kit (NEB). This system consists of fully recombinant proteins and ribosomes and tRNAs purified from *E. coli*, which makes it possible to work under nuclease-free conditions. However, in its current state of development, protein expression in this kit is much lower than in the RTS and recoveries in droplets were reduced about fivefold.

10. Keep reactions on ice whenever possible to avoid degradation by nucleases present in the RTS kit.

11. The addition of linear (23), non-coding DNA to the IVTT mix has been shown to improve the enrichment of specific binders, although the reason for this effect is not entirely clear. Possible explanations might be that the excess of DNA alleviates

template degradation by nucleases present in the IVTT mix, saturates the droplet interface with DNA (ensuring that templates are kept in solution) and/or reduces non-specific binding to beads.

12. It is possible to scale up the library size, e.g. by emulsifying three reactions and combining them in a 15-mL falcon tube before expression.

13. When recovering the aqueous phase after breaking the emulsions, hold the extraction tube and the tube containing BG close together in one hand to avoid leaking of the aqueous phase containing diethyl ether whilst pipetting. Hold extraction tube against the light to see phase border.

14. For model selections or monitoring of recovery separate from amplification of templates for the next selection round, we recommend using SensiMix SYBR NoRox kit for real time PCR, which contains all necessary reagents including SYBR green and a highly sensitive hot-start Taq polymerase and gives very reliable and accurate results. However, if the real time PCR product is to be used directly for next selection round, a proofreading polymerase might be more appropriate to avoid the accumulation of mutations that might compromise your library. If primers contain Uracil for USER assembly, use Pfu Turbo Cx Hotstart DNA polymerase (Agilent Technologies) as like Taq, this proofreading polymerase does not stall at Uracil residues.

Acknowledgements

MK was supported by a fellowship from the EU Marie-Curie ITN ProSA. FH is an ERC Starting Investigator.

References

1. Kaltenbach, M., Schaerli, S., and Hollfelder, F. (2009) Microdroplets – A Tool for Protein Engineering. *BIOforum* 13, 19–21.
2. Miller, O. J., Bernath, K., Agresti, J. J., Amitai, G., Kelly, B. T., Mastrobattista, E., Taly, V., Magdassi, S., Tawfik, D. S., and Griffiths, A. D. (2006) Directed evolution by in vitro compartmentalization. *Nature Methods* 3, 561–70.
3. Schaerli, Y., and Hollfelder, F. (2009) The potential of microfluidic water-in-oil droplets in experimental biology. *Molecular Biosystems* 5, 1392–404.
4. Tawfik, D. S., and Griffiths, A. D. (1998) Man-made cell-like compartments for molecular evolution. *Nature Biotechnology* 16, 652–56.
5. Gronemeyer, T., Godin, G., and Johnsson, K. (2005) Adding value to fusion proteins through covalent labelling. *Current Opinion in Biotechnology* 16, 453–58.
6. Keppler, A., Gendreizig, S., Gronemeyer, T., Pick, H., Vogel, H., and Johnsson, K. (2003) A general method for the covalent labeling of fusion proteins with small molecules in vivo. *Nat Biotechnol* 21, 86–9.
7. Keppler, A., Kindermann, M., Gendreizig, S., Pick, H., Vogel, H., and Johnsson, K. (2004) Labeling of fusion proteins of O-6-alkylguanine-DNA alkyltransferase with small molecules in vivo and in vitro. *Methods* 32, 437–44.

8. Williams, N. H., Takasaki, B., Wall, M., and Chin, J. (1999) Structure and nuclease activity of simple dinuclear metal complexes: Quantitative dissection of the role of metal ions. *Accounts of Chemical Research* 32, 485–93.

9. Schroeder, G. K., Lad, C., Wyman, P., Williams, N. H., and Wolfenden, R. (2006) The time required for water attack at the phosphorus atom of simple phosphodiesters and of DNA. *Proceedings of the National Academy of Sciences of the United States of America* 103, 4052–55.

10. Kaltenbach, M., Stein, V., Hollfelder, F. (2011) SNAP Dendrimers: Multivalent Protein Display on Dendrimer-Like DNA for Directed Evolution. *Chembiochem* 12, 2208–16.

11. Stein, V., Sielaff, I., Johnsson, K., and Hollfelder, F. (2007) A covalent chemical genotype-phenotype linkage for in vitro protein evolution. *Chembiochem* 8, 2191–94.

12. Bertschinger, J., Grabulovski, D., and Neri, D. (2007) Selection of single domain binding proteins by covalent DNA display. *Protein Engineering Design & Selection* 20, 57–68.

13. Bertschinger, J., and Neri, D. (2004) Covalent DNA display as a novel tool for directed evolution of proteins in vitro. *Protein Engineering Design & Selection* 17, 699–707.

14. Doi, N., and Yanagawa, H. (1999) STABLE: protein-DNA fusion system for screening of combinatorial protein libraries in vitro. *FEBS Lett* 457, 227–30.

15. Yonezawa, M., Doi, N., Higashinakagawa, T., and Yanagawa, H. (2004) DNA display of biologically active proteins for in vitro protein selection. *J Biochem* 135, 285–8.

16. Yonezawa, M., Doi, N., Kawahashi, Y., Higashinakagawa, T., and Yanagawa, H. (2003) DNA display for in vitro selection of diverse peptide libraries. *Nucleic Acids Res* 31, e118.

17. Stein, V., Sielaff, I., Johnsson, K., and Hollfelder, F. (2007) A covalent chemical genotype-phenotype linkage for in vitro protein evolution. *Chembiochem* 8, 2191–94.

18. Juillerat, A., Gronemeyer, T., Keppler, A., Gendreizig, S., Pick, H., Vogel, H., and Johnsson, K. (2003) Directed evolution of O-6-alkylguanine-DNA alkyltransferase for efficient labeling of fusion proteins with small molecules in vivo. *Chemistry & Biology* 10, 313–17.

19. Gronemeyer, T., Chidley, C., Juillerat, A., Heinis, C., and Johnsson, K. (2006) Directed evolution of O6-alkylguanine-DNA alkyltransferase for applications in protein labeling. *Protein Eng Des Sel* 19, 309–16.

20. Juillerat, A., Heinis, C., Sielaff, I., Barnikow, J., Jaccard, H., Kunz, B., Terskikh, A., and Johnsson, K. (2005) Engineering substrate specificity of O-6-alkylguanine-DNA alkyltransferase for specific protein labeling in living cells. *Chembiochem* 6, 1263–69.

21. Stein, V., and Hollfelder, F. (2009) An efficient method to assemble linear DNA templates for in vitro screening and selection systems. *Nucleic Acids Res* 37, e122.

22. Stein, V., Kaltenbach, M., and Hollfelder, F. (2011) in "Functional Genomics, Second Edition" (Kaufmann, M., and C., K., Eds.), Humana Press.

23. Stein, V. (2008) A protein display system based on human O^6-alkylguanine alkyltransferase and in vitro compartmentalisation (Doctoral Thesis), Cambridge.

Chapter 8

cDNA Display: Rapid Stabilization of mRNA Display

Shingo Ueno and Naoto Nemoto

Abstract

The cDNA display method is a robust in vitro display technology that converts an unstable mRNA–protein fusion (mRNA display) to a stable mRNA/cDNA–protein fusion (cDNA display) whose cDNA is covalently linked to its encoded protein using a well-designed puromycin linker. We provide technical details for preparing cDNA display molecules and for the synthesis of the puromycin linker for the purpose of screening the functional proteins and peptides.

Key words: cDNA display, mRNA display, In vitro virus, In vitro selection, Peptide aptamer, Puromycin, Directed evolution, Protein engineering, Antibody technology, Protein–protein interaction

1. Introduction

The linkage strategy of genotype (gene) to phenotype (its encoded protein) is indispensable for in vitro selection and evolution of proteins. "In vitro virus" (1) and mRNA display (2) were developed for generating an mRNA–protein fusion molecule by using a cell-free translation system in a test tube to screen functional proteins and peptides from vast-sized libraries ($>10^{12}$). The antibiotic puromycin has been utilized for the key molecule to connect an mRNA to its encoded protein during the translation reaction in both methods. One of the most different points between in vitro virus and mRNA display is the ligation method of an mRNA to a puromycin linker; that is, T4 RNA ligase for in vitro virus and T4 DNA ligase for mRNA display were used. From now on, we consider in vitro virus with mRNA display as one technique for simplicity, although some differences exist between these methods. Until recently, we have been confronted with the following two difficulties, which need to be overcome in mRNA display. First, the instability of mRNA restricts the conditions that can be used in selection procedure. It is very important for generating functional proteins and peptides to perform screens under varying conditions. Second, the performance of mRNA display is prone to demand laborious, troublesome, and time-consuming work, resulting in reductions in library size. Thus, rapid performance of mRNA display minimizes loss of library size and shortens the time of the selection cycle.

We believe that the above problems were elegantly solved by designing a versatile puromycin linker (see Fig. 1), which comprises four major portions: a "ligation site" for T4 RNA ligase, a "biotin site" for simple manipulation (streptavidin-coated magnetic bead handling), a "reverse transcription primer site" for the synthesis of mRNA/cDNA–protein fusion, and a "restriction enzyme site" for the release of the mRNA/cDNA–protein fusion from the bead. The preparation of the puromycin linker may be troublesome for a novice user at first. However, once the puromycin linker is synthesized, the user can prepare mRNA and cDNA display molecules smoothly and perform the in vitro selection safely and rapidly. We, therefore, recommend that, first of all, a user should master making the puromycin linker completely (described in Subheading 3.1). Second, we recommend that a user should perform a simple model experiment of in vitro selection using cDNA display [e.g., the enrichment experiment of B-domain of protein A (BDA) against immunoglobulin G (IgG) from a cDNA display library comprised of BDA and an equimolar POU-specific DNA binding domain of Oct-1 (PDO)] (described in Subheading 3.12) before attempting a selection against a novel target molecule. We believe that cDNA display can be also useful for any mRNA display-like method.

Fig. 1. (a) Schematic diagram of the puromycin-linker construct and hybridized mRNA (*lower case*). The puromycin linker comprises three regions: the ligation site for T4 RNA Ligase (used in Subheading 3.5), the primer region for reverse transcription (used in Subheading 3.8.2), and the restriction site of *Pvu*II for the release of the molecules from the solid surface (used in Subheading 3.9), and three moieties: biotin moiety for immobilization on streptavidin magnetic beads (used in Subheading 3.8.1), FITC moiety for detection and quantification (used in Subheadings 3.1, 3.5, 3.7 and 3.12), and puromycin moiety for the covalent linking of the translated protein to mRNA (used in Subheading 3.6). Nucleotides in the puromycin linker are written in *bold* and Spc18 is an abbreviation of Spacer 18. (b) Schematic diagram of the preparation route of puromycin linker. The puromycin linker is synthesized by coupling two fragments (puromycin segment and biotin segment) using a heterobifunctional cross-linking reagent EMCS (*N*-(6-Maleimidocaproyloxy) succinimide). Synthesis of the puromycin linker is composed of three steps: reduction of the puromycin segment (Subheading 3.1.1), the EMCS-modification of the biotin segment (Subheading 3.1.2), and the cross-linking of the reduced puromycin segment and the EMCS-modified biotin segment (Subheading 3.1.3). Nucleotides in the segments are written in *bold* and DMT is the abbreviation for the dimethoxytrityl group. (c) Schematic diagrams of the chemical structures of the puromycin segment (a) and biotin segment (b). Nucleotides in the structures are written in *bold*.

8 cDNA Display: Rapid Stabilization of mRNA Display 115

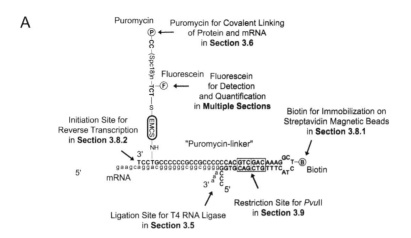

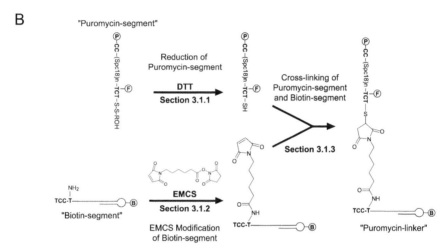

2. Materials

All reagents should be of molecular biology grade to avoid contamination of ribonuclease.

2.1. Synthesis of Puromycin Linker

1. Puromycin segment (see Fig. 1C-a): DNA oligomer composed of the following sequence, 5′-(5′-Thiol-Modifier C6S-S)-TC-(Fluorescein-dT)-(Spacer 18)n-CC-Puromycin-3′, $n = 1$–4 (see Note 1). This oligomer can be obtained from a custom DNA synthesis service with the phosphoramidite reagents of 5′-Thiol-Modifier C6, Amino-Modifier C6 dT, Spacer 18, Puromycin-CPG from Glen Research. The synthesis should be performed at 1.0 µmol scale and the purification should be done by high-performance liquid chromatography (HPLC). The final yield is about 100–200 nmol when it is synthesized at 1.0 µmol scale.

2. Biotin segment (see Fig. 1C-b): DNA oligomer composed of the following sequence, 5′-CCCGGTGCAGCTGTTTCATC-(Biotin-dT)-CGGAAACAGCTGCACCCCCGCCGCCC CCCG-(Amino-Modifier C6 dT)-CCT-3′. This oligomer can be obtained from a custom DNA synthesis service with the phosphoramidite reagents of Biotin-dT, Amino-Modifier C6 dT from Glen research. The synthesis should be performed at 1.0 µmol scale and the purification should be done by HPLC. The final yield is 30–60 nmol when it is synthesized at 1.0 µmol scale.

3. 1 M Na_2HPO_4, pH 9.0: 1 M Na_2HPO_4 in ddH_2O, pH comes to 9.0.

4. 200 mM phosphate buffer, pH 7.2.

5. 20 mM phosphate buffer, pH 7.2.

6. 1 M dithiothreitol (DTT): 1 M DTT in 10 mM sodium acetate, pH 5.2.

7. NAP-5 column (GE Healthcare).

8. 100 mM EMCS (*N*-(6-Maleimidocaproyloxy) succinimide): dissolve 2 mg of EMCS in 65 µL of *N,N*-dimethylformamide (see Note 2).

9. 3 M sodium acetate, pH adjusted to 5.2.

10. 99.5% Ethanol.

11. 70% Ethanol: prepared by dilution of 99.5% ethanol with ddH_2O.

12. HPLC column: Symmetry 300C18 Column, 5 µm, 4.6 × 250 mm, Waters.

13. 100 mM TEAA (triethylammonium acetate): prepare by dilution of 2 M TEAA (pH 7.0, HPLC grade, various suppliers) with ddH_2O (see Note 3).

14. 80% Acetonitrile: prepare by dilution of 100% acetonitrile (HPLC grade, various suppliers) with ddH$_2$O (see Note 3).
15. Quick-Precip™ Plus Solution (Edge Biosystems) for ethanol precipitation.
16. Vortex mixer (various suppliers).
17. Vacuum centrifuge (such as SpeedVac from Thermo Fisher Scientific).

2.2. Denaturing Urea Polyacrylamide Gel Electrophoresis

1. Temperature-controllable gel electrophoresis apparatus (such as AE-6510, ATTO) (see Note 4).
2. Glass plate: an example of size is 90 mm (W) × 80 mm (H) × 0.75–1.0 mm (D).
3. Spacer: 0.75–1.0 mm thick.
4. Comb: 0.75–1.0 mm thick.
5. 10× TBE: 890 mM Tris–HCl, 890 mM boric acid, 20 mM EDTA, pH 8.3.
6. 40% acrylamide/bis-solution (19:1): 40% acrylamide and 2% bis-acrylamide in ddH$_2$O. Store at 4 °C (see Note 5).
7. Ammonium persulfate (APS): prepare a 20% solution (w/v) in ddH$_2$O. Store at 4°C.
8. N,N,N',N'-tetramethylethylenediamine (TEMED). Store at 4°C.
9. 2× Loading buffer: mix 9.6 g of urea, 1 g of sucrose, 2 mg of bromophenol blue, 2 mg of xylene cyanol, 2 mL of 10× TBE, and ddH$_2$O up to 10 mL.
10. 10-bp DNA ladder (various suppliers).
11. SYBR® Gold nucleic acid gel stain (Invitrogen): other DNA staining reagents can be used.
12. Fluorescence scanner (e.g., Pharos FX from Bio-Rad, Typhoon from GE Healthcare).

2.3. Transcription

1. RiboMAX™ Large Scale RNA Production System-T7 (Promega), which contains T7 Enzyme Mix, 5× T7 transcription buffer, 100 mM each NTP, nuclease-free water and RNase-free DNase. In vitro transcription kit or T7 RNA polymerase supplied by other companies can be used.
2. 5× T7 transcription buffer: 400 mM HEPES–KOH (pH 7.5), 120 mM MgCl$_2$, 10 mM spermidine, 200 mM DTT.
3. 25 mM each rNTP Mix: mix equal volumes of four individual 100 mM rNTPs (rATP, rUTP, rGTP, and rCTP).
4. RQ1 RNase-Free DNase (Promega). DNases supplied by other companies can be used.

5. RNeasy Mini Kit (Qiagen). RNA purification columns supplied by other companies can be used.

6. Nuclease-free water.

2.4. Ligation

1. T4 RNA Ligase (Takara or other supplier).
2. 10× T4 RNA Ligase buffer: 500 mM Tris–HCl (pH 7.5), 100 mM $MgCl_2$, 100 mM DTT, 10 mM ATP.
3. T4 polynucleotide kinase (Takara or other supplier).
4. RNeasy Mini Kit (Qiagen). RNA purification columns supplied by other companies can be used.
5. Nuclease-free water.
6. Thermal cycler.

2.5. Translation

1. Retic Lysate IVT™ Kit (Ambion), which contains rabbit reticulocyte lysate, translation mix, and nuclease-free water. Store at −80°C. Rabbit reticulocyte lysates supplied by other companies can be used.
2. 3 M KCl prepared with nuclease-free water.
3. 1 M $MgCl_2$ prepared with nuclease-free water.
4. Nuclease-free water.

2.6. Denaturing Urea Sodium Dodecyl Sulfate–Polyacrylamide Gel Electrophoresis

1. Sodium dodecyl sulfate–polyacrylamide gel electrophoresis (SDS–PAGE) apparatus (various suppliers).
2. Glass plate: an example of size is 100 mm (W)×70 mm (H)×1.0 mm (D).
3. Comb: 1.0 mm thick.
4. 2× Sample buffer: mix 1 g of sucrose, 4 mL of 10% SDS, 2 mg of bromo-phenol blue, 2.5 mL of 500 mM Tris–HCl (pH 6.8), and ddH_2O up to 10 mL.
5. 2-Mercaptoethanol.
6. 10× Running buffer: 250 mM Tris, 1,920 mM glycine, 1.0% SDS.
7. 1.5 M Tris–HCl, pH 8.8.
8. 500 mM Tris–HCl, pH 6.8.
9. 40% Acrylamide/bis-solution (37.5:1): 40% acrylamide and 1% bis-acrylamide in ddH_2O. Store at 4°C (see Note 5).
10. 10% Sodium dodecyl sulfate (SDS): 10% SDS in ddH_2O.
11. Ammonium persulfate (APS): prepare a 20% solution in ddH_2O. Store at 4°C.
12. N,N,N',N'-tetramethylethylenediamine (TEMED). Store at 4°C.

2.7. Synthesis of mRNA/cDNA–Protein Fusion Library

1. Streptavidin-magnetic beads: MAGNOTEX-SA (Takara) or Dynabeads® M-270 Streptavidin (Invitrogen). Store at 4°C.
2. Solution A: diethylpyrocarbonate (DEPC)-treated 100 mM NaOH, DEPC-treated 50 mM NaCl.
3. Solution B: DEPC-treated 100 mM NaCl.
4. 2× Binding buffer: 20 mM Tris–HCl (pH 8.0), 2 mM EDTA, 2 M NaCl, 0.2% Triton X-100. This buffer is supplied with MAGNOTEX-SA.
5. Ribosome releasing buffer: 50 mM Tris–HCl (pH 8.0), 20 mM EDTA prepared with nuclease-free water.
6. 5× RT buffer: 250 mM Tris–HCl (pH 8.3), 375 mM KCl, 15 mM $MgCl_2$.
7. Reverse transcriptase: SuperScript III Reverse Transcriptase (Invitrogen), to which 5× RT buffer [250 mM Tris–HCl (pH 8.3), 375 mM KCl, 15 mM $MgCl_2$] and 100 mM DTT are attached. Store at –20°C. Reverse transcriptase supplied by other various companies can be used.
8. 2.5 mM each dNTP Mix (various suppliers). Store at –20°C.
9. Magnetic separator (such as DynaMag™ series from Invitrogen).
10. Temperature-controllable tube rotator (such as SNP-24B, NISSIN) (see Note 6).
11. Nuclease-free water.

2.8. Restriction Enzyme Treatment

1. *Pvu*II restriction enzyme (Takara or other supplier). Store at –20°C.
2. 10× *Pvu*II buffer: 100 mM Tris–HCl (pH 7.5), 100 mM $MgCl_2$, 10 mM DTT, 500 mM NaCl. Store at –20°C.
3. 0.1% BSA: supplied with *Pvu*II restriction enzyme from Takara. Store at –20°C.
4. Magnetic separator (such as DynaMag™ series from Invitrogen).
5. Temperature-controllable tube rotator (such as SNP-24B, NISSIN) (see Note 6).
6. Nuclease-free water.

2.9. His-tag Purification

1. Ni-NTA magnetic agarose beads (Qiagen). Store at 4°C.
2. Ni-NTA binding buffer: 50 mM NaH_2PO_4, 300 mM NaCl, 10 mM imidazole, 0.05% Tween 20, pH 8.0.
3. Ni-NTA wash buffer: 50 mM NaH_2PO_4, 300 mM NaCl, 20 mM imidazole, 0.05% Tween 20, pH 8.0.
4. Ni-NTA elution buffer: 50 mM NaH_2PO_4, 300 mM NaCl, 250 mM imidazole, 0.05% Tween 20, pH 8.0.

	5. Gel filtration column (such as Micro Bio-Spin 6 Column from Bio-Rad). Store at 4°C.
	6. Tube rotator (various suppliers).
2.10. Preparation of Biotinylated IgG for Test Screening	1. IgG from rabbit serum (I5006, Sigma). Store at 4°C.
	2. EZ-Link™ Sulfo-NHS-SS-Biotin (21331, Thermo Scientific). Store at 4°C.
	3. 100 mM phosphate buffer: 100 mM phosphate, 150 mM NaCl, pH 7.4.
	4. Micro Dializer (NIPPON Genetics or other suppliers).
	5. 1-L beaker.
	6. Magnetic stirrer.
	7. Stirrer bar.
2.11. Test Screening	1. Double-strand (ds) DNA coding for B domain of protein A (BDA): 5′-GATCCCGCGAAATTAATACGACTCACTATAGGGGAAGTATTTTTACAACAATTACCAACAACAACAACAAACAACAACAACATTACATTTTACATTCTACAACTACAAGCCACCatggataacaaattcaacaaagaacaacaaatgctttctatgaaatcttacatttacctaacttaaacgaagaacaacgcaatggtttcatccaaagcctaaaagatgacccaagccaaagcgctaaccttttagcagaagctaaaaagctaaatgatgctcaagcaccaaaagctgacaacaaattcaacGGGGGAGCAGCCATCATCATCATCATCACGGCGGAAGCAGGACGGGGGGCGGCGGGGAAA-3′, which is composed of T7 promoter, omega sequence, Kozak consensus sequence, the gene of BDA, Linker-hybridize region. The sequence coding for BDA is noted with small letters (see Fig. 3a and Note 7).
	2. dsDNA coding for POU-specific DNA binding domain of Oct-1 (PDO): 5′-GATCCCGCGAAATTAATACGACTCACTATAGGGGAAGTATTTTTACAACAATTACCAACAACAACAACAAACAACAACAACATTACATTTTACATTCTACAACTACAAGCCACCatggaccttgaggagcttgagcagtttgccaagaccttcaaacaaagacgaatcaaacttggattcactcagggtgatgttgggctcgctatggggaaactatatggaaatgacttcagccaaactaccatctctcgatttgaagccttgaacctcagctttaagaacatggctaagttgaagccacttttagagaagtggctaaatgatgcagagGGGGGAGGCAGCCATCATCATCATCATCACGGCGGAAGCAGGACGGGGGGCGGCGGGGAAA-3′, which is composed of T7 promoter, omega sequence, Kozak consensus sequence, the gene of PDO, Linker-hybridize region. The sequence coding for PDO is noted with small letters (see Fig. 3a and Note 7).
	3. 100 mM phosphate buffer: same as written in Subheading 2.10 (item 3).

4. 2× Binding buffer: same as written in Subheading 2.7 (item 4).
5. 10× Selection buffer: 500 mM Tris–HCl (pH 7.6), 10 mM EDTA, 5 M NaCl, 1.0% Tween 20. Prepare with nuclease-free water.
6. 1× Elution buffer: 500 mM Tris–HCl (pH 7.6), 10 mM EDTA, 5 M NaCl, 50 mM DTT, 1.0% Tween 20. Prepare with nuclease-free water.
7. Tube rotator (various suppliers).
8. 99.5% Ethanol.
9. 70% Ethanol: prepared by dilution of 99.5% Ethanol with ddH$_2$O.
10. Primer 1: 5′-(F)-CAACAACATTACATTTTACATTCTACAA CTACAAGCCACC-3′, where (F) is fluorescein isothiocyanate (FITC). Prepare a 20 µM solution in ddH$_2$O.
11. Primer 2: 5′-TTTCCCCGCCGCCCCCGTCCTGCTTCCG CCGTGATGAT-3′. Prepare a 20 µM solution in ddH$_2$O.
12. Taq DNA polymerase (various suppliers) or other thermostable polymerase. Store at −20°C.
13. 2.5 mM each dNTP Mix (various suppliers). Store at −20°C.
14. 10× PCR buffer (various suppliers, supplied with DNA polymerase in general). Store at −20°C.
15. Magnetic separator (such as DynaMag™ series from Invitrogen).
16. Temperature-controllable tube rotator (see Note 6).
17. Thermal cycler.

3. Methods

The procedure of in vitro selection of functional proteins and peptides in cDNA display method (see Fig. 2) is as follows: (1) Preparation of the dsDNA library against each binding target (Subheading 3.3), (2) in vitro transcription of the dsDNA library into the mRNA library (Subheading 3.4), (3) ligation of the mRNA library with puromycin linkers (Subheading 3.5), (4) in vitro translation of the linker-ligated mRNA library to form the mRNA–protein fusion library (Subheading 3.6), (5) purification of the mRNA–protein fusion library using streptavidin magnetic beads and reverse transcription of the mRNA–protein fusion library to form the mRNA/cDNA–protein fusion library (Subheading 3.8), (6) release of the mRNA/cDNA–protein fusion library from the beads by *Pvu*II treatment (Subheading 3.9), (7) purification of

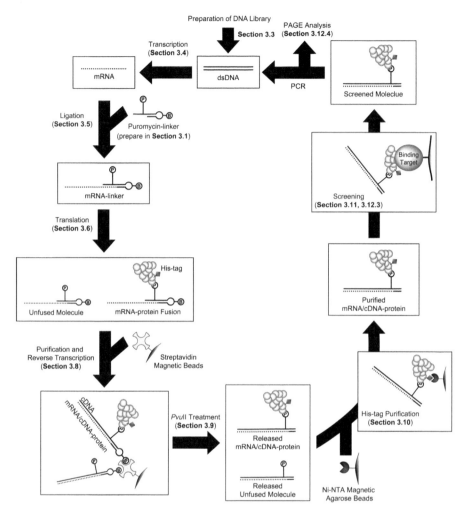

Fig. 2. The schematic diagram of the screening cycle of the cDNA display. The initial dsDNA library is prepared first (Subheading 3.3). The dsDNA library is transcribed into the mRNA library (Subheading 3.4). The mRNA library is ligated with the puromycin linker (prepared in Subheading 3.1) (Subheading 3.5). The ligated mRNA library is translated in vitro and the mRNA–protein fusion library is formed (Subheading 3.6). The mRNA–protein fusion library is purified on streptavidin magnetic beads and reverse-transcribed to form the mRNA/cDNA–protein fusion library (Subheading 3.8). The mRNA/cDNA–protein fusion library is released from the beads by PvuII treatment (Subheading 3.9). The mRNA/cDNA–protein fusion molecules are purified by Ni-NTA magnetic agarose beads (Subheading 3.10). The purified mRNA/cDNA–protein fusion library is exposed to the binding target and bound molecules are screened (Subheadings 3.11 and 3.12.3). The screened molecules are amplified by PCR to generate the next dsDNA library or analysis (Subheading 3.12.4).

mRNA/cDNA–protein fusion molecules using the affinity of the His-tag and Ni-NTA magnetic agarose beads (Subheading 3.10), (8) screening of the mRNA/cDNA–protein fusion molecules that possess a binding capacity to target molecules (Subheadings 3.11 and 3.12.3), and (9) PCR amplification (with or without mutation) of the screened molecules to make the dsDNA library of the next screening cycle. In this article, we do not show the details for the section of preparation of the initial dsDNA library and the section of screening (an example of these protocols is shown in Chapter 13), since the library design and screening conditions are varied and should be investigated for each target molecule.

The quality (e.g., randomness and diversity) of an initial library is essential for success in a selection. The initial library for in vitro selection can be classified into two categories: first, a synthetic random library, and second, a library derived from a natural source or a mutational library of a natural protein. In the construction of the synthetic random library, the NNK/NNS (N=A/T/G/C, K=G/T, S=C/G) codons are used when the random region is relatively short and a biased composition of the amino acids is not required. When the random region is relatively long or the biased composition of the amino acids is required, the advanced library construction methods such as Multiline split DNA synthesis (MLSDS) (3), Y-ligation-based block shuffling (YLBS) (4), and the use of trinucleotide phosphoramidites (5) should be used because the NNK/NNS codons include the amber termination codon (which cannot be used in cDNA display because the early termination of translation prevents linkage between the mRNA and its encoded protein) and the undesired bias of the amino acids composition when used in relatively longer random regions. cDNA libraries may be purchased from various suppliers or can be prepared afresh. Mutational libraries of natural proteins may be prepared using methods such as error-prone PCR, DNA shuffling, and so on (6).

The purity of the immobilized target molecule is also important to success in an affinity selection. The target molecule should be as pure as possible to avoid selecting undesired proteins, which bind to contaminants in the target sample. In general, the target molecule is immobilized on agarose or magnetic beads via chemical linkage or affinity binding between biotin and streptavidin with which the target molecule and the beads are derivatized, respectively.

We recommend performing the test screening described in Subheading 3.12 to confirm the entire procedure before proceeding to the actual screen.

3.1. Synthesis of Puromycin Linker

3.1.1. Reduction of Puromycin Segment

1. Prepare a 3 mM solution of the puromycin segment in ddH$_2$O.
2. Mix 3.3 µL of 3 mM puromycin segment (10 nmol) with 22.5 µL of 1 M Na$_2$HPO$_4$ (pH 9.0) and 2.5 µL of 1 M DTT in a 1.5-mL tube.

3. React for 1 h at room temperature with agitation using a vortex mixer.

4. Preequilibrate a NAP-5 column with 9 mL of 20 mM phosphate buffer (pH 7.2).

5. Apply the reacted mixture to the NAP-5 column and discard the flow-through.

6. Add 450 µL of 20 mM phosphate buffer (pH 7.2) onto the NAP-5 column and discard the flow-through.

7. Add 1 mL of 20 mM phosphate buffer (pH 7.2) onto the NAP-5 column and collect the fluorescent fraction (see Notes 8 and 9).

3.1.2. EMCS Modification of Biotin Segment

1. Prepare a 0.5 mM solution of the biotin segment in ddH$_2$O.

2. Mix 10 µL of 0.5 mM biotin segment (5 nmol) with 50 µL of 200 mM phosphate buffer (pH 7.2) and 10 µL of 100 mM EMCS in a 1.5-mL tube.

3. Incubate the mixture for 30 min at 37°C (see Note 10).

4. Add 10 µL of Quick-Precip™ Plus Solution and 2 volumes of 99.5% ethanol.

5. Centrifuge the mixture at maximum speed for 3 min at room temperature using a microcentrifuge.

6. Discard the supernatant and add 200 µL of 70% ethanol.

7. Centrifuge the mixture at maximum speed for 3 min at room temperature.

8. Remove the supernatant and proceed to the next step immediately (see Note 11).

3.1.3. Cross-Linking of the Puromycin Segment and the Biotin Segment

1. Dissolve the precipitated pellet from Subheading 3.1.2 in the whole collected solution from Subheading 3.1.1 to cross-link the puromycin segment and the biotin segment.

2. Incubate the mixture overnight at 4°C.

3. Add 1/20 volumes of 1 M DTT and react for 30 min at room temperature using a vortex mixer.

4. Add 10 µL of Quick-Precip™ Plus Solution and 2 volumes of 99.5% ethanol.

5. Centrifuge the mixture at maximum speed for 15 min at room temperature using microcentrifuge.

6. Discard the supernatant and add 200 µL of 70% ethanol.

7. Centrifuge the mixture at maximum speed for 5 min at room temperature.

8. Remove the supernatant and dissolve the pellet with 30 µL of ddH$_2$O (see Note 12).

9. Mix 0.2–0.5 μL of the product with an equal volume of 2× loading buffer and run on the 8 M urea 16% denaturing polyacrylamide gel electrophoresis at 250 V for 45 min at 60°C in a 0.5× TBE buffer as in Subheading 3.2. As a reference, run the puromycin segment, biotin segment, and 10-bp DNA ladder on the same gel. After running, image the gel by fluorescence of FITC and SYBR® Gold stain using the fluorescence scanner (see Note 13).

3.1.4. HPLC Purification of Puromycin Linker

1. Perform the HPLC purification of the product to remove the unreacted biotin segment. HPLC conditions are as follows: column, Waters Symmetry 300C18, 4.6×250 mm, particle sizes 5 μm; solvent A, 100 mM TEAA; solvent B, acetonitrile/water (80:20, v/v); gradient, B/A (15–35%, 30 min); flow rate, 0.5 mL/min; detection, absorbance at 260 and 490 nm.

2. Dry up the collected sample using vacuum centrifuge and dissolve it with appropriate volume (100–200 μL) of nuclease-free water (see Note 14) and add 0.1 volume of Quick-Precip™ Plus Solution and 3 volumes of 99.5% ethanol.

3. Centrifuge the mixture at maximum speed at room temperature for 3 min using microcentrifuge.

4. Discard the supernatant and rinse the precipitated pellet with 70% ethanol.

5. Remove residual alcohol and dissolve the pellet in 50 μL of nuclease-free water.

6. Determine the concentration of the product by absorbance at 260 nm. Usually the yield is about 1 nmol (see Note 15).

7. Store the puromycin linker at −20°C.

3.2. Denaturing Urea-PAGE

1. Prepare 10 mL of the appropriate percentage 8 M urea polyacrylamide gel: appropriate volume of 40% acrylamide/bis-solution (19:1), 4.8 g of urea, 0.5 mL of 10× TBE and ddH$_2$O up to 10 mL. Add 25 μL of 20% APS and 10 μL of TEMED and mix quickly. Immediately load the solution into a previously assembled 0.75–1.0 mm thick glass plate, set the comb, and allow the gel to polymerize (see Notes 5 and 16).

2. When the gel is polymerized, remove the comb and the spacer at the bottom of the gel and wash the gel plate with water.

3. Mount the gel plate on the apparatus and fill the bottom and upper tank with 0.5× TBE.

4. Prewarm the running buffer and prerun the gel for 15–20 min.

5. Before applying the samples, wash the well of the gel with 0.5× TBE by pipetting to remove the sinking urea.

6. Apply the samples, which are mixed with an equal volume of 2× loading buffer, in the well.

7. Run the electrophoresis at the appropriate condition.

8. After running, remove the glass plate and stain the gel with SYBR® Gold nucleic acid gel stain if required, place the gel on the nonluminescence glass, and image the fluorescence of the gel using a fluorescence scanner.

3.3. Preparation of DNA Library

Our DNA library for cDNA display is composed of the following base sequences (see Fig. 3a): T7-promoter, omega sequence, which is the translation enhancer of tobacco mosaic virus (7), Kozak consensus sequence, which is consensus sequence of the near upstream region of the initiation codon in eukaryotes (8), the library sequence without a stop codon, which encodes diverse proteins or peptides, the spacer sequence coding for the four amino acids "Gly-Gly-Gly-Ser", the hexahistidine tag, the spacer sequence coding for three amino acids "Gly-Gly-Ser," and the hybridization region, which is hybridized with the DNA sequence of the puromycin linker. These sequences are written from 5′ to 3′. An example of the detailed protocol of preparation of the DNA library is written in Chapter 13.

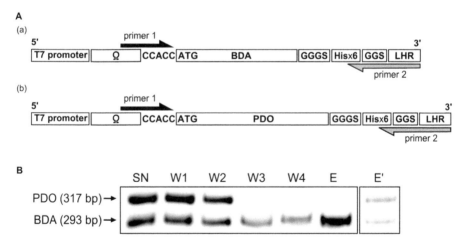

Fig. 3. (**A**) DNA constructs for test screening. (**a**) DNA construct coding for BDA. (**b**) DNA construct coding for PDO. T7 promoter, promoter sequence for T7 polymerase; Ω translation enhancer of tobacco mosaic virus, *CCACC* Kozak consensus sequence, *ATG* initiation codon, *BDA* coding region of BDA, *PDO* coding region of PDO, *GGGS* spacer sequence composed of four amino acids Gly-Gly-Gly-Ser, *His × 6* Histidine-tag, *GGS* spacer sequence composed of three amino acids Gly-Gly-Ser, *LHR* linker hybridized region on which puromycin linker is hybridized. *Arrows* indicate hybridizing region of primer 1 and 2 used for amplification of screened molecules in Subheading 3.12.4. (**B**) An example of a result for test screening. The equimolar binary mixture of cDNA–protein coding for BDA and PDO was exposed to a single affinity selection against IgG. The PCR products of the cDNA-display molecules sampled during and after the screening procedure were analyzed by 8 M urea denaturing 5% PAGE. PCR products derived from BDA and PDO are shown by *arrows*. *SN* supernatant of the initially added mixture, *W1–4* supernatant after each wash procedure, *E* after the screening sample, *E'* after the screening sample of the screening procedure against the streptavidin magnetic beads without IgG.

3.4. In Vitro Transcription

1. The amount of the template DNA used for in vitro transcription varies depending on the diversity of the library. The diversity of the library is dependent on the number of randomized or mutated amino acids. In general, the diversity of the library varies from 1×10^{10} to 1×10^{14}, which corresponds to a library size from 20 fmol to 200 pmol if it is a single copy. In this protocol, we describe the in vitro selection using a protein or peptide library with eight randomized amino acids as an example. The diversity of the library with eight amino acids is $20^8 = 2.56 \times 10^{10}$, which corresponds to the library size of 42.5 fmol if it is a single copy. Therefore, we use more than 0.5 pmol of the template DNA for in vitro transcription to give a copy number of about 10 for every molecular variant in the library. It is necessary to vary the reaction volumes given in the following steps depending on the diversity of the library.

2. Prepare the transcription solution: mix more than 0.5 pmol of the dsDNA from Subheading 3.3, 30 μL of 25 mM each rNTP Mix, 20 μL of 5× T7 transcription buffer, 4 μL of T7 Enzyme-Mix, and nuclease-free water up to 100 μL (see Notes 17 and 18).

3. Incubate the mixture at 37°C for 2–4 h. Then, add 4 μL of RNase-free DNase (1 U/μL) and incubate at 37°C for 15 min.

4. Analyze the product by denatured polyacrylamide gel electrophoresis as in Subheading 3.2 (see Note 19) and purify the mixture using a RNA purification silica-membrane column, such as RNeasy Mini Kit, and elute the transcript from the column with nuclease-free water. Determine the concentration of the transcript by absorbance at 260 nm.

5. Store the transcript at −20 or −80°C until the ligation reaction.

3.5. Ligation of the Transcript and Puromycin Linker

1. Prepare the following solution in a PCR tube: 100 pmol of the transcript from Subheading 3.4, 200 pmol of the puromycin linker from Subheading 3.1, 20 μL of 10× T4 RNA ligase buffer, and nuclease-free water up to 188 μL.

2. Heat the solution at 95°C and cool to 25°C gradually over 15 min.

3. After cooling to 25°C, add 6 μL of T4 polynucleotide kinase (10 U/μL) and 6 μL of T4 RNA Ligase (30 U/μL) and incubate at 25°C for 1–2 h (see Note 20).

4. Analyze 1 μL of the reaction mixture by 8 M urea denatured PAGE running with the unligated transcript as a reference as in Subheading 3.2 (see Note 19). After running, image the gel by the fluorescence of FITC and SYBR® Gold stain using the fluorescence scanner. If ligation is achieved a band shifted to higher molecular mass is observed. In general, >90% of the transcript is ligated with puromycin linker.

5. Purify the ligated product using the RNA purification silica-membrane column, such as RNeasy Mini Kit, and elute the ligated product with nuclease-free water.

6. Determine the concentration of the ligated product by absorbance at 260 nm and estimate the copy number of the library.

7. Store the ligated product at −20 or −80°C until the translation reaction step.

3.6. In Vitro Translation (Synthesis of mRNA–Protein Fusion Library)

1. The final yield of the mRNA/cDNA–protein fusion library, which will be used in the screening, is about 1% of the initial library, which was input into the translation reaction. The puromycin linker ligated mRNA library, which is input into the translation reaction, should have a copy number of at least 100. We use 50 pmol of the puromycin linker ligated mRNA library, which has about 1,000 copies because the diversity of the library is 2.56×10^{10}, which corresponds to 42.5 fmol if it is a single copy.

2. Prepare the in vitro translation solution as follows: mix on ice 50 pmol of puromycin linker ligated RNA from Subheading 3.5, 20 μL of 20× translation mix (−Leu), 20 μL of 20× translation mix (−Met), 544 μL of rabbit reticulocyte lysate, and nuclease-free water up to 800 μL (see Note 20).

3. Incubate the mixture at 30°C for 20 min (see Note 21).

4. Add 320 μL of 3 M KCl and 96 μL of 1 M $MgCl_2$ and incubate at 37°C for 90 min.

5. Analyze the formation of the mRNA–protein fusion formation by using the denaturing urea–SDS–PAGE in the next section.

3.7. Denaturing Urea–SDS–PAGE

1. Prepare the 6% separation gel as follows: mix 2.5 mL of 1.5 M Tris–HCl (pH 8.8), 4.8 g of urea, 1.5 mL of 40% acrylamide/bis-solution (37.5:1), 100 μL of 10% SDS and ddH_2O up to 10 mL. Add 25 μL of 20% APS and 5 μL of TEMED and mix. Load the solution into the previously assembled glass-plate until the solution reaches 1–2 cm below the teeth of the comb. Overlay water-saturated *n*-butanol or ddH_2O on the separation gel and allow the gel to polymerize (see Note 22).

2. Prepare the 4% stacking gel as follows: mix 1.25 mL of 500 mM Tris–HCl (pH 6.8), 2.4 g of urea, 0.5 mL of 40% acrylamide/bis-solution (37.5:1), 50 μL of 10% SDS, and ddH_2O up to 5 mL. Add 12.5 μL of 20% APS and 5 μL of TEMED and mix. Remove the water-saturated *n*-butanol or ddH_2O overlaid on the separation gel and load the stacking gel solution onto the polymerized separation gel, set the comb and allow the gel to polymerize (see Note 22).

3. Mix 12 μL of the translated mixture from Subheading 3.6 (corresponding to 0.5 pmol of mRNA in the reaction) with an

equal volume of 2× sample buffer and 1/20 volume of 2-mercaptethanol (see Note 23). Apply the sample and run on the denaturing urea–SDS–PAGE prepared above. As a reference, run puromycin linker ligated RNA from Subheading 3.5 on the same gel. Run at 10 mA on the stacking gel, 20 mA on the separating gel until the bromophenol blue reaches the bottom of the gel in a 1× running buffer. After running, image the fluorescence of the FITC derived from the puromycin linker using the fluorescence scanner. If mRNA–protein formation is achieved, an upper-shifted band is observed. In general, 20–30% of fusion formation is observed.

4. Although the translated sample can be stored at −20°C, proceed to the next step as soon as possible.

3.8. Synthesis of mRNA/cDNA–Protein Fusion Library

3.8.1. Purification of mRNA–Protein Fusion from the Lysate by Streptavidin Magnetic Beads

1. Discard the supernatant of 500 μL of the streptavidin-magnetic beads (see Notes 24 and 25) and wash the beads two times with 500 μL of solution A and one time with 500 μL of solution B.

2. Suspend the beads in 1,204 μL of the translated sample and 1,204 μL of the 2× binding buffer. Prepare this solution in several tubes because the volume is too large for one tube.

3. Incubate the mixture at 25°C for 30 min with agitation using a tube rotator (see Notes 26 and 27).

4. Wash the beads three times with 2 mL of 1× binding buffer.

5. Suspend the beads in 250 μL of the ribosome releasing buffer and incubate for 10 min at room temperature with agitation using the tube rotator (see Notes 26 and 27).

6. Wash the beads three times with 2 mL of 1× binding buffer.

7. Wash the beads one time with 2 mL of 1× RT buffer and discard the supernatant.

3.8.2. Reverse Transcription

1. Prepare the following reaction solution: 50 μL of 5× RT buffer, 50 μL of 2.5 mM each dNTP Mix, 12.5 μL of 100 mM DTT, 5 μL of SuperScript® III Reverse Transcriptase (200 U/μL), and nuclease-free water up to 250 μL (see Note 20).

2. Suspend the beads, on which the mRNA–protein fusions are captured, in the preprepared reaction solution from step 1 and incubate at 45°C for 30 min with agitation using a temperature-controllable tube rotator (see Notes 6, 26 and 27).

3.9. Restriction Enzyme Treatment (Release of mRNA/cDNA–Protein Fusion Library from Magnetic Beads)

1. Prepare the following reaction solution: 25 μL of 10× *Pvu*II buffer, 25 μL of 0.1% BSA, 5 μL of *Pvu*II (10 U/μL), and nuclease-free water up to 250 μL.

2. Discard the supernatant of the reverse transcription solution from Subheading 3.8.2 and wash three times with 1,500 μL of 1× binding buffer and then one time with 1,500 μL of 1× *Pvu*II buffer.

3. Discard the supernatant and suspend the beads, on which the mRNA/cDNA–protein fusions are captured, in the prepared *Pvu*II reaction solution from step 1 and incubate at 37°C for 1 h using the temperature-controllable tube rotator (see Notes 6, 26 and 27).

4. Collect the supernatant in which the mRNA/cDNA–protein fusions are dissolved.

3.10. His-tag Purification (Purification of mRNA/cDNA–Protein Molecules)

1. Discard the supernatant of 160 μL of Ni-NTA magnetic agarose beads (see Note 25) and suspend the beads with 250 μL of Ni-NTA binding buffer.

2. Add 250 μL of the collected sample from Subheading 3.9 (step 4) to the Ni-NTA magnetic agarose beads solution from step 1 and incubate for 60 min at room temperature with agitation using the tube rotator (see Notes 26 and 27).

3. Wash the beads two times with 800 μL of Ni-NTA wash buffer.

4. Add 75 μL of Ni-NTA elution buffer to the beads and incubate for 10 min at room temperature with agitation using the tube rotator.

5. Collect the supernatant and remove the excessive imidazole using the gel filtration column, which is equilibrated with the buffer used in the screening, according to the manufacturer's directions.

6. Store the purified mRNA/cDNA–protein fusion library at 4°C until the screening step and use it as soon as possible.

3.11. Screening

The protocol of the screening varies depending on the binding target. It is necessary to set and investigate the screening condition (such as the buffer composition, incubation time, wash condition, etc.) depending on each experiment. We show an example of the actual screening in Chapter 13. When a user performs an actual screening, refer to the protocol in Chapter 13.

3.12. Test Screening

In this section, we show the protocol of the test screening to confirm the screening procedures and one's skill, and we recommend performing this test screening before the actual screening. In the test screening, the B domain of protein A (BDA), which specifically binds to the Fc fragment of the immunoglobulin G (IgG), is used as a model molecule and should be enriched.

3.12.1. Preparation of Biotinylated IgG

1. Dissolve 1 mg of IgG in 1 mL of 100 mM phosphate buffer.
2. Dissolve 1 mg of sulfo-NHS-SS-Biotin in 166.7 μL of nuclease-free water (see Note 28).
3. Mix 100 μL of IgG solution and 13.45 μL of sulfo-NHS-SS-Biotin solution.

4. Incubate the mixture for 2 h at 4°C.
5. Perform a dialysis of the sample with 1 L of 100 mM phosphate buffer over night at 4°C using Micro Dializer under stirring.
6. Store the dialyzed sample at 4°C.

3.12.2. Preparation of mRNA/cDNA–Protein Fusion Mixture

1. Prepare two kinds of puromycin linker ligated mRNA coding for BDA and PDO, respectively, according to the protocol described above using two kinds of dsDNA coding for each gene [written in Subheading 2.11 (items 1 and 2), see Fig. 3a].
2. Prepare the in vitro translation solution as follows: mix on ice 4.5 pmol of puromycin linker ligated mRNA coding for BDA, 4.5 pmol of puromycin linker ligated mRNA coding for PDO, 3.6 µL of 20× translation mix (−Leu), 3.6 µL of 20× translation mix (−Met), 98 µL of rabbit reticulocyte lysate, and nuclease-free water up to 144 µL.
3. Perform in vitro translation reaction and proceed until the His-tag purification step according to the protocols written above with a reduced volume.

3.12.3. Screening Against Biotinylated IgG

1. Wash 40 µL of streptavidin-magnetic beads three times with 100 mM phosphate buffer.
2. Suspend the beads in the following mixture: 15 µL of biotinylated IgG from Subheading 3.12.1, 5 µL of 100 mM phosphate buffer and 20 µL of 2× binding buffer.
3. Incubate for 20 min at room temperature with agitation using the tube rotator.
4. Wash the beads three times with 1× binding buffer and one time with 1× selection buffer.
5. Suspend the beads in the following mixture: half of the volume of the mRNA/cDNA–protein fusion mixture from Subheading 3.12.2, 10 µL of 5× selection buffer, and nuclease-free water up to 50 µL.
6. Incubate for 30 min at 25°C with agitation using the tube rotator and collect the supernatant (SN).
7. Wash the beads as follows: add 300 µL of 1× selection buffer to the beads and incubate for 3 min in suspension using the tube rotator. Then collect the supernatant (W1).
8. Repeat step 7 three times and collect each of the supernatants (W2–4) (see Note 29).
9. Add 50 µL of 1× Elution buffer and incubate for 10 min at 37°C using the temperature-controllable tube rotator (see Notes 6, 26, 27 and 30).
10. Collect the supernatant (E).

11. Add 10 μL (for SN and E) or 20 μL (for W1–4) of Quick-Precip™ Plus Solution and 2 volumes of 99.5% ethanol to the collected samples.

12. Centrifuge the mixtures at maximum speed for 3 min at room temperature using a microcentrifuge.

13. Discard the supernatants and add 200 μL of 70% ethanol.

14. Centrifuge the mixtures at maximum speed for 3 min at room temperature.

15. Remove the supernatant and dissolve the pellets in 12 μL of ddH$_2$O.

3.12.4. PCR Amplification and PAGE Analysis

1. Perform PCR amplification on each collected sample (SN, W1–4, E) from Subheading 3.12.3 and the mRNA/cDNA–protein fusion mixture before screening from Subheading 3.12.2 as follows: Mix 4 μL of each collected sample, 0.5 μL of 20 μM primer 1, 0.5 μL of 20 μM primer 2, 2 μL of 25 mM each dNTP Mix, 2.5 μL of 10× PCR buffer, 0.125 μL of DNA polymerase (5 U/μL), and ddH$_2$O up to 25 μL in the PCR tubes. The PCR program is as follows: 95°C for 2 min, repeat 30 cycles of three steps, 95°C for 25 s, 69°C for 20 s, 72°C for 30 s, followed by 72°C for 2 min.

2. Analyze 3 μL of the PCR products by denatured polyacrylamide gel electrophoresis as in Subheading 3.2. The PAGE condition is as follows: 8 M urea, 5% polyacrylamide gel, run at 200 V for 50 min at 60°C. After running, image the fluorescence of FITC derived from primer 1. If the enrichment of the BDA gene is achieved, the 293 bp band derived from the BDA gene will be the majority in the sample "E" whereas the 317 bp band derived from the PDO gene and the 293 bp band derived from the BDA gene are equally strong before screening. An example of the test-screening result is shown in Fig. 3b.

4. Notes

1. The number of spacer 18 U has no effect on mRNA–protein formation when $n = 1$–4 in our experiments.

2. Prepare the EMCS solution just before use because the reactive groups, maleimide and NHS ester, are hydrolyzed easily in aqueous solution. To prevent condensate formation inside the vial, which causes hydrolysis of the reactive groups, open the cap of the vial stored at 4°C after the temperature of the vial is back to room temperature.

3. Perform the degassing of 100 mM TEAA and 80% acetonitrile before use to prevent air-bubble formation in the HPLC system.

4. PAGE at 60°C using temperature-controllable PAGE apparatus is recommended to denature the polynucleotide completely.

5. Because unpolymerized acrylamide is a neurotoxin, wear gloves and avoid exposure.

6. If there is not a temperature-controllable tube rotator, use a normal tube rotator in a temperature-controllable chromatography chamber or tap the tube frequently during the reaction to keep the beads in suspension.

7. Although the protocol for the preparation of this dsDNA is described in ref. 9, it requires laborious and troublesome procedures. Therefore, we recommend synthesizing the dsDNA using a custom artificial gene synthesis service.

8. Set the NAP-5 column on a clamp stand and put three 1.5-mL tubes under the outlet of the column. Add 1 mL of phosphate buffer onto the column and collect the first no-fluorescence drops in the first 1.5-mL tube. When the fluorescence has become observed in drops, collect the drops in the second 1.5-mL tube. When the fluorescence passed away, collect the drops in the third 1.5-mL tube.

9. Proceed to Subheading 3.1 (step 3) as soon as possible, because the reduced thiols are able to form disulfide-bonds with each other.

10. Do not allow to react for over 30 min because a longer reaction causes hydrolysis of the maleimide, which has to react with the thiol group of the reduced puromycin segment in the next section.

11. Proceed to the next step as soon as possible to prevent hydrolysis of the maleimide.

12. Remove the supernatant completely, because the unreacted puromycin segment, which cannot be removed by HPLC purification at Subheading 3.1.4, is removed in this step.

13. The cross-linked product will migrate to the same point as 80–90-bp double-stranded DNA in a denaturing polyacrylamide gel.

14. Confirm the purity of the sample before ethanol precipitation using denaturing polyacrylamide gel electrophoresis.

15. 1 nmol of puromycin linker is sufficient for a screening for a single binding target when the diversity of a library is lower than 1×10^{14}.

16. The polyacrylamide gel wrapped in plastic wrap is able to be stored at 4°C.

17. The capping analog is not required. It has almost no effect on the yield of mRNA–protein fusion formation. However, it is desirable to use the capping analog if degradation of mRNA is observed.

18. The RNase inhibitor is mixed in the T7 Enzyme-Mix.
19. The percentage of acrylamide and the electrophoresis condition varies depending on the length of the nucleotide sequence. Please investigate these in each case.
20. RNase inhibitors supplied from various companies can be used if necessary, even though RNA degradation is not observed without an RNase inhibitor in our experiments.
21. An incubation for longer than 20 min may cause degradation of mRNA. An incubation time of 20 min is sufficient for mRNA–protein fusion formation.
22. In preparation of the gel solution, it is important to mix completely to polymerize the gel uniformly, but do not agitate vigorously because oxygen taken into the solution during the agitation inhibits polymerization. A microwave oven can be used to dissolve urea. After setting the comb, incubate for at least 1 h to polymerize the gel completely. The 8 M urea–SDS–polyacrylamide-gel wrapped in plastic wrap can be stored at 4°C.
23. Heat denaturing is not necessary after adding 2× sample buffer.
24. The required volume of the streptavidin-magnetic beads to capture the biotinylated nucleotide entirely may vary depending on the length and sequence of the nucleotide. We recommend investigating the volume of the beads for each nucleotide.
25. Magnetic beads can be gathered and a supernatant can be discarded or collected easily using a magnetic separator.
26. Adjust the speed of rotation to keep the beads in suspension.
27. If liquid adheres to the cap of the tube during agitation, divide the solution into multiple tubes. If the solution slips into the space between the cap and the tube, it stays there and cannot be mixed efficiently.
28. Prepare the solution just before use, because sulfo-NHS ester is hydrolyzed easily in aqueous solution. To prevent condensate formation inside the vial, which causes hydrolysis of the reactive groups, open the cap of the vial stored at 4°C after the temperature of the vial is back to room temperature.
29. We recommend exchanging the tubes in each of the washing steps to avoid contaminants that bind to the wall of the tube nonspecifically.
30. The disulfide bond in the linker between IgG and biotin is cleaved by DTT in the elution buffer and mRNA/cDNA–protein fusion molecules, which bind to IgG, are released from the beads.

Acknowledgments

Much of the contents of this work comprises the collected wisdom of a number of colleagues. The authors would like to thank Prof. Yuzuru Husimi for providing helpful comments to the manuscript.

References

1. Nemoto N, Miyamoto-Sato E, Husimi Y and Yanagawa H (1997) In vitro virus: Bonding of mRNA bearing puromycin at the 3′-terminal end to the C-terminal end of its encoded protein on the ribosome in vitro. FEBS Lett 414:405–408
2. Roberts RW and Szostak JW (1997) RNA-peptide fusions for the in vitro selection of peptides and proteins. Proc Natl Acad Sci USA. 94:12297–12302
3. Tabuchi I, Soramoto S, Ueno S and Husimi Y (2004) Multi-line split DNA synthesis: a novel combinatorial method to make high quality peptide libraries. BMC Biotechnol 4:19
4. Kitamura K, Kinoshita Y, Narasaki S, Nemoto N, Husimi Y and Nishigaki K (2002) Construction of block-shuffled libraries of DNA for evolutionary protein engineering: Y-ligation-based block shuffling. Protein Eng 15:843–853.
5. Yáñez J, Argüello M, Osuna J, Soberón X and Gaytán P (2004) Combinatorial codon-based amino acid substitutions. Nucl Acids Res 32:e158
6. Neylon C (2004) Chemical and biochemical strategies for the randomization of protein encoding DNA sequences: library construction methods for directed evolution. Nucl Acids Res 32:1448–1459
7. Sawasaki T, Ogasawara T, Morishita R and Endo Y (2002) A cell-free protein synthesis system for high-throughput proteomics. Proc Natl Acad Sci USA. 99:14652–14657
8. Kozak M (1997) Recognition of AUG and alternative initiator codons is augmented by G in position +4 but is not generally affected by the nucleotides in positions +5 and +6. EMBO J 16:2482–2492
9. Yamaguchi J, Naimuddin M, Biyani M, Sasaki T, Machida M, Kubo T, Funatsu T, Husimi Y and Nemoto N (2009) cDNA display: a novel screening method for functional disulfide-rich peptides by solid-phase synthesis and stabilization of mRNA-protein fusions. Nucl Acids Res 37:e108

Part IV

Applications of Ribosome Display Methods Using Natural Amino Acids

Chapter 9

Optimisation of Antibody Affinity by Ribosome Display Using Error-Prone or Site-Directed Mutagenesis

Leeanne Lewis and Chris Lloyd

Abstract

Affinity optimisation of antibodies can be achieved with great success by using directed evolution approaches, that is, the creation of and selection from diverse libraries. Here, we describe in detail methods to optimise antibody affinity for an antigen through directed evolution using ribosome display. Diversification of antibody single chain variable (scFv) domains is carried out by error-prone PCR and oligonucleotide-directed mutagenesis to generate random and targeted libraries respectively. Subsequent libraries are converted to ribosome display format and taken through cycles of transcription, translation, and selection. Since the starting point and the recovered product are linear DNA, this can easily be manipulated further to allow accumulation of beneficial mutations through iterative cycles of selection.

Key words: Ribosome display, scFv, Antibody, Antibody library, Directed evolution, Random mutagenesis, Directed mutagenesis, In vitro selection, Affinity maturation

1. Introduction

Increasing the binding affinity of a protein for a given target (for example improving an antibody–antigen interaction) is a frequent objective undertaken by protein engineering, and is one in which directed evolution approaches have had considerable success (1). Phage (2–4), bacterial (5) and yeast display (6) have been show to be valuable techniques to select improved proteins from diverse protein libraries. However, as each of these approaches requires living cells to display or propagate the protein libraries, they are typically limited in size by the transformation efficiency of the host cell. Additionally, directed evolution with these techniques requires frequent and labour intensive switching between *in vitro* and *in vivo* steps. Ribosome display overcomes these limitations as it is performed entirely in vitro, therefore allowing simultaneous selection and evolution of proteins from very large mutagenic libraries (>10^{12} variants) (7, 8). Ribosome display has been shown to be highly effective for the generation of antibodies with high affinity (10), for the improvement of other properties such as protein stability (9) and for the selection of peptides (8).

Affinity maturation by ribosome display starts with a linear DNA library, making it compatible with all in vitro methods of mutagenesis. During a cycle of ribosome display (see Fig. 1), DNA is transcribed to mRNA for use in in vitro translation under conditions that promote formation of stable ternary complexes consisting of mRNA, ribosome and translated protein. These complexes are then subjected to antigen selection with non-specific or conformationally disrupted protein complexes being removed by wash steps. mRNA from specifically bound complexes is then isolated for reverse transcription into cDNA and then amplification by PCR. Additional cycles of selection can be carried out to enrich for high affinity binders.

This protocol describes the use of ribosome display for isolating affinity matured single chain Fv (scFv) antibody fragments from random and targeted libraries, but is equally applicable to the display and selection of peptides and other binding proteins.

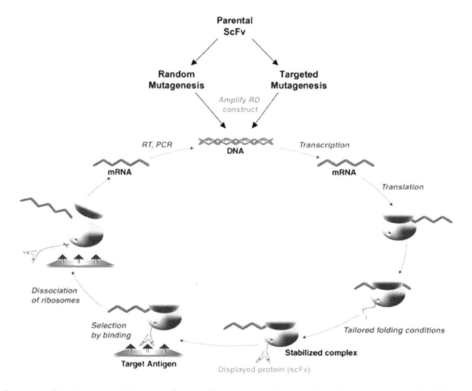

Fig. 1. Summary of the ribosome display cycle for the affinity optimisation of scFv antibody fragments. The DNA sequence encoding a scFv of interest is subjected to diversification by random (error-prone PCR) or targeted (site-directed) mutagenesis. The resulting library is converted to ribosome display format for transcription to mRNA, translation, and selection.

The complementarity-determining region (CDR) loops, in particular the CDR3 domains, of an antibody form the major interaction sites with antigen and are, therefore, the main areas to target for site-directed mutagenesis. Owing to the capacity of ribosome display to accommodate libraries of variants of greater than 10^{12}, up to seven amino acid residues can be randomised simultaneously in a single library whilst ensuring all possible diversity is maintained. Random mutagenesis libraries can also be made, in this case by subjecting scFv sequences to error-prone (EP)-PCR. The improvements in affinity achieved are generally more modest compared to the targeted approach, but random mutagenesis does offer other advantages in terms of speed and technical ease. Random mutagenesis can also provide information about non-predictable positions, such as those within the antibody framework regions that may significantly contribute to an affinity improvement. Mutational frequency can be tailored during EP-PCR to control the amount of nucleotide mutations in a given sequence. Additionally, EP-PCR can be performed between rounds of selection and coupled with site-directed libraries to allow further exploration of sequence space. Both these mutagenesis approaches require scFv with retained binding and improved affinity to be captured and enriched from a background of clones with impaired binding specificity and affinity. This is achieved through multiple rounds of affinity selection which involves reducing the antigen concentration round-by-round and/or the use of competitors during selection to favour recovery of clones with slower dissociation rates (9–12).

2. Materials

2.1. Preparation of the Parental scFv DNA Construct for Ribosome Display

1. DNA encoding the scFv of interest, for example purified plasmid DNA. Store at −20°C.
2. 2× Taq PCR master mix (or separate Taq polymerase plus a suitable reaction buffer and dNTP mix). Store at −20°C.
3. SDCAT V_H specific primer: 5′AGACCACAACGGTTTCCCTCTAGAAATAATTTTGTTTAACTTTAAGAAGGAGATATATCCATGGCC+≥10 nucleotides matching the start of the scFv coding sequence (ending in ≥1 C or G).
4. HisMycRev V_L specific primer (see Note 1): 5′ATTCAGATCCTCTTCTGAGATGAGATGGTGATGATGATGATGTGCGGCCGCACC+≥10 complementary nucleotides to the end of the scFv coding sequence (ending in ≥1 C or G).
5. Nuclease-free water. Store at 4°C.
6. 1× TAE gel electrophoresis buffer: 40 mM Tris-base, 1.14% (v/v) glacial acetic acid, 1 mM ethylenediaminetetraacetic acid (EDTA), pH 7.6.

7. 1% and 1.5% (w/v) agarose, 1× TAE gels with a DNA staining agent (see Note 2). Store at 4°C.
8. DNA gel loading dye.
9. DNA ladder (200 bp to 2 kb). Store at 4°C.
10. DNA gel purification kit.
11. 10 mM Tris-HCl pH8.0.
12. T7te primer: 5′GTAGCACCATTACCATTAGCAAG.
13. MycgeneIIIshortadapt primer (see Note 1): 5′ATCTCAG AAGAGGATCTGAATGGTGGCGGCTCCGGTTCCG GTGAT.
14. Gene III tether DNA – optionally prepared by gene synthesis, or by PCR as described in Subheading 3.1 step 4 (see Note 3). Store at −20°C.
15. Miniprep purification kit for plasmid DNA.
16. T7B primer: 5′ATACGAAATTAATACGACTCACTATAGGG AGACCACAACGG.
17. T6te primer: 5′CCGCACACCAGTAAGGTGTGCGGTATC ACCAGTAGCACCATTACCATTAGCAAG.

2.2. Construction of Error–Prone Mutagenesis Libraries

1. DNA encoding the scFv of interest, see Subheading 2.1 item 1.
2. Diversify™ PCR Random Mutagenesis Kit (Clontech) (see Note 4).
3. SDCAT V_H specific primer, see Subheading 2.1 item 3.
4. HisMycRev V_L specific primer, see Subheading 2.1 item 4.
5. Nuclease-free water, see Subheading 2.1 item 5.
6. 1× TAE gel electrophoresis buffer, see Subheading 2.1 item 6.
7. 1% and 1.5% (w/v) agarose, 1× TAE gels, see Subheading 2.1 item 7.
8. DNA gel loading dye.
9. DNA ladder, see Subheading 2.1 item 9.
10. DNA gel purification kit.
11. 2× Taq PCR master mix, see Subheading 2.1 item 2.
12. T7te primer, see Subheading 2.1 item 12.
13. MycgeneIIIshortadapt primer, see Subheading 2.1 item 13.
14. Gene III tether DNA, see Subheading 2.1 item 14.
15. 10 mM Tris-HCl pH 8.0.
16. T7B primer, see Subheading 2.1 item 16.
17. T6te primer, see Subheading 2.1 item 17.

2.3. Construction of CDR-Directed Mutagenesis Libraries

1. DNA encoding the scFv of interest, see Subheading 2.1 item 1.
2. QuikChange Site-Directed Mutagenesis Kit (Agilent) comprising: *Pfu* Turbo DNA polymerase supplied with 10× reaction buffer, 12.5 mM each dNTP PCR nucleotide mix, *Dpn*I restriction enzyme (see Note 5).
3. Forward and reverse oligonucleotides for stop template production (see Note 6).
4. Forward and reverse oligonucleotides for CDR randomisation (see Note 7).
5. Nuclease-free water, see Subheading 2.1 item 5.
6. 1× TAE gel electrophoresis buffer, see Subheading 2.1 item 6.
7. 1% and 1.5% (w/v) agarose, 1× TAE gels, see Subheading 2.1 item 7.
8. DNA gel loading dye.
9. DNA ladder, see Subheading 2.1 item 9.
10. Chemically competent *E. coli* cells. Store at −70°C (see Note 8).
11. T7te primer, see Subheading 2.1 item 12.
12. 2× Taq PCR master mix, see Subheading 2.1 item 2.
13. SDCAT V_H specific primer, see Subheading 2.1 item 3.
14. DNA gel purification kit.
15. HisMycRev V_L specific primer, see Subheading 2.1 item 4.
16. MycgeneIIIshortadapt primer, see Subheading 2.1 item 13.
17. Gene III tether DNA, see Subheading 2.1 item 14.
18. 10 mM Tris-HCl pH 8.0.
19. T7B primer, see Subheading 2.1 item 16.
20. T6te primer, see Subheading 2.1 item 17.

2.4. In Vitro Transcription and mRNA Translation

1. Non-stick, RNase-free, 1.5 ml microcentrifuge tubes (Sarstedt) (see Note 9).
2. Ribomax Large Scale RNA Production Kit (Promega) comprising the following: T7 transcription buffer, 25 mM rNTP mix (prepared by pooling equal volumes of 100 mM rUTP, rATP, rGTP, and rCTP), and T7 RNA polymerase (see Note 10).
3. Illustra ProbeQuant G-50 micro column (GE Healthcare).
4. 200 mg/ml heparin. Store at 4°C.
5. 10× *E. coli* wash buffer: 0.5 M Tris–acetate pH 7.5, 1.5 M sodium chloride, 0.5 M magnesium acetate, 1% (v/v) Tween-20. Store at 4°C.

6. Premix X: 250 mM Tris–acetate pH 7.5, 1.75 mM of each amino acid, 10 mM ATP, 2.5 mM GTP, 5 mM cAMP, 150 mM acetylphosphate, 2.5 mg/ml *E. coli* tRNA, 0.1 mg/ml folinic acid, 7.5% (w/v) PEG 8000. Store at –20°C.
7. 5 mg/ml protein disulphide isomerase (PDI). Store at –20°C.
8. 2 M potassium glutamate. Store at 4°C.
9. 0.1 M magnesium acetate. Store at 4°C.
10. S30 *E. coli* extract. Store at –80°C (see Note 11).
11. Microcentrifuge with cooling capacity.

2.5. Selection and Capture of Specifically Bound scFv-Ribosome-mRNA Complexes

1. Non-stick, RNase-free, 1.5-ml and 2.0-ml microcentrifuge tubes, see Subheading 2.4 item 1.
2. 200 mg/ml heparin. Store at 4°C.
3. 10× *E. coli* wash buffer, see Subheading 2.4 item 5.
4. Dynabeads M-280 streptavidin beads (Invitrogen). Store at 4°C.
5. 10% (w/v) non-fat dried milk in water, autoclaved at 121°C for 15 min and cooled to 50°C. Store at 4°C.
6. EB20: 50 mM Tris–acetate pH 7.5, 150 mM sodium chloride, 20 mM EDTA. Store at 4°C.
7. 10 mg/ml *S. cerevisiae* RNA in nuclease-free water. Store at –20°C.
8. High Pure RNA Isolation Kit (Roche).

2.6. Converting Selection Output mRNA Back to DNA in Ribosome Display Format

1. Non-stick, RNase-free, 1.5-ml microcentrifuge tubes, see Subheading 2.4 item 1.
2. 5× First Strand Buffer (Invitrogen). Store at –20°C.
3. 100 mM dithiothreitol (DTT). Store at –20°C.
4. T7te primer, see Subheading 2.1 item 12 or,
5. MycRestore primer: ATTCAGATCCTCTTCTGAGATGAG (see Note 1).
6. dNTP mix (25 mM each dATP, dTTP, dCTP, dGTP). Store at –20°C.
7. 40 U/μl RNasin. Store at –20°C.
8. 200 U/μl Superscript II Reverse Transcriptase (Invitrogen). Store at –20°C.
9. Nuclease-free water, see Subheading 2.1 item 5.
10. 2× Taq PCR master mix, see Subheading 2.1 item 2.
11. T7B primer, see Subheading 2.1 item 16.
12. Dimethylsulfoxide (DMSO).
13. 1× TAE gel electrophoresis buffer, see Subheading 2.1 item 6.

14. 1% and 1.5% (w/v) agarose, 1× TAE, gels, see Subheading 2.1 item 7.
15. DNA gel loading dye.
16. DNA ladder, see Subheading 2.1 item 9.
17. DNA gel purification kit.
18. T6te primer, see Subheading 2.1 item 17.

2.7. Real-Time PCR Analysis of Selection Outputs for Relative Quantitation of cDNA (See Note 12)

1. 2× Taqman PCR master mix. Store at 4°C.
2. Forward primer: 5′CTTGATTCTGTCGCTACTGATTA.
3. Reverse primer: 5′CCATTAGCAAGGCCGGAA.
4. Taqman probe: 5′FAM-GTCACCAATGAAACCATCGATAGCAGCA-TAMRA.
5. Nuclease-free water, see Subheading 2.1 item 5.

3. Methods

This method describes the conversion of scFv DNA to the format required for ribosome display, two different types of library construction and then a single round of ribosome display selection. The result is an enriched pool of DNA that can be used for additional ribosome display cycles, mutagenesis or cloning for further analysis of individual selected clones.

3.1. Preparation of the Parental scFv DNA Construct for Ribosome Display

Before starting the affinity maturation process, it is essential that parent scFv molecules are assessed for their capacity to express and fold correctly under ribosome display conditions. This step can also be helpful in determining the optimal starting concentration of antigen for the actual library selections (see Note 13). For the first time user, it is advised that this be carried out more than once to become comfortable with the method and ensure consistent results. The protocol describes how a scFv DNA construct is prepared for ribosome display by overlapping extension PCR to add the required flanking sequences; however, the whole construct may also be generated by gene synthesis (see Fig. 2). The construct contains *Nco*I and *Not*I sites for cloning into expression vectors after selection; it is, therefore, important to check the scFv sequence for these sites and remove them by site-directed mutagenesis if necessary. Alternatively, these cloning sites can be modified if to suit other vectors without adversely affecting the process. The conversion process described below is also used to convert scFv libraries to ribosome display format.

1. Set up the following PCR reaction to amplify the scFv coding sequence (see Note 14); 50 μl of 2× PCR master mix, 2 μl of

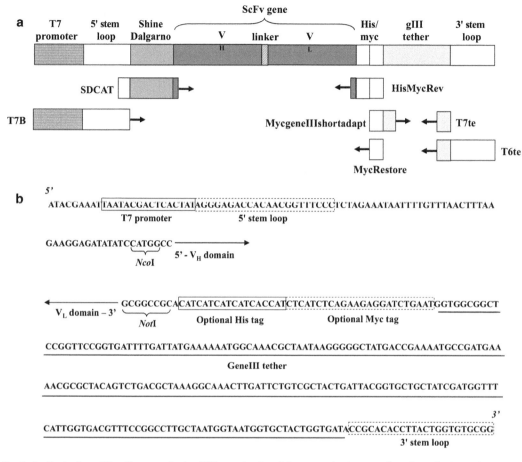

Fig. 2. An illustration of the ribosome display DNA construct and the approximate annealing sites of primers (**a**). The T7 promoter is the start signal for transcription. The Shine-Dalgarno motif is the initiator sequence of prokaryotic translation. The gene III tether is fused in-frame to the scFv coding sequence to facilitate display and folding of the parent scFv beyond the ribosome tunnel. 5′ and 3′ stem-loop sequences are used in the mRNA construct to reduce mRNA degradation and facilitate ribosome stalling during translation. Stop codons are absent to enhance ribosome stalling and the production of antibody–ribosome–mRNA complexes. The 5′ and 3′ nucleotide sequences required to adapt a scFv coding DNA to ribosome display format (**b**).

10 μM SDCAT V_H specific primer, 2 μl of 10 μM HisMycRev V_L specific primer (see Note 1), 10 ng of scFv plasmid DNA and nuclease-free water to a final volume of 100 μl.

2. Amplify using the following conditions: 94°C for 3 min, 25 cycles of (94°C for 30 s, 55°C for 30 s and 72°C for 105 s), 72°C for 5 min.

3. Separate the PCR products by electrophoresis on a 1% agarose, 1× TAE gel. The amplified fragment should be ~900 bp in length. Excise the appropriate band from the gel and purify using a DNA gel purification kit.

4. Set up the following PCR reaction to amplify the gene III tether (this can be done at the same time as step 2) (see Notes 3 and 15): 50 μl of 2× PCR master mix, 2 μl of 10 μM T7te primer, 2 μl of 10 μM MycgeneIIIshortadapt primer (see Note 1), 10 ng of gene III tether DNA and nuclease-free water to a final volume of 100 μl.

5. Amplify using the conditions in Subheading 3.1 step 2.

6. Separate the PCR products by electrophoresis on a 1.5% agarose, 1× TAE gel. The amplified fragment is 221 bp in length.

7. Excise the appropriate band from the gel and purify (see Note 16).

8. Set up the following PCR reaction to assemble the scFv with the gene III tether: 50 μl of 2× PCR master mix, 50–200 ng of gel-purified SDCAT/HisMycRev product from Subheading 3.1 step 3, 50 ng–200 ng of gel-purified gene III tether from Subheading 3.1 step 7 and nuclease-free water to a final volume of 96 μl (leaving 4 μl to allow for later primer additions).

9. Amplify using the following conditions: 94°C for 3 min, 5 cycles of (94°C for 30 s, 50°C for 30 s, 72°C for 105 s). During the 5th annealing step at 50°C, pause the PCR block and add 4 μl of a mix of SDCAT V_H specific primer and T7te primer (at 10 μM each). Resume the programme as follows: three cycles of (94°C for 30 s, 35°C for 30 s, 72°C for 105 s) and 15 cycles of (94°C for 30 s, 50°C for 30 s, 72°C for 105 s), 72°C for 5 min.

10. Separate the PCR products by electrophoresis on a 1% agarose, 1× TAE gel. The amplified fragment is ~1,100 bp in length (see Note 17). Cut the appropriate band from the gel and purify.

11. Check a 5 μl PCR sample on a 1% agarose, 1× TAE gel for size and purity and quantify the purified recombined product.

12. Proceed directly to the next step or store the purified DNA at −20°C.

13. Set up the following PCR reaction to add the T7 sequence, 5′ and 3′ stem loops and amplify the final ribosome display construct: 50 μl of 2× PCR master mix, 2 μl of 10 μM T7B primer, 2 μl of 10 μM T6te primer, 50–200 ng of gel-purified product from Subheading 3.1 step 10, and nuclease-free water to a final volume of 100 μl.

14. Amplify using the conditions in Subheading 3.1 step 2.

15. Check a 5 μl PCR sample on a 1% agarose, 1× TAE gel for size and purity (see Notes 18 and 19). Store the PCR product at −20°C.

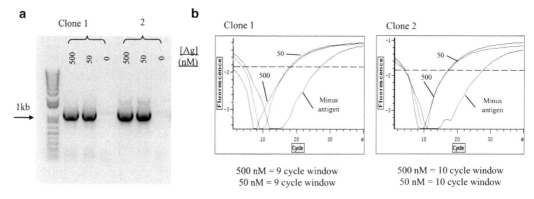

Fig. 3. Example selection data for two parent scFv test selections. 25 cycles of endpoint PCR (**a**) and real-time PCR analysis (**b**) are shown. Both clones show strong specific amplification at both antigen concentrations relative to their no antigen control.

16. If the size of the construct is correct and there are no non-specific products visible, proceed to in vitro transcription and translation (Subheading 3.4), and then selection (Subheading 3.5). See Fig. 3 for examples of typical scFv parent selections.

3.2. Construction of Error-Prone Mutagenesis Libraries

Generation of error-prone (EP) libraries from a single clone or a population of clones is relatively straight forward and rapid. It is important to ensure that only the scFv sequence is subjected to EP-PCR, and not the gene III tether or any other elements of the construct, as this could have deleterious effects on display efficiency and downstream processing.

EP-PCR is performed using the Diversify™ PCR Random Mutagenesis Kit from Clontech. The error rate can be adjusted by the use of different manganese sulphate and/or dGTP concentrations. The conditions in the following protocol are for the highest error rate (8.1 nucleotide changes per 1,000 bp according to the manufacturer).

1. Set up the following PCR reaction to introduce mutations into the scFv construct: 5 μl of 10× Titanium Taq buffer, 4 μl of 8 mM MnSO$_4$, 5 μl of 2 mM dGTP, 1 μl of 50× Diversify dNTP mix, 1 μl of 10 μM SDCAT V$_H$ specific primer, 1 μl of 10 μM HisMycRev V$_L$ specific primer, 1 μl of Titanium Taq DNA polymerase, 10 ng of purified scFv plasmid DNA template, and nuclease-free water to a final volume of 50 μl.

2. Amplify using the following conditions: 94°C for 3 min, 25 cycles of (94°C for 30 s, 68°C for 120 s).

3. Separate the PCR products by electrophoresis on a 1% agarose, 1× TAE gel (see Note 20). The amplified fragment should be ~900 bp in length. Cut the appropriate band from the gel and purify.

4. Additional EP-PCR may be carried out at this point using 1 ng of purified DNA from the first error-prone reaction as the template for a second EP-PCR reaction by repeating Subheading 3.2 steps 1–3.

5. Convert the purified scFv EP library to ribosome display format using the product from Subheading 3.2 step 3 or 4 and the gene III tether DNA from Subheading 3.1 steps 4–7 for assembly PCR as described in Subheading 3.1 steps 8–16. See Note 21 for library QC.

3.3. Construction of CDR-Directed Mutagenesis Libraries

3.3.1. Stop Template Construction

Generation of targeted mutagenesis libraries requires the preparation of "stop template", that is a DNA sequence encoding the scFv sequence of interest, where stop codons have been introduced into the region that is to be randomised. This will prevent display of any non-mutated scFv sequences that are present in the library and therefore prevent their preferential enrichment over individual library members.

1. Set up a mutagenesis PCR with the following components: 5–50 ng of purified parental plasmid DNA template, 1 μl of 125 ng/μl forward stop primer, 1 μl of 125 ng/μl reverse stop primer, 5 μl of 10× reaction buffer, 1 μl of 12.5 mM dNTP mix, 1 μl of 2.5 U Pfu Turbo DNA Polymerase, and nuclease-free water to a final volume of 50 μl.

2. Amplify using the following conditions: 95°C for 30 s, 17 cycles of (95°C for 30 s, 55°C for 60 s, 68°C for 300 s) 37°C for 120 s.

3. Add 10 U of *Dpn*I restriction enzyme to the PCR and mix by pipetting. Incubate for 1 h at 37°C to digest the parental (i.e. non-mutated) supercoiled dsDNA.

4. Transform a sample, e.g. 1–10 μl of the *Dpn*I digested reaction into chemically competent *E. coli* cells.

5. DNA sequence several colonies to identify and confirm the desired scFv stop template(s). Prepare plasmid DNA from appropriate clones.

3.3.2. CDR Randomisation

PCR cassette mutagenesis is employed to fully randomise blocks of up to seven amino acid residues per library in any chosen CDR. If the CDR is longer than seven amino acids, the whole CDR can be covered by using separate, possibly overlapping randomisation blocks.

1. Amplify the parent stop template plasmid DNA from Subheading 3.3.1 step 5 and add the gene III tether as described in Subheading 3.1 steps 1–11.

2. For amplification of the mutated region, set up the following PCR reaction: 50 μl of 2× PCR master mix, 2 μl of 10 μM

SDCAT V_H specific forward primer, 2 µl of 10 µM reverse mutagenic primer (see Note 7), 50–200 ng of purified DNA from Subheading 3.3.2 step 1, and nuclease-free water to a final volume of 100 µl.

3. Amplify using the conditions in Subheading 3.1 step 2.

4. Separate the PCR products by electrophoresis on a 1% agarose, 1× TAE gel (see Note 20). The PCR product size will vary according to the region being targeted within the scFv sequence. Cut the appropriate band from the gel and purify (see Note 22).

5. For amplification of the non-mutated region, set up the following PCR reaction:

6. 50 µl of 2× PCR master mix, 2 µl of 10 µM T7te primer, 2 µl of 10 µM forward non-mutagenic primer (see Note 7), 50–200 ng of purified DNA from step 1 and nuclease-free water to a final volume of 100 µl.

7. Amplify using the conditions in Subheading 3.1 step 2.

8. Separate the PCR products by electrophoresis on a 1% agarose, 1× TAE gel. Cut the appropriate band from the gel and purify.

9. Set-up the following PCR reaction to combine mutated and non-mutated variable domain regions: 50 µl of 2× PCR master mix, 50–200 ng gel-purified mutated PCR product from Subheading 3.3.2 step 4, 50–200 ng of purified non-mutated PCR product from Subheading 3.3.2 step 8 (a 1:1 M ratio of both PCR products should be used) and nuclease-free water to a final volume of 96 µl.

10. Amplify using the conditions in Subheading 3.1 step 9.

11. Separate the PCR products by electrophoresis on a 1% agarose, TAE gel. Cut the appropriate band from the gel and purify (see Note 17).

12. Amplify to generate the final ribosome display construct as described in Subheading 3.1 steps 13–16. See Note 23 for library QC.

3.4. In Vitro Transcription and mRNA Translation (See Note 11)

1. Assemble a transcription reaction as follows in a non-stick, RNase-free microcentrifuge tube in the order listed: 10 µl of 5× T7 Transcription Buffer, 15 µl of 25 mM rNTP mix, 20 µl of linear DNA template (PCR reaction), and 5 µl of T7 polymerase mix.

2. Mix well by pipetting or gentle vortexing. Pulse-spin to ensure all liquid is at the bottom of the tube. Incubate at 37°C for 2–3 h.

3. Pulse-spin to collect condensation from the tube lid.

4. Purify mRNA using a ProbeQuant G50 micro column: Vortex the column to resuspend the matrix, break off the tip and place into a micro centrifuge tube. Loosen cap ¼ turn and centrifuge for 1 min at $735 \times g$ to remove the storage buffer. Discard the flow-through, shake off any residual liquid at the tip of the column and place the column into a fresh, RNase-free, microcentrifuge tube. Carefully pipette 50 μl of transcription reaction to the centre of the surface of the resin in the column. Replace the cap and loosen ¼ turn, then centrifuge for 2 min at $735 \times g$ to elute the purified mRNA.

5. Transfer the mRNA immediately to ice.

6. Quantify an mRNA sample by measuring OD A260 in a spectrophotometer (see Note 24).

7. Proceed immediately with in vitro translation or snap-freeze the mRNA on dry ice for storage at −70°C.

8. Prepare heparin-block (HB) buffer on ice by combining: 5 ml of 10× *E. coli* wash buffer, 45 ml of sterile-filtered MilliQ water and 625 μl of 200 mg/ml heparin. Store on ice or in the fridge at all times and make fresh for each experiment.

9. Thaw the Premix X and PDI on ice. Begin to thaw the S30 extract on ice and start the following procedure immediately (see Note 25).

10. Prepare a master mix for the total number of translation reactions required. A single translation reaction is needed for each library and this is then divided between three selections. 330 μl translation reactions comprising 300 μl of translation mix and 30 μl of mRNA are used. For each library translation, combine the following in the order listed in an RNase-free microcentrifuge tube on ice: 52.2 μl of nuclease-free water, 33 μl of 2 M potassium glutamate, 22.8 μl of 0.1 M magnesium acetate, 6 μl of 5 mg/ml PDI and 66 μl of Premix X. This master mix is not yet complete, S30 *E. coli* extract will be added after the mRNA has been thawed and diluted. Pipette gently to mix.

11. If mRNA has been produced on a previous day and stored at −70°C, defrost it quickly by holding between fingertips and place immediately on ice when thawed. For each library or clone, add 30 μl of mRNA at 1 μg/μl (approximately 6×10^{13} molecules) to the bottom of a pre-chilled 1.5 ml RNase-free microcentrifuge tube on ice. Immediately refreeze any unused mRNA.

12. Complete the translation master mix by adding 120 μl of S30 *E. coli* extract per translation to the master mix from Subheading 3.4 step 10 and mix by gentle pipetting. Do not vortex.

13. Add 300 µl of translation master mix to the 30 µl of mRNA and mix gently by pipetting up and down. Immediately transfer the tube to a 37°C heating block for 6–10 min (see Notes 26 and 27).

14. After translation, immediately pipette each 330 µl translation reaction into a pre-chilled 2 ml RNase-free microcentrifuge tube containing 1,320 µl of HB buffer on ice to stabilise the scFv-ribosome-mRNA complexes. Mix thoroughly but gently.

15. Centrifuge the samples at maximum speed for 5 min at 4°C in a pre-chilled bench-top microcentrifuge and replace on ice. Proceed to Subheading 3.5 immediately.

3.5. Selection and Capture of Specifically Bound scFv-Ribosome-mRNA Complexes

Enrichment for the highest affinity scFv within a population is achieved through affinity selection, which can be performed by decreasing antigen concentration from round to round or by adding unlabelled antigen as a competitor to remove scFv with the fastest dissociation rates. This is best achieved in solution-phase using a biotinylated antigen with streptavidin-coated magnetic beads for capture, but other formats can be used. Antigen quality is critical for the success of in vitro selection and should be as pure as possible. Selection quality is assessed by end-point and real-time PCR, where a successful selection is deemed to have a sufficiently high cDNA recovery relative to a selection performed without antigen, termed "selection window". Multiple rounds of selection are usually required to enrich for the highest affinity clones, often resulting in selections being performed at very low concentrations of antigen in later rounds. Selections are no longer effective and have therefore reached their limit when there is no longer a selection window present. If further gains in affinity are needed at this point, additional mutagenesis may be considered, such as (further) error-prone PCR or recombination of different targeted regions for synergy (see Note 28).

The following protocol describes the process for carrying out three selections per library; two containing antigen and one without antigen (see Note 29). Unless stated all steps should be performed at 4°C (see Note 30).

1. For each library, pre-chill three 1.5-ml RNase-free, microcentrifuge tubes on ice and prepare an additional 2 ml RNase-free microcentrifuge tube on ice containing 1,320 µl ice-cold HB buffer.

2. Prepare de-biotinylated milk for selection (see Note 31). Add 100 µl of streptavidin beads to 1 ml of 10% autoclaved non-fat dried milk in water, in an RNase-free, microcentrifuge tube. Incubate with end-over-end mixing for at least 10 min. Collect the beads using a magnetic particle concentrator and transfer the milk to a fresh tube. Store on ice until required.

3. Transfer 500 µl of stabilised scFv-ribosome-mRNA complexes from Subheading 3.4 step 15 to each pre-chilled RNase-free selection tube. Add 50 µl of de-biotinylated sterile milk. Add biotinylated antigen at the required concentration (see Note 13) to the positive antigen selections only.

4. Incubate the selections at 4°C for the required time (2 h to overnight) with gentle end-over-end rotation (see Note 32).

5. Prepare streptavidin beads (see Note 33). Wash the beads four times in HB buffer in RNase-free, microcentrifuge tubes and resuspend in HB to the original volume. Store on ice until required.

6. Capture antigen-bound complexes by addition of washed streptavidin beads to each selection. Incubate for 1–2 min at 4°C.

7. Wash the beads five times with 800 µl of HB buffer at 4°C to remove non-specifically bound complexes. Resuspend the beads in 220 µl of EB20 containing *S. cerevisiae* mRNA (at 10 µg/ml final concentration) at 4°C. Incubate for 10 min at 4°C (see Note 34).

8. Transfer 200 µl of EB20, which now contains the eluted mRNA, to 400 µl of lysis buffer from the High Pure RNA Isolation Kit in a fresh, RNase-free, microcentrifuge tube. Vortex immediately to mix well.

9. Isolate mRNA using the High Pure RNA Isolation Kit and a bench-top microcentrifuge pre-chilled at 4°C (see Note 35).

10. Transfer samples to the upper chamber of the filter tubes, centrifuge for 30 s at $10,000 \times g$ and discard the flow-through.

11. Dilute 10 µl of DNase I in 90 µl DNase incubation buffer for each sample. Pipette 100 µl of DNase I solution onto each filter. Incubate for 10 min at 15°C (see Note 36).

12. Follow the manufacturer's instructions for washing the captured mRNA.

13. Check filter tubes for residual liquid. If present, centrifuge for an additional 15 s at $10,000 \times g$ and discard the flow-through.

14. Discard the collection tubes and insert filter tubes into fresh, pre-chilled RNase-free, microcentrifuge tubes.

15. For elution of RNA, add 40 µl of elution buffer (or nuclease-free water) to the centre of each filter and centrifuge at 4°C for 1 min at $10,000 \times g$.

16. Discard the filter columns and immediately place the eluted mRNA on ice. Proceed directly to Subheading 3.6.

3.6. Converting Selection Output mRNA Back to DNA in Ribosome Display Format

1. Prepare a master mix for the required number of reverse transcription (RT) reactions to be performed (one per selection plus a negative control). Mix on ice in a fresh RNase-free microcentrifuge tube, per selection: 4 µl of 5× First Strand buffer, 2 µl of 100 mM DTT, 0.25 µl of 100 µM reverse primer (see Note 37), 0.5 µl of 25 mM dNTP mix, 0.5 µl of 40 U/µl RNasin, and 0.5 µl of 200 U/µl Superscript II Reverse Transcriptase.

2. Mix well by pipetting and aliquot 7.75 µl of RT master mix into the bottom of a thin-walled 0.2 ml PCR tube per reaction.

3. Mix eluted mRNA from Subheading 3.5 step 16 by pipetting up and down and add 12.25 µl of mRNA to the RT reaction. Mix well. Unused mRNA should be frozen immediately and stored at −70°C.

4. Incubate RT reactions in a PCR block at 50°C for 30 min (see Note 38).

5. Transfer completed RT reactions onto ice or store at −20°C if PCR is to be performed at a later time.

6. Prepare a master mix for the required number of PCR reactions to be performed in a fresh RNase-free microcentrifuge tube, on ice. Per RT reaction: 34.5 µl of nuclease-free water, 50 µl of 2× PCR master mix, 0.25 µl of 100 µM reverse (RT) primer (e.g. T7te), 0.25 µl of 100 µM T7B, and 5 µl of DMSO.

7. Mix well by pipetting and aliquot 90 µl into the bottom of a thin-walled 0.2 ml PCR tube per reaction. Keep on ice.

8. Add 10 µl of cDNA from Subheading 3.6 step 5 and mix well. Store unused cDNA at −20°C.

9. Amplify using the following conditions: 94°C for 3 min, 25–40 cycles of (94°C for 30 s, 55°C for 30 s and 72°C for 105 s), 72°C for 5 min (see Note 39 for cycle number information).

10. Compare 5 µl of each PCR sample on a 1% agarose, 1× TAE gel. Figure 4 shows an example of a single round of selection performed on two different libraries (see Notes 40 and 41).

11. To process appropriate selection outputs for further rounds of selection or for vector sub-cloning, run the remaining PCR reaction on a large 1% agarose, 1× TAE gel. Cut the appropriate band from the gel and purify. Store purified DNA at −20°C. This DNA can be used for sub-cloning to analyse the sequence (see Note 42) and function of individual clones in this selection output as desired. To carry out further rounds of selection on an output recovered with T7te (see Note 41), proceed to the next step. If use of an upstream primer such as MycRestore was required to produce a good quality product, rebuild the

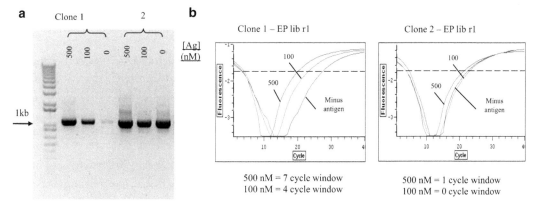

Fig. 4. Round 1 selection of mutagenic libraries. 25 cycles of endpoint PCR (**a**) and real-time PCR analysis (**b**). Good selection windows are present for clone 1 at both 500 and 50 nM antigen. Clone 2 only has a 1 cycle window at 500 nM antigen concentration. This is not uncommon for a first round of selection with a mutagenic library, and proceeding to a round 2 with this product often yields a sufficiently large selection window.

full length construct by addition of the gene III tether as described in Subheading 3.1 steps 4–16.

12. Re-amplify the output DNA by PCR as follows: 50 μl of 2× PCR master mix, 2 μl of 10 μM T7B primer, 2 μl of 10 μM reverse primer T6te, 5 μl of gel-purified product from Subheading 3.6 step 11, and 41 μl nuclease-free water.

13. Amplify DNA using the following conditions: 94°C for 3 min, 20 cycles of (94°C for 30 s, 55°C for 30 s and 72°C for 105 s), 72°C for 5 min.

14. Check a 5 μl PCR sample on a 1% agarose, 1× TAE electrophoresis gel for size and purity. Store DNA at −20°C.

15. Begin the next round of selection as described in Subheading 3.4.

3.7. Real-Time PCR Analysis of Selection Outputs for Relative Quantitation of cDNA

1. Relative quantitation of cDNA yields can be assessed by real-time PCR by adding the following to a PCR plate (see Note 12): 12.5 μl of 2× Taqman Universal PCR master mix, 5 μl of cDNA from Subheading 3.6 step 5 (this may need to be diluted prior to real-time PCR), 2 μl of 10 μM forward primer, 2 μl of 10 μM reverse primer, 1 μl of 5 μM Taqman probe, and 15 μl of nuclease-free water.

2. Mix by pipetting several times, briefly centrifuge to remove air bubbles, and amplify DNA using the following conditions: 50°C for 2 min, 95°C for 10 min, 39 cycles of (95°C for 15 s, plate read, 60°C for 1 min).

3. Compare CT values (cycle number to threshold) to determine relative cDNA levels in positive versus negative selections.

4. Notes

1. The HisMycRev primer adds a His and Myc tag to the coding sequence. If this sequence is already present, the MycRestore primer (Subheading 2.6 item 5) may be used instead. The construct can be made with other or no tags present, but will require alternative primers to substitute the HisMycRev, MycgeneIIIshortadapt, and MycRestore primers.

2. For DNA visualisation, we use SYBR Safe™ (Invitrogen) with blue light, although other DNA stains and visualising aids may be used.

3. The gene III tether sequence is derived from the gene III sequence of M13 filamentous phage, spanning amino acids 269–336 (SwissProt P69168). The tether is used to provide a non-structured portion at the C-terminus of the displayed protein, such that the ribosomal tunnel can cover at least 20–30 amino acid residues of the emerging polypeptide without interfering with the folding of the protein of interest. This sequence can be obtained by gene synthesis (see Fig. 2) or amplified from the M13 phagemid vector pCantab6 (13). Other tether sequences may also be employed.

4. This method describes the use of the Diversify™ PCR Random Mutagenesis Kit from Clontech for the error-prone PCR reactions; however, alternatives are available.

5. The QuikChange Site-Directed Mutagenesis kit is available from Agilent; however, the individual components can also be purchased separately.

6. Stop template preparation by site-directed mutagenesis requires a pair of complementary oligonucleotides to replace two adjacent amino acids with stop codons (e.g. TGATAA) within the area to be randomised. Design the oligonucleotides with a complementary region of at least 15 nucleotides either side of the amino acid codons that are to be replaced by stop codons. Include at least one G or C at the 3′ end of each oligonucleotide.

7. Saturation mutagenesis is achieved by using a degenerate oligonucleotide annealing to (and therefore complementary to) the coding strand to incorporate NNS codons at the positions of the amino acid codons that are to be randomised. The NNS primer should extend from the targeted amino acid block by approximately 24 nucleotides in each direction, and there should be at least one G or C at the 3′ end and a G/C content of at least 40%. This oligonucleotide is used for PCR with the upstream SDCAT V_H specific primer to produce the mutated/library scFv fragment. The non-mutated fragment comprising the rest of the scFv is prepared with the downstream primer,

T7te, and a primer annealing to the non-coding strand 3' to the NNS-mutagenic site (i.e. complementary to the flanking region of the NNS oligonucleotide to enable assembly PCR). This non-mutagenic primer should be approximately 20 nucleotides long.

8. The QuikChange Site-Directed Mutagenesis Kit from Agilent is supplied with chemically competent *E. coli* cells. If using individual components for site-directed mutagenesis rather than this kit, alternative competent cells can be used.

9. Non-stick, RNase-free microcentrifuge tubes are preferred for steps involving pure mRNA, i.e. transcription, cell-free translation and selection steps, and for elution of purified mRNA. The lids of these particular tubes tend to snap less in a microcentrifuge without an internal lid, when eluting from columns.

10. The Ribomax Large Scale RNA Production Kit from Promega is recommended; however, other kits are available.

11. S30 *E. coli* extract is a bacterial cell extract containing ribosomes and translation factors that is commonly used for cell-free translation. We use an *E. coli* MRE600 S30 extract prepared in-house; however, S30 extract can also be purchased. Preparation of the S30 extract is a reasonably significant undertaking so for occasional users we recommend using a commercially available cell-free translation reagent. Purified recombinant *E. coli* translation systems that are suitable for ribosome display are also available, for example PURExpress (New England Biolabs). We recommend testing the chosen translation reagent for suitability by performing a trial selection of the parent scFv, and in particular the optimal translation time should be determined empirically. The ribosome display method described here is only applicable to prokaryotic cell-free translation systems since eukaryotic ribosomes require different buffers for optimal performance.

12. Real-time PCR using the primers described here will only amplify cDNA products which have the gene III tether present.

13. A concentration of biotinylated antigen should be chosen above the K_D of the parent antibody for the first round of selection, and then lowered over subsequent rounds to select for variants with improved affinity (e.g. a two- to tenfold reduction in antigen concentration per round).

14. With each PCR step, it is advisable to add the samples to the block when the temperature has reached 94°C and to include a no template control. For assembly PCR, a negative control with only one of the templates is also advised.

15. The generation of tether via PCR amplification can be performed in triplicate so large quantities are available for subsequent assembly PCRs.

16. Use of Tris-based elution buffers helps to protect against deterioration of the tether DNA.

17. There may be some non-recombined products present in the sample. A careful gel extraction should eliminate these from the desired product prior to amplification of the full-length construct for transcription.

18. There should be one strong band at ~1,200 bp. Non-specific bands of lower molecular weight are not acceptable because these will be amplified preferentially in subsequent rounds.

19. PCR products should not be purified for transcription.

20. When constructing multiple libraries, it is advised to separate the PCR products on different gels, or with an empty lane between samples to help reduce the risk of cross contamination.

21. Successful error-prone library construction should be verified by cloning (e.g. using *Nco*I and *Not*I digested scFv library and ligation into a suitable vector) and DNA sequencing of a representative number of clones (e.g. 88).

22. There may be additional PCR products present, as well as a high background of template DNA. It is, therefore, crucial to check the correct size of the bands to proceed with and work precisely when removing them from the gels.

23. Successful targeted randomised library construction should be verified by cloning and sequencing of a representative number of clones (e.g. 88). Ideally at least 70% of the targeted region should be randomised with no contaminating clones or original parent present.

24. For accurate mRNA quantitation, ensure the dilution made gives a solution in the linear range of the spectrophotometer (0.1–1.0 units); typically 1 in 150 is suitable. The concentration of the transcribed mRNA should be higher than 2 μg/μl. If the concentration is lower, the transcription reaction should be repeated. If mRNA concentration is consistently lower, repeat the amplification of the PCR product template and ensure a very strong band is visible.

25. All reagents except the *E. coli* S30 extract should be defrosted and/or placed on ice prior to setting up the translation reaction. The *E. coli* S30 extract should be thawed at the last minute immediately prior to use. Do not warm or vigorously mix the *E. coli* S30 extract.

26. The translation time needs to be optimised for different batches of S30 lysate. Refer to manufacturer's instructions if a commercial purified translation system is being used.

27. The start of the in vitro translation reactions can be staggered in convenient intervals, e.g. every 15 s, to allow accurate timing of multiple reactions.

28. To achieve additive or synergistic gains in affinity, recombination of the individual randomised regions can be carried out. For example, separate affinity selected V_H and V_L targeted library outputs can be recombined to make a new library by amplifying the separate domains with overlapping primers within the linker region and an appropriate N- or C-terminal primer already described. These amplified domains can then be recombined by assembly PCR.

29. Performing a no antigen control is crucial to determine selection efficiency.

30. For best results, the whole selection process should be kept cold; working on ice at all times to reduce ternary complex dissociation and mRNA degradation. Speed, co-ordination and efficiency are key to uniformity and consistency of selections. It helps to have all tubes for this section pre-chilled on ice, and labelled prior to starting.

31. It is possible that the biotin found naturally in milk could interfere with the selection process when using streptavidin-biotinylated antigen capture. For this reason, biotin is removed from the autoclaved milk using streptavidin beads prior to using the milk in selections.

32. A sufficient equilibration time should be used to allow even the highest affinity interactions to reach equilibrium e.g. below 10 pM, equilibration times of over 2 h may be required. Typically it is convenient to allow low pM selections to proceed overnight.

33. Use 1 μl beads per "nM" of biotinylated antigen, down to a minimum of 50 μl, i.e. use 50 μl of beads for 50 nM antigen and less.

34. Automated washing and elution (e.g. Kingfisher ML, Thermo) is highly recommended to improve the efficiency and reproducibility of the process, as well as being more user friendly for bead washing at 4°C.

35. The use of a vacuum aspirator is recommended to remove column flow-through wastes as this helps increase speed and consistency and reduces sample cross-contamination.

36. It is advisable to defrost the reverse transcription reagents and make the master mix for Subheading 3.6 step 1 during the 10-min DNase incubation step.

37. It is recommended to initially use T7te as the RT primer, as it allows a single step reamplification of the full RD construct with T6te. However, it may be beneficial to use more upstream primers (e.g. MycRestore) if quality or yields are poor with T7te (see *also* Notes 40 and 41).

38. It is advisable to defrost the reagents and prepare the PCR master mix for Subheading 3.6 step 6 during the 30 min incubation time of the RT reactions.

39. The optimal number of PCR cycles varies depending on the yield and quality of cDNA. Guidance can be obtained from the model selection of the parent clone and by performing real-time PCR before the endpoint PCR, see Subheading 3.7. Alternatively, trial PCRs can be performed to determine a suitable number of cycles. In general 25 PCR cycles is a good starting point, as further cycles can be carried out if necessary.

40. The two outputs from selections with antigen should be similar and yield clearly more DNA than the selection without antigen (see Fig. 3 for parent selections and Fig. 4 for library selections). If this is the case, the selection has been successful and the appropriate outputs can be processed for further application. This can be confirmed by real-time PCR, see Subheading 3.7.

41. There should only be one strong band per lane at about 1,200 bp (900 bp for MycRestore amplified outputs). Non-specific bands of lower molecular weight will be amplified preferentially at the next PCR and thus eliminate the specific product. Lower molecular weight smears suggest degradation of the specific product. Repeating the RT and PCR with different combinations of primers can often result in these problems being rectified. If problems persist, however, the initial construct may need to be made again and selections repeated.

42. Ideally, the selections should contain no non-mutated parent sequence and there should be good sequence diversity in early rounds of selection.

References

1. Hida K, Hanes J and Ostermeier M. (2007) Directed evolution for drug and nucleic acid delivery, Adv Drug Deliv Rev 59: 1562–1578.
2. Thom G, Cockroft A C, Buchanan A G, *et al.* (2006) Probing a protein-protein interaction by in vitro evolution, Proc Natl Acad Sci U S A 103: 7619–7624.
3. Persson H, Wallmark H, Ljungars A, *et al.* (2008) In vitro evolution of an antibody fragment population to find high-affinity hapten binders, Protein Eng Des Sel 21: 485–493.
4. Friedman M, Orlova A, Johansson E, *et al.* (2008) Directed evolution to low nanomolar affinity of a tumour-targeting epidermal growth factor receptor-binding affibody molecule, J Mol Biol 376: 1388–1402.
5. Daugherty P S. (2007) Protein engineering with bacterial display, Curr Opin Struct Biol 17: 474–480.
6. Boder E T, Midelfort K S and Wittrup K D. (2000) Directed evolution of antibody fragments with monovalent femtomolar antigen-binding affinity, Proc Natl Acad Sci U S A 97: 10701–10705.
7. Douthwaite J and Jermutus L. (2006) Exploiting directed evolution for the discovery of biologicals, Curr Opin Drug Discov Devel 9: 269–275.
8. Rothe A, Hosse R J and Power B E. (2006) Ribosome display for improved biotherapeutic molecules, Expert Opin Biol Ther 6: 177–187.
9. Hawkins R E, Russell SJ and Winter G. (1992) Selection of phage antibodies by binding affinity.

Mimicking affinity maturation, J Mol Biol 226: 889–896.

10. Schier R, Bye J, Apell G, *et al.* (1996) Isolation of high-affinity monomeric human anti-c-erbB-2 single chain Fv using affinity-driven selection, J Mol Biol 255: 28–43.

11. Zahnd C, Spinelli S, Luginbuhl B, *et al.* (2004) Directed in vitro evolution and crystallographic analysis of a peptide-binding single chain antibody fragment (scFv) with low picomolar affinity, J Biol Chem 279: 18870–18877.

12. Zahnd C, Sarkar C A and Pluckthun A. Computational analysis of off-rate selection experiments to optimize affinity maturation by directed evolution, Protein Eng Des Sel 23: 175–184.

13. McCafferty J, Fitzgerald KJ, Earnshaw J, *et al.* (1994) Selection and rapid purification of murine antibody fragments that bind a transition-state analog by phage display, Appl Biochem Biotechnol 47: 157–171; discussion 171–153.

Chapter 10

Affinity Maturation of Phage Display Antibody Populations Using Ribosome Display

Maria A. Groves and Adrian A. Nickson

Abstract

Ribsosome display is a PCR-based in vitro display technology that it well suited for the selection and evolution of high-affinity antibodies. In particular, ribosome display lends itself to the evolution of functional characteristics, such as potency, and thereby facilitates the production of therapeutic antibodies from lead candidates. In this chapter, we describe how to mature large phage display antibody populations (>10^7) by performing increasingly stringent selections with decreasing antigen concentration. This process takes advantage of ribosome display's intrinsic ability to evolve sequence during selection. Ribosome display can also be used as a complementary tool to phage display for isolating high-affinity antibodies from naïve libraries. Ultimately, maturation of large antibody populations by ribosome display will help to speed up the process of generating antibody therapeutics.

Key words: Ribosome display, Affinity maturation, scFv library, In vitro selections, Antibody therapeutics

1. Introduction

Phage display has been extensively used since its development (1, 2) for the isolation of antibodies for research purposes and for the development of therapeutics (3, 4), such as the antibody against Tumour Necrosis Factor α – adalimumab. Commonly, naïve antibody phage display libraries, derived from non-immunised humans, are used as a basis for the isolation of lead single-chain variable fragments (scFvs) with promising binding characteristics. Initial IC_{50} values in biological assays for these scFvs are typically in the range 0.2–200 nM (5–7). In some cases, these antibody fragments have been converted directly to antibody (IgG) therapeutics without optimisation of affinity, for instance in the development of HGS-ETR1 (7). In other cases, to obtain antibodies of sufficiently high affinity, the initial sequences have been optimised using display techniques, which can lead to affinity enhancements of up to 1,000-fold. For example, phage display was used to optimise a human scFv that bound to BlyS, producing an approximately 20-fold improvement in the IC_{50} value for the inhibition of binding of BlyS to its receptor (LymphoStat-B) (8).

The development of ribosome display provides a complementary PCR-based in vitro technology to phage display, which is well suited for the selection and evolution of antibody functional characteristics, such as potency (9–16). In this chapter, we describe the methodology for the affinity maturation of phage display antibody populations using ribosome display. Unlike phage display, this technology is able to evolve the antibody libraries by introducing random mutation during the selection cycles. This evolution, coupled with increasingly stringent selection conditions, leads to the generation of higher affinity antibodies from already enriched phage display antibody populations.

We begin by describing how to convert a phage display library (or similar) into a format that is amenable to ribosome display, and list the features that are essential to produce the ternary ribosome–mRNA–protein complex. We then detail how the antibody fragments are matured using several rounds of ribosome display, and show how the selection outputs can be sub-cloned and analysed to identify protein sequences with high affinity for the desired target.

2. Materials

2.1. Conversion of Phage Display Outputs to Ribosome Display Format

1. SDCAT-DP47 forward primer (AGACCACAACGGTTTCCC TCTAGAAATAATTTTGTTTAACTTTAAGAAGGA GATATATCCATGGCCGAGGTGCAGC) (see Note 1).

2. Myc-Restore reverse primer (ATTCAGATCCTCTTCTGAGA TGAG) (see Note 2).

3. Myc-Forward primer (ATCTCAGAAGAGGATCTGAATGG TGGCGGCTCCGGTTCCGGTGAT) (see Note 3).

4. GeneIII-Reverse primer (CCGTCACCGACTTGAGCC) (see Note 4).

5. T7B forward primer (ATACGAAATTAATACGACTCACTA TAGGGAGACCACAACGG) (see Note 5).

6. T6te reverse primer (CCGCACACCAGTAAGGTGTGCGGT ATCACCAGTAGCACCATTACCATTAGCAAG) (see Note 6).

7. T7te reverse primer (GTAGCACCATTACCATTAGCAAG) (see Note 7).

8. Nuclease-free water (Promega, or alternative supplier).

9. Quantitative DNA ladder (100 bp to 10 kbp).
10. Plasmid midiprep kit.
11. DNA gel extraction kit.
12. 2× PCR master mix (or separate *Taq* polymerase with suitable reaction buffer and dNTPs).
13. Agarose gel electrophoresis sample loading buffer.
14. TAE gel running buffer: 40 mM Tris-base, 1.14% (v/v) glacial acetic acid, 1 mM EDTA, pH 7.6.
15. 2×TY growth media: 16 g/l tryptone, 10 g/l yeast extract, 5 g/l NaCl.
16. Agarose gels: 1% (w/v) agarose (electrophoresis grade) and appropriate DNA gel stain (Invitrogen, see Note 8) in 1×TAE buffer. These should be freshly made on the day of use.
17. 2×TYAG selective growth media: 2% (w/v) glucose and 100 μg/ml ampicillin in 2×TY media.
18. At least 100 μl of phagemid glycerol stock (see Note 9).

2.2. Optional Error-Prone Mutagenesis

1. Primers (see Subheading 2.1, items 1–7).
2. Nuclease-free water (Promega, or alternative supplier).
3. Quantitative DNA ladder (see Subheading 2.1, item 9).
4. Plasmid midiprep kit.
5. DNA gel extraction kit.
6. PCR master mix (see Subheading 2.1, item 12).
7. TAE gel running buffer (see Subheading 2.1, item 14).
8. Agarose gel electrophoresis sample loading buffer.
9. Agarose gels (see Subheading 2.1, item 16).
10. 2×TY growth media (see Subheading 2.1, item 15).
11. Diversify® PCR Random Mutagenesis Kit (Clontech) or alternative.

2.3. Transcription of Ribosome Display Libraries

1. Non-stick, RNAse-free 1.5-ml microfuge tubes (Sarstedt, see Note 10).
2. Ribomax™ Large Scale RNA production system – T7 (Promega).
3. Illustra ProbeQuant™ G-50 Micro Columns (GE Healthcare) or equivalent.

2.4. Ribosome Display Selections

1. SDCAT-DP47 forward primer (see Note 1).
2. T7B forward primer (see Note 5).
3. T6te reverse primer (see Note 6).
4. T7te reverse primer (see Note 7).
5. Real-time PCR detection system.

6. Non-stick, RNAse-free 1.5-ml and 2.0-ml microfuge tubes (Sarstedt, see Note 10).
7. Nuclease-free water (Promega, or alternative supplier).
8. Heparin at 200 mg/ml.
9. Quantitative DNA ladder (see Subheading 2.1, item 9).
10. 10% (w/v) non-fat dried milk in water, autoclaved on liquids cycle.
11. Streptavidin-coated magnetic beads, for example M280 Dynabeads® (Invitrogen).
12. Biotinylated target protein.
13. High Pure RNA Isolation Kit (Roche) or equivalent.
14. *Saccharomyces cerevisiae* RNA at 10 μg/ml.
15. 10× *E. coli* wash (HB) buffer: 0.5 M Tris–acetate pH 7.5, 1.5 M NaCl, 0.5 M magnesium acetate, 1% (v/v) Tween-20.
16. EB20 buffer: 50 mM Tris–acetate pH 7.5, 150 mM NaCl, 20 mM EDTA.
17. Premix X: 250 mM Tris–acetate pH 7.5, 1.75 mM of each standard amino acid, 10 mM ATP, 2.5 mM GTP, 5 mM cAMP, 150 mM acetylphosphate, 2.5 mg/ml *E. coli* tRNA, 0.1 mg/ml folinic acid, 7.5% (w/v) PEG-8000.
18. Protein disulphide isomerase (PDI) at 5 mg/ml.
19. Potassium glutamate at 2 M.
20. Magnesium acetate at 0.1 M.
21. S30 *E. coli* extract (see Note 11).
22. Superscript II reverse transcriptase and 5× first strand buffer (Invitrogen) or equivalent.
23. DTT at 100 mM.
24. dNTP mix containing 25 mM each of dATP, dTTP, dCTP, and dGTP.
25. RNasin® (Promega) at 40 U/μl.
26. PCR master mix (see Subheading 2.1, item 12).
27. DMSO (dimethylsulphoxide).
28. DNA gel extraction kit.
29. TAE gel running buffer (see Subheading 2.1, item 14).
30. Agarose gel electrophoresis sample loading buffer.
31. Agarose gels (see Subheading 2.1, item 16).
32. 2× TaqMan® Universal PCR Master Mix (Applied Biosystems).
33. Real-time PCR primers and probe (see Note 12).

2.5. Cloning of Enriched Ribosome Display Outputs

1. Sterile 0.5-ml and 1.5-ml microfuge tubes.
2. *Not*I restriction enzyme.
3. *Nco*I restriction enzyme.
4. Appropriate restriction buffer (supplied with restriction enzyme).
5. Bovine serum albumin at 10 mg/ml (supplied with restriction enzyme).
6. T4 DNA ligase.
7. Nuclease-free water (Promega, or alternative supplier).
8. PCR master mix (see Subheading 2.1, item 12).
9. DNA gel extraction kit.
10. Chemically competent *E. coli* cells (e.g. TG1, XL1-blue).
11. TAE gel running buffer (see Subheading 2.1, item 14).
12. Agarose gel electrophoresis sample loading buffer.
13. Agarose gels (see Subheading 2.1, item 16).
14. 2×TY growth media (see Subheading 2.1, item 15).
15. 2×TYG growth media: 2% (w/v) glucose in 2×TY media.
16. 2×TYAG selective agar plates: 2% (w/v) glucose, 100 µg/ml ampicillin and 15 g/l agar in 2×TY media.
17. Suitable phage or phagemid vector (see Note 13).

2.6. Identification of scFvs with Required Specificity by Phage ELISA

1. M13 KO7 helper phage at 3×10^{13} p.f.u./ml (see Note 14).
2. Anti-M13 HRP-conjugate antibody.
3. TMB substrate (3,3′,5,5′-Tetramethylbenzidine).
4. Stop reagent: 0.5 M H_2SO_4.
5. 100% Ethanol.
6. 2×TY growth media (see Subheading 2.1, item 15).
7. 2×TYG growth media (see Subheading 2.5, item 15).
8. 2×TYAG selective growth media (see Subheading 2.5, item 16).
9. 2×TYAK selective growth media: 100 µg/ml ampicillin and 50 µg/ml kanamycin in 2×TY media.
10. 1× PBS: 137 mM NaCl, 2.7 mM KCl, 10 mM Na_2HPO_4, 1.76 mM KH_2PO_4, pH 7.4.
11. 2× PBS (twice the concentration of 1×PBS, see item 10).
12. PBST: 0.1% (v/v) Tween-20 in 1×PBS (see item 10).
13. MPBS: 3% (w/v) skimmed milk powder in 1× PBS (see item 10).
14. 2×MPBS: 6% (w/v) skimmed milk powder in 2×PBS (see item 11).
15. Purified target antigen, (5–25 µg per plate).

3. Methods

3.1. Conversion of Phage Display Outputs to Ribosome Display Format

This protocol describes the conversion of a phage display library into a format that is compatible with ribosome display. First, the phage library DNA is isolated and quantified (see Note 15), along with the pCANTAB6 plasmid (or alternative) containing a suitable tether sequence (see Note 16). Next, the scFv portion of the phage display output is amplified and purified, as is the tether sequence. The two sequences are linked using a pull-through (recombinatorial) PCR and, after purification, a final amplification is performed using T7B and T6te primers to add the mRNA stem-loops and promoter sequence (see Note 17).

3.1.1. Isolation of Plasmid DNA from Phage Display Outputs

1. Inoculate 500 ml of 2×TYAG with 100 µl of the required output glycerol stock and grow at 37°C for 8 h, shaking at 300 rpm.
2. Take 150 ml of each culture and pellet the cells by centrifugation at $1,500 \times g$ for 10 min at 4°C. Discard the supernatant. At this point, the pellets can be stored at −20°C if required.
3. Purify the plasmid DNA from each sample using a plasmid midiprep kit, following the manufacturer's instructions.
4. Quantify the plasmid yields by UV spectroscopy (A_{260} reading).

3.1.2. Amplification of scFv Sequences by PCR

1. Set up the following PCR reaction for each library: 50 µl of PCR master mix (2×), 3 µl of Myc-Restore primer (10 µM), 3 µl of SDCAT-DP47 primer (10 µM), 3 µl of template DNA (100 ng/µl), and 41 µl of nuclease-free water. Also include a no-template control.
2. Amplify the sequences using the following PCR conditions: 94°C for 3 min; 25 cycles of 94°C for 30 s, 55°C for 30 s and 72°C for 105 s; 72°C for 5 min.
3. Separate the PCR products on a preparative 1% agarose/TAE gel. Cut out the appropriate band from the gel (see Note 18) and purify using a gel extraction kit according to the manufacturer's instructions.
4. Store the purified scFv DNA library at −20°C.

3.1.3. Production of the GeneIII Tether

1. Set up the following PCR reaction to amplify the tether from pCANTAB6: 50 µl of PCR master mix (2×), 2 µl of GeneIII-Reverse primer (10 µM), 2 µl of Myc-Forward primer (10 µM), 1 µl of pCANTAB6 vector (10 ng/µl), and 45 µl of nuclease-free water. Also include a no-template control.
2. Amplify the sequences using the conditions described in Subheading 3.1.2, step 2.

3. Gel-purify the PCR product as described in Subheading 3.1.2, step 3.

4. Store the purified tether DNA at −20°C.

3.1.4. Pull-Through Reaction to Link the GeneIII Tether to the scFv Library

1. Run a 5 µl sample of both the purified scFv and tether products on the same analytical 1% agarose/TAE gel. Verify their relative sizes and determine the DNA concentrations. If both products look pure, then set up the following PCR reaction for each library: 50 µl of PCR master mix (2×), 10 µl of purified scFv library (10–20 ng/µl), 3 µl of purified geneIII tether (50 ng/µl), and 35 µl of nuclease-free water.

2. Generate the full-length sequences using the following PCR conditions: 94°C for 3 min; 5 cycles of 94°C for 30 s, 50°C for 30 s and 72°C for 105 s.

3. During the fifth annealing step, (i.e. at 50°C), pause the PCR block and add 2 µl of mixed primers (SDCAT-DP47 and T7te, each at 10 µM) to the reactions. Resume the PCR program and, once the 72°C extension step is complete, proceed with the following reaction conditions: 3 cycles of 94°C for 30 s, 35°C for 30 s and 72°C for 105 s; 15 cycles of 94°C for 30 s, 50°C for 30 s and 72°C for 105 s; 72°C for 5 min.

4. Run a 5 µl sample of each pull-through product on a 1% agarose/TAE gel and compare to 2 µl of the initial purified scFv library. The pull-through product should be noticeably larger at ~1,100 base pairs.

5. Gel-purify the PCR product as described in Subheading 3.1.2, step 3 (see Note 19).

6. Store the full-length construct at −20°C.

3.1.5. Amplification of the Full-Length Transcription Template

1. Set up the following PCR reaction for each sample: 50 µl of PCR master mix (2×), 3 µl of T7B forward primer (10 µM), 3 µl of T6te reverse primer (10 µM), 10 µl of purified pull-through product (10–20 ng/µl), and 34 µl of nuclease-free water.

2. Amplify the sequences using the following PCR conditions: 94°C for 3 min; 20 cycles of 94°C for 30 s, 55°C for 30 s and 72°C for 105 s; 72°C for 5 min.

3. Check 5 µl of each PCR sample on a 1% agarose/TAE gel for size and purity (see Note 20). If the size is correct, and there are no non-specific products visible, then the DNA is ready for transcription (see Subheading 3.3). *Do not gel-purify the product at this stage!*

4. Store the transcription template at −20°C.

3.2. Optional Error-Prone Mutagenesis

This mutagenesis protocol provides a method to generate increased diversity in a selection output by exploiting amino-acid sequence space (see Note 21). First, a round of error-prone mutagenesis is performed using SDCAT V_H specific and Myc-Restore primers, which generates the increased sequence diversity. A recombinatorial PCR is then used to add the geneIII tether to the 3′ end of the scFv library DNA, and this is followed by a re-amplification with T7B and T6te primers to add the promoter sequences and stem-loops. Finally, a subset of the library is cloned and sequenced for QC purposes.

1. Gel-purify the selection output as described in Subheading 3.1.2, step 3. Use a different agarose gel for each output. Estimate the concentration of this product by running a 5 μl sample on a 1% agarose/TAE gel alongside a quantitative DNA marker.

2. Store the purified DNA at −20°C.

3. Set up the following PCR reaction to introduce mutations into the library constructs (see Note 22): 5 μl of Titanium *Taq* buffer (10×), 4 μl of $MnSO_4$ (8 mM), 5 μl of dGTP (2 mM), 1 μl of Diversify® dNTP mix (50×), 1 μl of SDCAT V_H specific primer (10 μM), 1 μl of Myc-Restore primer (10 μM), 1 μl of Titanium *Taq* DNA polymerase, 1 μl of purified DNA template (10 ng/μl), and 31 μl of nuclease-free water. Also include a no-template control.

4. Amplify the sequences using the following PCR conditions: 94°C for 3 min; 25 cycles of 94°C for 30 s, 68°C for 2 min.

5. Gel-purify the PCR product as described in Subheading 3.1.2, step 3 (see Note 23). Keep the purified DNA on ice until the recombinatorial PCR (see Note 24).

6. Produce the geneIII tether as described in Subheading 3.1.3, with the modification that the PCR should be performed in triplicate so that large quantities of tether are available for the recombinatorial PCR (see Note 25).

7. Store the gel-extracted PCR products on ice until the recombinatorial PCR.

8. Perform a pull-through reaction as described in Subheading 3.1.4. Include an additional control reaction that contains the geneIII tether but no scFv template (see Note 26).

9. Perform a final amplification as described in Subheading 3.1.5. The PCR products should *NOT* be gel-purified before transcription.

3.3. Transcription of Ribosome Display Libraries

This protocol describes how mRNA transcripts are produced from a T7B-T6te amplified linear PCR product. Briefly, the reaction components are mixed (see Note 27) and are then incubated at 37°C for 2 h. The mRNA is purified, quantified, and stored at −80°C or used immediately for selections.

1. Thaw the transcription buffer and rNTP mix at room temperature. Use quickly, and refreeze immediately. Keep the T7 polymerase at −20°C until needed.

2. Assemble the transcription components in the order listed in a non-stick nuclease-free microcentrifuge tube: 10 μl of transcription buffer (5×), 15 μl of rNTP mix (25 mM ATP, CTP, GTP, UTP), 20 μl of linear DNA template (T7B-T6te PCR reaction), 5 μl of T7 polymerase. Prepare a master mix if performing multiple reactions.

3. Mix well by pipetting or gentle vortexing. If necessary, pulse-spin to ensure that all liquid has collected at the bottom of the tube.

4. Incubate at 37°C for at least 2 h and no more than 3 h.

5. Pulse spin to collect condensation from the tube lid.

6. Isolate mRNA using a rapid purification spin-column according to the manufacturer's instructions. Transfer purified mRNA immediately onto ice.

7. Quantify the mRNA sample by measuring the A_{260} of a 1 in 500 dilution in nuclease-free water (see Note 28).

8. Use the mRNA immediately for ribosome display selection, or snap freeze in aliquots on dry ice and store at −70°C until needed.

3.4. Ribosome Display Selections

This protocol details a single round of affinity maturation that can be repeated as required. A standard affinity optimisation campaign could consist of between two and ten cycles of ribosome display selection, with the target antigen concentration being reduced at each round of selection (see Note 29). It is strongly advised to perform a positive control selection in parallel. This can be a previous successful selection or a single scFv clone that is known to bind to another suitable antigen.

Firstly, the mRNA is translated in vitro and the resulting scFv–ribosome–mRNA complexes are stabilised. These ternary complexes are incubated with a biotin-tagged target antigen at 4°C for several hours, and bound individuals are captured using streptavidin-coated magnetic beads, (although this protocol is amenable to other methods of tagging and capture as required). Any non-specific binders are washed away. The output mRNA is eluted, purified, amplified by RT-PCR, and then used as a template for both real-time PCR and end-point PCR. This final cDNA output is gel-purified and re-amplified for subsequent rounds of selection, or for sub-cloning into an appropriate expression vector.

3.4.1. Initial Preparation for Ribosome Display Selections

1. Pre-chill a bench-top microcentrifuge to 4°C.

2. Prepare heparin-block (HB) buffer on ice by combining the following in a 50 ml falcon: 5 ml of 10× *E. coli* wash buffer; 45 ml of sterile, filtered milliQ water and 625 μl of heparin at

200 mg/ml. Prepare sufficient buffer for the number of selections being performed (~15 ml per translation reaction, see Note 30). Store on ice at all times and prepare fresh buffer each day.

3. For each translation, pre-chill four 1.5-ml RNAse-free microfuge tubes on ice (one for the translation reaction, three for the selections) and prepare an additional 2-ml RNAse-free microfuge tube on ice, containing 1,320 μl of ice-cold HB buffer.

4. Prepare the strepatavidin-coated magnetic beads. Use 50 μl of beads per selection (plus an extra 100 μl of beads if blocking the selections – see Note 31). Wash the beads four times with chilled HB buffer in RNAse-free microfuge tubes, and resuspend in HB buffer to the original volume. Store on ice until required.

5. Optional: If blocking the selections, prepare de-biotinylated, sterile milk (see Note 31). Add 100 μl of streptavidin-coated magnetic beads to 1 ml of autoclaved 10% (w/v) skimmed milk powder in water in an RNAse-free microfuge tube. Incubate with end-over-end rotation for at least 10 min. Collect the beads using a magnetic particle separator and transfer the milk to a fresh tube. Store on ice until required.

6. Thaw premix X, protein disulphide isomerase and S30 extract on ice and start the following procedure immediately. These components must be used as soon as possible once thawed!

3.4.2. Translation of scFv DNA and Stabilisation of Ribosome Complexes

All steps in this section are to be performed on ice unless otherwise specified.

A 300 μl in vitro translation reaction is performed for each library, population or single clone to be selected. This reaction is then split between three selections: two with antigen and one without as a negative control.

1. Prepare a translation master mix for the total number of translation reactions required (plus one extra volume to account for pipetting errors) as follows. For one translation reaction, combine the following reagents in the order listed in an RNAse-free microfuge tube: 60.9 μl of nuclease-free water, 38.5 μl of potassium glutamate (2 M), 26.6 μl of magnesium acetate (0.1 M), 7 μl of protein disulphide isomerase (5 mg/ml), and 77 μl of premix X. Mix gently by pipetting up and down several times.

2. If mRNA has been produced on a previous day and stored at −70°C, thaw quickly by holding between fingertips and place immediately on ice when thawed. For each translation of library, control, or parent add 30 μl of diluted mRNA at 1 μg/ml (approximately 6×10^{13} molecules) to the bottom of a pre-chilled 1.5 ml RNAse-free microfuge tube on ice (from Subheading 3.4.1, step 3).

3. Very gently mix the thawed S30 extract by pipetting, but do not vortex or mix vigorously. Add 140 μl per translation to the mastermix and mix all reagents by pipetting up and down five times. Proceed to the next step *immediately*.

4. Add 300 μl of translation master mix to each 30 μl stock of mRNA, and mix gently by pipetting up and down five times. Immediately transfer one of the tubes to a 37°C heat-block for exactly 9 min (see Note 32). Repeat for all other selections at suitable intervals (10–30 s).

5. After translation, immediately pipette each 330 μl translation reaction into a pre-chilled 2 ml RNAse-free microfuge tube containing 1,320 μl of chilled HB buffer (from Subheading 3.4.1, step 3) and mix by gentle pipetting. This buffer stabilises the ternary scFv–ribosome–mRNA complexes.

6. Once all samples have been removed from the hot-block and stabilised, centrifuge at $17,000 \times g$ for 5 min at 4°C in a pre-chilled bench-top centrifuge (from Subheading 3.4.1, step 1) and place on ice.

3.4.3. Selection and Capture of Specifically Bound Ternary Complexes

1. For each translation reaction, transfer 500 μl of supernatant to three of the pre-chilled 1.5-ml RNAse-free selection tubes (from Subheading 3.4.1, step 3). If required, add 50 μl of chilled, de-biotinylated, sterile milk (from Subheading 3.4.1, step 5) to each tube. Cap the negative selection to prevent accidental addition of antigen.

2. For each positive selection, add the appropriate biotinylated antigen, at the required concentration. Do not exceed 30 μl total volume for antigen addition.

3. Incubate all selections at 4°C for the required time (see Note 33) with gentle end-over-end rotation.

4. Capture the selected complexes by the addition of 50 μl HB-washed streptavidin-coated magnetic beads (from Subheading 3.4.1, step 4) to each selection. Incubate for 5 min at 4°C with gentle end-over-end rotation.

5. Wash each selection five times with 800 μl pre-chilled HB buffer and transfer the beads to 220 μl of EB20 containing a 1/1,000 dilution of *S. cerevisiae* mRNA stock and mix well. Allow 10 min for the ternary complexes to dissociate and then remove the magnetic beads from the elution buffer (see Note 34).

6. For each selection, pre-chill a fresh RNAse-free microfuge tube on ice and add 400 μl of lysis buffer from the High Pure RNA Isolation Kit (see Note 35). This step should be performed during the 10 min mRNA elution.

7. Transfer 200 μl of the EB20 solutions (containing the eluted mRNA) into the pre-chilled RNAse-free microfuge tubes containing lysis buffer. Vortex each tube immediately to mix.

3.4.4. Purification of the Output mRNA and Reverse Transcription

1. Isolate mRNA using the High Pure RNA Isolation Kit, (see Note 35), and a bench-top microcentrifuge pre-chilled to 4°C (see Note 36). Use the kit according to the manufacturer's instructions, but elute in 40 μl of nuclease-free water. Proceed immediately to the next step.

2. Prepare a master mix for the required number of reverse transcription (RT) reactions to be performed (include two extra reaction volumes to allow for a no-template control and for pipetting error). For one sample, mix the following components in a fresh RNAse-free microfuge tube on ice: 4 μl of First Strand buffer (5×), 2 μl of DTT (0.1 M), 0.25 μl of T7te reverse primer (100 μM), 0.5 μl of dNTP mix (25 mM each oligonucleotide), 0.5 μl of RNasin® (40 U/μl), and 0.5 μl of Superscript II Reverse Transcriptase (200 U/μl).

3. Mix the RT master mix well and for each sample (plus the no-template control) aliquot 7.75 μl into the bottom of a thin-walled 0.2-ml PCR tube. Keep on ice.

4. Mix the eluted mRNA from step 1 by pipetting up and down five times, and add 12.25 μl of each sample to the 7.75 μl of RT mix. Mix well, but gently, by pipetting up and down five times. Immediately freeze any unused mRNA and store at −80°C.

5. Incubate the reactions in a PCR block at 50°C from 30 min. Prepare the PCR master mix during this time (see Subheading 3.4.5, step 1).

6. Transfer completed RT reactions onto ice.

3.4.5. End-Point PCR

1. Prepare a master mix for the required number of PCR reactions to be performed (include three extra reaction volumes to allow for the RT no-template control, a PCR no-template control and to account for pipetting error). For one reaction, mix the following components in a fresh RNAse-free microfuge tube on ice: 34.5 μl of nuclease-free water, 50 μl of PCR master mix (2×), 0.25 μl of T7te reverse primer (100 μM), 0.25 μl of SDCAT V_H specific forward primer (100 μM), and 5 μl of DMSO.

2. Mix well by pipetting. For each sample, (plus the RT no-template control and the PCR no-template control), aliquot 90 μl of master mix into the bottom of a thin-walled 0.2-ml PCR tube. Keep on ice.

3. Add 10 μl of each cDNA (from Subheading 3.4.4, step 6) to a separate PCR tube. Mix well, but gently, by pipetting up and down five times. Freeze any unused cDNA and store at −80°C.

4. Amplify the cDNA samples using the following conditions: 94°C for 3 min, 25–35 cycles (see Note 37) of 94°C for 30 s, 55°C for 30 s, 72°C for 105 s, 72°C for 5 min.

5. Run 5 μl of each PCR sample on a 1% agarose/TAE gel alongside a quantitative DNA ladder. The two positive selections should be similar, and should yield clearly more DNA than the selection without antigen (see Note 38). If this is the case, then the selection appears successful, and the positive outputs can be reamplified for a further processing (Subheading 3.4.7).

3.4.6. Optional: Real-Time PCR to Quantify Output cDNA Yields (see Note 39)

1. Dilute the selection output cDNA, (use PCR tube strips for convenience). Transfer 5 μl of each cDNA sample (from Subheading 3.4.4, step 6) to a PCR tube containing 15 μl of nuclease-free water and mix by pipetting several times.

2. Prepare a master mix for the required number of real-time reactions. Perform each reaction in duplicate and allow six extra volumes for the RT no-template control, a real-time no-template control and to allow for pipetting error. For one reaction, mix the following components in a fresh RNAse-free microfuge tube on ice: 12.25 μl of TaqMan Universal PCR Master Mix (2×), 2 μl of GeneIIIst-115F forward primer (10 μM), 2 μl of GeneIII-188R reverse primer (10 μM), 1 μl of GeneIIIst-141T probe (5 μM), and 2.5 μl of nuclease-free water.

3. Mix well by pipetting. For each sample, (plus RT no-template control, plus real-time no-template control), aliquot 20 μl of master mix into two wells of a 96-well PCR plate. Add 5 μl of each diluted cDNA sample to duplicate wells, and then seal the plate with a thermostable, optically transparent, adhesive cover. Plates can be prepared in advance and stored at 4°C for up to 4 h.

4. Centrifuge the plate for a few seconds in a bench-top centrifuge, (supporting the PCR plate in a 96-well, round-bottomed, microtitre plate for example), to ensure that all samples are at the bottom of the wells and that any air bubbles are expelled.

5. Perform the real-time PCR using the following conditions: 50°C for 2 min; 95°C for 10 min; 40 cycles of 95°C for 15 s, read fluorescence at 520 nm, 60°C for 60 s (see Note 40).

3.4.7. Re-amplification of Selection Outputs for Further Processing

1. To process successful outputs for a further round of selection or for sub-cloning into pCANTAB6, gel-purify the DNA as described in Subheading 3.1.2, step 3. Store purified DNA at −20°C. Estimate the concentration of each product by running a 5 μl sample on a 1% agarose/TAE gel alongside a quantitative DNA ladder.

2. Re-amplify the output DNA using the T7B and T6te primers as described in Subheading 3.1.5. Check 5 μl of each PCR sample on a 1% agarose/TAE gel for size and purity (see Note 41). Store DNA at −20°C.

3.5. Cloning of Enriched Ribosome Display Outputs

One of the major advantages of ribosome display is the fact that it is an entirely in vitro technology. This allows for fast library construction and selection, large library sizes, and easy manipulation of pools between rounds of selection. However, if individual clones have to be analysed by sequencing, or in a screening assay, then the DNA has to be ligated into an appropriate vector and transformed into cells. This protocol describes the cloning of scFv constructs into the pCANTAB6 vector; however, the same procedure can be applied to other antibody formats and vectors. First, the selection output is purified alongside a fresh batch of pCANTAB6 and then both are digested with *Nco*I and *Not*I restriction enzymes. The scFv insert and phagemid vector are ligated and transformed into chemically competent TG1 cells. Finally, the cells are grown overnight at 30°C and a representative number of colonies are picked for sequencing and screening.

3.5.1. Preparation of the scFv Insert and pCANTAB6 Vector

1. Separate the PCR products (see Note 42) on a 1% agarose/TAE gel, leaving at least one empty lane between each sample to reduce the chance of cross-contamination. Gel-purify as described in Subheading 3.1.2, step 3 (see Note 43). Elute the sample in 30 µl of nuclease-free water (*not Buffer TE*) and store purified DNA at −20°C. Estimate the concentration of each product by running a 5 µl sample on a 1% agarose/TAE gel alongside a quantitative DNA marker.

2. For pCANTAB6, use a standard plasmid miniprep kit according to the manufacturer's instructions. Elute the vector in 50 µl Buffer EB or nuclease-free water and quantify the DNA yield by measuring A_{260} of a 1 in 20 dilution in nuclease-free water.

3.5.2. Digestion of scFv Insert DNA with NotI and NcoI

1. Combine the following in a sterile eppendorf tube: 10 µl of gel-purified selection output DNA in water (30–100 ng/µl), 1 µl of *Not*I enzyme (10 U/µl), 1 µl of *Nco*I enzyme (10 U/µl), 5 µl of restriction buffer 3 (10×), 0.5 µl of bovine serum albumin (10 mg/ml), and 32.5 µl of nuclease-free water. Perform the reaction in triplicate for each sample.

2. Incubate at 37°C for 90 min.

3. Separate the PCR products on a 1% agarose/TAE gel, running the three triplicate samples in adjacent lanes. Cut the appropriate bands from the gel (see Note 44) and co-purify the three replicates as described in Subheading 3.1.2, step 3. Store the purified DNA at −20°C. Estimate the concentration of each product by running a 5 µl sample on a 1% agarose/TAE gel alongside a quantitative DNA marker.

3.5.3. Digestion of Vector DNA with NotI and NcoI (see Note 45)

1. Combine the following in an eppendorf tube: 10 µl of pCANTAB6 plasmid (100 ng/µl), 1 µl of *Not*I enzyme (10 U/µl), 5 µl of restriction buffer 3 (10×), 0.5 µl of bovine serum albumin (10 mg/ml), and 35.5 µl of nuclease-free water.

2. Incubate at 37°C for 90 min.

3. Heat-denature the samples at 65°C for 20 min to remove *Not*I enzyme from the DNA cleavage site.

4. Spin the samples to collect condensation and vortex gently. Remove a 5 μl aliquot and then add 1 μl of *Nco*I restriction enzyme (10 U/μl).

5. Incubate at 37°C for 90 min.

6. Run 5 μl of each sample on a 1% agarose/TAE gel alongside the singly digested sample (from step 4) and 5 μl of undigested vector (see Note 46).

7. If digestion is complete, gel-purify as described in Subheading 3.1.2, step 3. Up to four replicate samples can be pooled and applied to a single column. Elute the digested vector in 30 μl of nuclease-free water or Buffer TE and store the purified DNA at −20°C. Estimate the concentration of each product by running a 5 μl sample on a 1% agarose/TAE gel alongside a quantitative DNA ladder.

3.5.4. Ligation of Insert and Vector (see Note 47)

1. For each ligation, combine the following in a microfuge tube on ice: 2 μl of digested pCANTAB6 vector (~50 ng/μl), 3 μl digested scFv insert (~15 ng/μl), 3 μl of T4 DNA ligase buffer (10×), 1 μl of T4 DNA ligase (6,000 Weiss units/ml), and 21 μl of nuclease-free water. Include a negative control reaction containing vector and ligase but no insert to monitor the efficiency of the ligation reactions.

2. Incubate at room temperature for 1 h, and then leave on ice until ready to transform.

3.5.5. Transformation of Ligated DNA into E. coli Cells (see Note 48)

1. Transform 10 μl of ligation reaction into 100 μl of TG1 cells. Perform in triplicate for each selection output, and include both a positive and negative transformation control.

2. Incubate all aliquots on water–ice for 30 min, transfer to a 42°C water bath for 45 s, and return immediately to water–ice for a further 2 min.

3. Add 900 μl of 2×TYG (no ampicillin) growth media to each aliquot, and shake cells for 1 h at 37°C and 150 rpm.

4. Plate out transformations onto 2×TYAG agar plates, leave to dry for ~10 min and incubate overnight at 30°C (see Note 49).

5. Prepare colony source plates for further analysis by picking a representative number of colonies from each selection output into 2×TYAG media in a standard 96-well plate. Grow the cells overnight at 30°C and 150 rpm, and then add glycerol to a final concentration of 15% (v/v). Freeze the cultures at −80°C.

3.6. Identification of scFvs with Required Specificity by Phage ELISA

This protocol details how to perform a specificity ELISA when the DNA of interest has been cloned into the pCANTAB6 phagemid (see Note 50). If other expression vectors and/or display formats are being used then these protocols may need to be substantially altered to produce an effective ELISA screen. First, the source plates, (prepared in Subheading 3.5.5, step 5), are replicated and the daughter cultures grown to mid-logarithmic phase. The bacteria are then infected with helper phage and cultured overnight to produce a high density of phage particles. An appropriate number of microtitre plates are coated with antigen, and then both the phage cultures and the plates are blocked to reduce non-specific binding. The antigen-coated plates are first incubated with the phage cultures, and then incubated with a secondary (detection) antibody. Finally, the reaction is developed and positive binding clones are identified.

3.6.1. Preparation of Blocked Phage Particles

1. Replicate each source plate into a 96 deep-well block, (500 μl 2×TYAG per well), and grow for about 5 h at 37°C, 280 rpm. We recommend that control colonies are added to each daughter block to act as an internal reference (see Note 51).

2. Remove the daughter blocks from the incubator and check that growth is visible in all wells – the media should be visibly turbid. If this is the case, dispense 100 μl of diluted helper phage into each well, (5 μl of M13 KO7 helper phage per 10 ml 2×TYAG media).

3. Incubate the blocks for 1 h at 37°C and 150 rpm. The lower rotation speed is necessary to allow the phage to infect the TG1 cells.

4. Centrifuge the blocks for 10 min at 2,000×g, room temperature.

5. Remove the supernatant, dispense 500 μl of 2×TYAK media into all wells of the daughter blocks (see Note 52), and culture overnight at 25°C and 280 rpm.

6. Add an equal volume (500 μl) of 2×MPBS to each phage culture and incubate the blocks at room temperature for 1 h.

7. Centrifuge the blocks for 5 min at 2,000×g, room temperature.

3.6.2. Preparation of Antigen-Coated Plates

1. For each source plate, prepare one antigen-coated plate by adding 50 μl of antigen solution, (at the appropriate concentration – see Note 53), into each well of a "sticky" microtitre plate (e.g. Nunc Maxisorp™ or Immobiliser™). Alternatively, a biotinylated form of the target antigen can be specifically bound to streptavidin-coated microtitre plates. In either case, coat another set of plates with a negative control (e.g. BSA) at the same concentration.

2. Cover the plates with a plate-seal and incubate overnight at 4°C (alternatively, coat for 2 h at room temperature or 1 h at 37°C).

3. Wash the antigen-coated plates three times with 1×PBS to remove unbound antigen and add 300 μl of 1×MPBS to each well.

4. Incubate for 1 h at room temperature and then wash three times with 1×PBS.

3.6.3. Incubations and Detection

1. Transfer 50 μl of blocked phage from each culture into the corresponding wells in both a positive antigen-coated plate and a negative control plate.

2. Incubate the plates with the phage cultures for 1 h at room temperature and then wash all plates three times with PBST.

3. Add 50 μl per well of secondary antibody, diluted to an appropriate concentration in 1×MPBS (see Note 54).

4. Incubate for 1 h at room temperature and then wash all plates three times with PBST.

5. Develop the reactions and identify positive wells (see Note 55).

3.6.4. Data Analysis (see Note 56)

1. *The negative control plates*: Ideally, the readings from the positive and negative control clones should be comparable, and should be only marginally higher than for the blank wells. In addition, where more than one source plate was screened, the control clones should give similar readings on all negative plates. Use the readings from the control clones to set the negative cut-off (see Note 57). Any phage clone that exhibits a binding signal above this value should be considered "non-specific" and should be discarded.

2. *The positive assay plates*: Ideally, the positive cut-off should be set at least fourfold higher than the negative cut-off (see Note 57). The readings from the positive control clones should be consistent between antigen plates, and should be well above the positive cut-off. The signals from the negative controls should be below the negative cut-off, as determined from the negative plates. Any phage clone that exhibits a binding signal above the positive cut-off should be considered a hit, provided that it is specific for the target antigen.

3. Once all hits have been identified, they can be tracked back to the original source colony plate stored at −80°C. Interesting colonies can be picked into fresh 2×TYAG media, grown up, and sequenced.

3.7. Characterisation of Purified scFv Sequences by Biacore

Output clones that produce a strong signal in the phage ELISA should be sequenced to confirm that the corresponding scFv is a "novel" hit. The most interesting of these hits can then be analysed by BIAcore (or similar) to assess the affinity improvements relative to parent protein. We recommend that the kinetic analysis is performed on purified scFv proteins, (see Note 58), since other cellular components can affect the BIAcore results.

4. Notes

1. The SDCAT-DP47 forward primer has the sequence **AGACCACAACGGTTTCCC**TCTAGAAATAATTTTGTTAACTTTA*AGAAGGAGA*TATATCC<u>ATGGCCGAGGTGCAGC</u>. This particular primer is specific to the start of the DP47 V_H framework, but all SDCAT forward primers have the general form shown above. The first 18 bases (shown in bold) are part of a 5′ mRNA stem-loop, which is required to prevent mRNA degradation during the antigen selections. The Shine-Dalgarno (SD) sequence, which promotes ribosome binding, is shown in bold and italicised. The variable region of the primer, underlined, is engineered to match the start of the V_H domain. This region should be approximately 15 nucleotides in length and should terminate with one or more G/C bases.

2. The Myc-Restore reverse primer is generic, and recognises the myc-tag that is common to all pCANTAB6 scFv sequences.

3. The Myc-Forward primer binds to the generic myc-tag. It is complementary to the reverse primer described above (Myc-Restore).

4. The GeneIII-Reverse primer is complementary to a nucleotide sequence within the geneIII protein.

5. The T7B forward primer has the sequence ATACGAAAT<u>TAATACGACTCACTATA</u>**GGGAGACCACAACGG**. It recognises and completes the 5′ stem-loop of the SDCAT primer (bold), and also adds a T7 promoter region to initiate transcription (underlined).

6. The T6te reverse primer has the sequence <u>CCGCACACCAGTAAGGTGTGCGG</u>**TATCACCAGTAGCACCATTACCATTAGCAAG**. It recognises the 3′ end of the geneIII tether sequence (bold) and contains a stem-loop (underlined) to help prevent degradation of the mRNA transcripts.

7. The T7te reverse primer is a short form of the T6te primer (above) and is used when the stem-loop is not required, since it generally results in better yields and a higher purity PCR product.

8. For DNA visualisation, we use SYBR® Safe from Invitrogen. This stain is excited with blue light and is therefore much safer than alternative stains that require excitation with ultraviolet light. In addition, the blue light is less destructive to the DNA samples and gives better ligation of selection outputs.

9. The phagemid glycerol stock can either be a naïve library (17) or a phage display selection output. The conversion process can also be performed with a single lead clone, in which case a much smaller volume is required.

10. Our preference is for Sarstedt microcentrifuge tubes, since they are non-stick and produce the best selection windows. In addition, their lids are less prone to snapping when eluting from mRNA purification columns.

11. We use a homemade system, prepared according to the literature (18). However, commercial *E. coli* translation systems are available.

12. We perform real-time PCR using primers and a probe that bind to the geneIII tether, since this sequence is present in all of our scFv constructs. If an alternative tether sequence is being used, then the real-time primers and probe will need to be redesigned. The forward primer (GeneIIIST-115F) has the sequence CTTGATTCTGTCGCTACTGATTAC, and the reverse primer (GeneIIIST-188R) has the sequence CCATTAGCAAGGCCGGAAG. The probe (GeneIIIST-141T) is synthesised as **FAM**-GTCACCAATGAAACCATCGATAGCAGCA-**TAMRA**, where FAM and TAMRA are donor and quencher molecules respectively.

13. The ribosome display output will typically be cloned back into the same vector that contained the initial phage display population. However, if the original format was not scFv then we recommend cloning the output into a scFv vector such as pCANTAB6, which is necessary to maintain the synergy between the V_H and V_L domains. In addition, this facilitates conversion back to ribosome display format if necessary.

14. We use homemade M13 K07 helper phage, prepared according to the literature (19). However, commercial helper phages are available.

15. The antibody format used for ribosome display (RD) selections is predominantly scFv, and this protocol describes the conversion of scFv phage display libraries to scFv ribosome display libraries. When converting a phage library from the pCANTAB6 vector, the V_H and V_L domains are already concatenated and can be amplified together using the SDCAT specific and Myc-Restore primers. When converting from other formats, (e.g. Fab), the V_H and V_L domains can be amplified separately with two extra complementary primers that produce a flexible linker between the two domains. In this chapter we exemplify our construction and selection protocols with the "DP47" phage library, which comprises the heavy chain sequence of the DP47 germ line, combined with V_H CDR3 and V_L sequences from a large antibody repertoire (over 8×10^8 unique members). For the standard scFv format, the V_H and V_L domains are linked by a 15 amino-acid $(Gly_4Ser)_3$ linker and each construct contains a generic section of DNA after the light-chain sequence that encodes a poly-histidine (His)-tag and c-Myc (myc)-tag to facilitate

detection and purification of isolated scFvs post-selection. However, these conversion protocols can be adapted to any phage display library composed of naïve, synthetic or single framework libraries. The standard SDCAT primers described above (see Note 1) allow for conversion of naïve or single framework phage display outputs to ribosome display scFv library format. For synthetic libraries, the appropriate primer design can be deduced from the SDCAT primers.

16. For ribosome display, a tether sequence (polypeptide spacer) is necessary. This tether occupies the ribosome tunnel in the stalled ribosome complex and thereby allows the scFv library sequences to protrude from the ribosome and fold. In our constructs, the tether is produced from the pCANTAB6 vector using the Myc-Forward and GeneIII-Reverse primers. However, this particular tether can be substituted for other nucleotide sequences of ~300 base pairs, provided that the 5′ tether primer contains a sequence that is complementary to the 3′ scFv primer.

17. There are several features present in the ribosome display scFv construct that are not present in the corresponding phage display construct. The SDCAT primers are used to add the Shine-Dalgarno (SD) sequence to the construct; however, the remaining features are only added in the final PCR before transcription using the T7B and T6te primers.

18. If a scFv construct is being amplified, the PCR product should be approximately 850 bp in length. The tether should be approximately 300 bp.

19. When separating the PCR products on a preparative 1% agarose/TAE gel, make sure that the gel is run for sufficient time that the full-length construct is clearly separate from any un-recombined scFv that may be present.

20. Following T7B-T6te amplification, there should only be one strong band at ~1,200 bp. Non-specific bands of lower molecular weight will be amplified preferentially at the next PCR and will thus eliminate the specific product. Lower molecular weight smears suggest degradation of the specific product. In either case, the conversion has not been successful and should be repeated.

21. This protocol uses a process known as error-prone mutagenesis, which can introduce up to eight nucleotide changes per 1,000 bp of template DNA. It has been adapted from the Diversify® PCR Random Mutagenesis Kit (Clontech). The average number of mutations that are introduced can be controlled by the concentrations of manganese sulphate and/or dGTP that are present in the PCR reaction. If necessary, template DNA can be subjected to several rounds of error-prone PCR to

generate a higher mutation frequency. Alternatively, mutations can be introduced at each selection round.

22. These conditions are for the highest error rate (8.1 nucleotide changes per 1,000 bp according to the manufacturer). For low or medium error rates, refer to the Diversify® manual.

23. The PCR products generated under error-prone mutagenesis conditions are generally of poorer quality than those seen from standard PCR reactions. For this step only, we recommend that if any PCR product is visible at ~850 bp then it should be extracted and processed, regardless of the purity of the sample.

24. The recombinatorial PCR should be performed on the same day as the primary gel extractions.

25. The geneIII tether should be freshly produced alongside the error-prone libraries, since the use of old tether can compromise the subsequent pull-through reaction. This is due to degradation of the complementary ends of the PCR products.

26. If this control reaction produces a full-length product (~1,100 bp) then the geneIII tether is contaminated and should be re-made.

27. We prepare mRNA in vitro from ribosome display libraries under the control of the T7 promoter using the Ribomax Large Scale RNA production system [T7] from Promega. The resulting mRNA can (and should) be used immediately for Ribosome Display selections. The template for transcription should be a non-purified PCR product (i.e. a T7B-T6te re-amplification reaction) as we have previously found that the use of purified PCR products can compromise translation efficiency, possibly due to contaminants from the spin columns. The template can be a single scFv sequence, (e.g. to verify that a parent clone is compatible with Ribosome Display), or a population such as an error-prone library or a previous selection output.

28. The concentration should be higher than 2 μg/μl. If the concentration is lower than this value, the transcription should be repeated.

29. The first round of ribosome display selection is typically performed at an antigen concentration that matches the preceding phage display selection. For naïve libraries, 100 nM is a suitable starting point. This concentration is typically reduced tenfold with each subsequent round of RD selection. However, if the selection outputs contain clones with lower affinity than the selection input, or if there is no discernable difference between selections with and without antigen, then the concentration drops may be too severe. In these cases, return to the last successful output and try dropping the antigen concentration two- to fivefold instead.

30. The protocol is designed for three selections per library, population or single clone: two on antigen and one without (+ + −).

31. Ribosome display selections are often blocked using a final concentration of 1% skimmed dried milk powder. However, many selections perform better in the absence of milk, especially when binding to the antigen is allowed to proceed overnight. In the event of a poor selection, consider either adding or omitting milk from the selection mix.

32. Although we have found between 7 and 9 min to be best for our in-house translation system, alternative systems may require optimisation of this incubation step.

33. The selections should be performed for sufficient time that even the highest affinity interactions are allowed to approach equilibrium. As a rule-of-thumb, the selections should be left for at least one dissociation half-life ($t_{1/2}$), which can be calculated using $\ln(2)/k_{off}$. For example, a clone with an estimated off-rate of 10^{-4} s^{-1} has a $t_{1/2}$ of ~2 h, whereas a clone with an off rate of 10^{-5} s^{-1} has a $t_{1/2}$ of ~20 h. The actual equilibration rates are impossible to calculate since (a) the on- and off-rates are usually determined for the isolated scFv, and not for the mRNA–ribosome–scFv complex, and (b) the concentration of ternary complex is unknown. For the early stages of affinity maturation, 2 h is usually sufficient time for equilibration. However, when dealing with picomolar affinities, the selections should be left to equilibrate overnight.

34. For the wash and elution steps we strongly recommend the use of an automated system, (e.g. a Kingfisher mL) that has been pre-chilled to 4°C.

35. Other RNA purification kits are available, and the volume of lysis buffer will depend on the manufacturer's instructions.

36. We recommend the use of a vacuum aspirator to remove the flow through from each column. This increases the speed and consistency of the process, and helps to reduce cross-contamination between samples. In addition, when transferring the columns to a new microcentrifuge tube prior to the final elution step, be careful not to let residual ethanol filtrate in the collection tube touch the bottom of the column since ethanol inhibits the reverse transcription step. If in doubt, re-spin the samples.

37. The number of amplification cycles used in the end-point PCR depends on the parent, the library type and the degree of enrichment that has occurred. Typically, 25 cycles are required for our "model" selections, whilst 30 cycles is standard for a parent test selection. Early rounds of ribosome display selection typically require 35 cycles, whereas later rounds require fewer (~30) cycles of amplification. Too few cycles will result in DNA yields that are too low to be quantified/purified, whereas too many cycles will reduce the selection window between the +ve

and −ve antigen runs and will favour the generation of non-specific PCR products. Guidance can be obtained by performing real-time PCR before the end-point PCR.

38. There should only be one strong band at ~1,100 bp, and an absence of non-specific bands of lower molecular weight. If present, these smaller products will be preferentially amplified in the next PCR step, causing deterioration in input quality and leading to selection failure. If the positive runs produce weak bands, with insufficient DNA for subsequent processing, the number of cycles in the end-point PCR can be increased up to a maximum of 35 cycles.

39. Real-time PCR is a technique that accurately measures cDNA levels, and it can be used in the context of ribosome display to assess the success of selections up to the end of the reverse transcription process. It is often used for troubleshooting purposes, to confirm unexpected end-point results or when no end-point PCR products are obtained, and the accurate cDNA quantification is used for method development and when testing new batches of reagent. Real-time PCR is also used to determine the appropriate number of cycles for the end-point PCR, allowing for sufficient amplification of the scFv product without excessive background amplification. The real-time reactions are based around an oligonucleotide probe that contains a 5′ reporter dye (e.g. FAM) and a 3′ quencher (e.g. TAMRA).

40. Decide on a reference sample, (most usually a positive run of the control selection), and calculate the cycle window (ΔC_T) for each sample ($= C_{T,sample} - C_{T,reference}$). The control selection should show excellent agreement between the two positive selection runs, and should produce a large cycle window between the positive and negative runs. Successful selections will show good agreement between the two positive selection runs ($\Delta C_T < 1$ cycle), with a cycle window of at least three cycles between the positive and negative runs (~10-fold difference in cDNA). In the vast majority of cases, real-time and end-point PCR data are consistent. This assumes that a sufficient, but not excess, number of end-point PCR cycles have been performed to visualise the products, without over amplifying the cDNA. Occasionally, real-time PCR data may indicate successful selections with good enrichment in cases where this is not reproduced by end-point PCR. The most likely explanation for this discrepancy is that the cDNA quality is poor, i.e. is not full-length. Good real-time data can still be obtained because only an internal ~100 bp region of the cDNA is amplified in the real-time PCR assay, whereas the end-point PCR uses primers that bind to the 5′ and 3′ termini of the scFv construct. In these cases the selections should be repeated and, if the problem persists, the libraries may need to be rebuilt using internal primers (e.g. Myc-Restore).

41. After the final amplification of a selection output there should only be one strong band at ~1,200 bp, and no visible bands of lower molecular weight. If these conditions are satisfied, then the T7B-T6te amplified cDNA is ready for another round of selection, (starting with transcription, Subheading 3.3), or subcloning into pCANTAB6 (Subheading 3.5). It is recommended that the functionality of relevant selection outputs is confirmed by subcloning the cDNA into pCANTAB6, followed by a phage ELISA screen with relevant positive and negative controls. A representative number of clones from each output (~88) should be sequenced for quality control purposes. If this selection has been based on an error-prone library, the outputs should contain four to six base changes per scFv if a high error rate has been used.

42. The insert DNA should be a fresh PCR product that has been amplified with T7B or SDCAT forward primers, together with Myc-Restore, T6te or T7te reverse primers. The SDCAT primer is advisable, since it spans the *Nco*I restriction site and can restore the sequence if it has been unintentionally mutated during the PCR steps.

43. The amplified scFv fragment should be ~900–1,200 bp in length, depending on the primers used for amplification and the size of the scFv.

44. The *Not*I/*Nco*I digested insert should be ~900 bp in length. Depending on the initial construct, it may be possible to see lower molecular weight bands that have been cleaved from the scFv sequence (e.g. the geneIII tether).

45. In general, 1 μg of pCANTAB6 plasmid is digested with *Not*I and *Nco*I in a 50 μl sample, which should be sufficient for cloning about eight inserts. If the *Not*I and *Nco*I restriction sites are more than 100 bp apart, (for example when cleaving vector that already contains a scFv insert), then a double digestion can be performed as described for the insert. However, if the two restriction sites are less than 100 bp apart, then a double digestion is not recommended and the two enzymes should be added sequentially.

46. The pCANTAB6 should be completely linearised after the first digestion and should therefore run as a single band of larger molecular weight than undigested vector. Depending on the size of the *Not*I–*Nco*I fragment, there may also be a noticeable decrease in size after the second digestion.

47. In our experience, ligations are most successful when using a "vector ends" to "insert ends" ratio of between 1:3 and 1:5. In addition, we try to minimise the total volume of insert plus vector added to each reaction, (<5 μl), as there are concerns that contaminants from the purification process are able to

inhibit the DNA ligase. This is one of the reasons that we suggest performing digestions in duplicate/triplicate followed by co-purification through a single column. Since 1 bp is approximately 660 g/mol, 1 μg of pCANTAB6 vector (4.6 kb) is ~330 fmol and 1 μg of insert (850 bp) is ~ 1,800 fmol. In a typical reaction, ~100 ng of vector and ~50 ng of insert (about a 3:1 ratio of insert ends to vector ends) are ligated together.

48. Ligated DNA is normally transformed into chemically competent TG1 cells by heat-shock. This particular cell-type is favoured since it exhibits high expression levels of both phage and isolated scFv fragments; however, other cell types (e.g. XL1-blue) and transformation methods (e.g. electroporation) can be used.

49. Successful ligation reactions should give 100-fold more colonies (or better) than either of the negative controls. Ideally, the transformation efficiency should be >10^8 colony forming units (cfu) per μg DNA.

50. The pCANTAB6 vector has been designed so that a *Not*I–*Nco*I insert is ligated upstream (5′) of the M13 filamentous phage geneIII sequence (20). When expressed at low levels, the product of this vector is a scFv that is linked, via a His-tag and myc-tag, to the N-terminus of a fully functional geneIII minor coat protein. Since pCANTAB6 is a phagemid vector, mature phage particles are not released from the host cells until those cells are also infected with M13 K07 helper phage. The genes within the helper phage enable the packaging of the viral DNA, and promote the assembly of mature virions that express a scFv antibody fragment on their surface. A standard phage ELISA is performed by immobilising the desired target antigen on the surface of a 96-well plate. Antibody fragments that interact strongly with the target will remain bound after several washing cycles, whereas those with poor binding characteristics will be removed. A secondary antibody is then applied which binds to a specific epitope of the scFv-phage construct, (e.g. His-tag, myc-tag or the phage particle itself). In addition, the secondary antibody is normally conjugated to a label (e.g. europium cryptate) or enzyme (e.g. horse-radish peroxidase – HRP) that produces a detectable signal. Any wells that exhibit a strong signal can be tracked back to the original clone that produced the phage particle, and the DNA sequence of the particular scFv insert can be determined. If other expression vectors and/or display formats are being used then these protocols may need to be substantially altered to produce an effective ELISA screen.

51. For the ELISA controls, we recommend at least two positive wells, (scFv sequences that are known to give a positive signal in the phage ELISA), and two negative wells, (scFv sequences that produce a background signal). Leave at least two wells completely empty to determine a "true" background signal.

52. Remove as much of the 2×TYAG media as possible, since trace amounts of glucose will act as a repressor of the scFv-geneIII contruct. The 2×TYAK media contains ampicillin to select cells with the pCANTAB6 phagemid, whilst the kanamycin is used to select cells that have also been infected with the helper phage.

53. Optimal coating conditions vary between different target antigens. If a suitable detection antibody is available, (either a commercial antibody, or a parent clone that has been expressed as a full-length IgG), then we recommend performing a presentation ELISA at the start of the project. In this experiment, a serial dilution of antigen is prepared (e.g. 10 μg/ml down to 10 ng/μl) and each dilution is coated onto a different well of the plate. Each well is then incubated with a standard concentration of detection antibody (e.g. 1 μg/ml), followed by secondary antibody, and then finally the signal from each well is determined. The optimal coating condition is the lowest dilution that still produces a maximal signal. It should be noted that this initial experiment, using full-length antibodies, only gives a rough guide to the optimal coating conditions. Once ELISA-positive phage clones are available to further optimise conditions, the antigen concentration may need to be altered. If a presentation ELISA is not feasible, then a starting antigen concentration of between 1 and 5 μg/ml should be used.

54. When using the anti-M13 HRP-conjugate antibody from GE Healthcare, the recommended dilution is 1 in 5,000. For other secondary antibodies, refer to the manufacturer's instructions.

55. For an HRP conjugate antibody, add 50 μl per well of detection reagent (TMB substrate) and develop for 5–20 min at room temperature. Stop the reaction by adding 50 μl per well of 0.5 M H_2SO_4. Use a plate reader to determine the absorbance of all wells at 450 nm. For other detection systems, (e.g. Europium cryptate), refer to the manufacturer's instructions.

56. The exact method of data analysis will be dependent on the antigen, ELISA format and control clones. In addition, the cut-offs, (used to designate whether a clone is a hit or not), are also likely to vary between repeats due to slight differences in phage yields. Nevertheless, these guidelines provide a good basis for data analysis.

57. Typically, we find that blank wells give a reading of ~0.05 AU and non-binding clones give a reading of 0.05–0.10 AU. We usually set the negative cut-off to be ~0.05 AU above the non-binding clones, i.e. 0.10–0.15. As a result, the positive cut-off is normally set at between 0.40 and 0.60 AU.

58. The pCANTAB6 vector has been designed to facilitate purification, since it appends both a Myc-tag and a His-tag to the

C-terminus of the scFv sequence. In addition, the vector contains an amber-stop codon between the scFv sequence and the geneIII sequence. When the proteins are expressed at low levels, (e.g. in a phage ELISA), then the amber-stop codon is suppressed and the scFv sequences are covalently linked to the geneIII minor coat protein of M13 filamentous phage. However, when expression is driven at high levels by the addition of IPTG, the majority of scFv sequences are truncated at the amber-stop codon leading to isolated scFv proteins. Typically, the clones of interest are grown up in a 500 ml culture to mid-log phase and then induced by the addition of 1 mM IPTG. After an incubation of 4 h, the cells are lysed and the scFv protein is purified using a nickel-affinity column.

References

1. McCafferty J, Griffiths A D, Winter G, *et al.* (1990) Phage antibodies: filamentous phage displaying antibody variable domains, Nature 348: 552–554.
2. Smith G P. (1985) Filamentous fusion phage: novel expression vectors that display cloned antigens on the virion surface, Science 228: 1315–1317.
3. Hudson PJ and Souriau C. (2003) Engineered antibodies, Nat Med 9: 129–134.
4. Osbourn J, Groves M and Vaughan T. (2005) From rodent reagents to human therapeutics using antibody guided selection, Methods 36: 61–68.
5. Edwards BM, Barash SC, Main SH, *et al.* (2003) The remarkable flexibility of the human antibody repertoire; isolation of over one thousand different antibodies to a single protein, BLyS, J Mol Biol 334: 103–118.
6. Hoet R M, Cohen E H, Kent R B, *et al.* (2005) Generation of high-affinity human antibodies by combining donor-derived and synthetic complementarity-determining-region diversity, Nat Biotechnol 23: 344–348.
7. Pukac L, Kanakaraj P, Humphreys R, *et al.* (2005) HGS-ETR1, a fully human TRAIL-receptor 1 monoclonal antibody, induces cell death in multiple tumour types in vitro and in vivo, Br J Cancer 92: 1430–1441.
8. Baker K P, Edwards B M, Main SH, *et al.* (2003) Generation and characterization of LymphoStat-B, a human monoclonal antibody that antagonizes the bioactivities of B lymphocyte stimulator, Arthritis Rheum 48: 3253–3265.
9. Groves MA and Osbourn JK. (2005) Applications of ribosome display to antibody drug discovery, Expert Opin Biol Ther 5: 125–135.
10. Hanes J, Jermutus L, Weber-Bornhauser S, *et al.* (1998) Ribosome display efficiently selects and evolves high-affinity antibodies in vitro from immune libraries, Proc Natl Acad Sci USA 95: 14130–14135.
11. Hanes J and Pluckthun A. (1997) In vitro selection and evolution of functional proteins by using ribosome display, Proc Natl Acad Sci USA 94: 4937–4942.
12. Hanes J, Schaffitzel C, Knappik A, *et al.* (2000) Picomolar affinity antibodies from a fully synthetic naive library selected and evolved by ribosome display, Nat Biotechnol 18: 1287–1292.
13. He M, Menges M, Groves M A, *et al.* (1999) Selection of a human anti-progesterone antibody fragment from a transgenic mouse library by ARM ribosome display, J Immunol Methods 231: 105–117.
14. Jermutus L, Honegger A, Schwesinger F, *et al.* (2001) Tailoring in vitro evolution for protein affinity or stability, Proc Natl Acad Sci USA 98: 75–80.
15. Yau KY, Groves MA, Li S, *et al.* (2003) Selection of hapten-specific single-domain antibodies from a non-immunized llama ribosome display library, J Immunol Methods 281: 161–175.
16. Zahnd C, Spinelli S, Luginbuhl B, *et al.* (2004) Directed in vitro evolution and crystallographic analysis of a peptide-binding single chain antibody fragment (scFv) with low picomolar affinity, J Biol Chem 279: 18870–18877.
17. Vaughan TJ, Williams A J, Pritchard K, *et al.* (1996) Human antibodies with sub-nanomolar affinities isolated from a large non-immunized phage display library, Nat Biotechnol 14: 309–314.

18. Chen H Z and Zubay G. (1983) Prokaryotic coupled transcription-translation, Methods Enzymol 101: 674–690.
19. Sambrook J and Russell D W. (2001) Molecular Cloning: A laboratory Manual, 3rd ed., pp 3.17–13.32.
20. McCafferty J, Fitzgerald KJ, Earnshaw J, *et al.* (1994) Selection and rapid purification of murine antibody fragments that bind a transition-state analog by phage display, Appl Biochem Biotechnol 47: 157–171; discussion 171–173.

Chapter 11

Evolution of Protein Stability Using Ribosome Display

Andrew Buchanan

Abstract

The opportunity to enhance protein stability has a number of potential benefits for biological therapeutics – for example extending in vivo half-life, enabling a longer shelf life, reducing the propensity to aggregate, or enabling soluble expression. Engineering protein stability has been attempted empirically, rationally, and using directed evolution based on phage display. Ribosome display is a powerful in vitro technology for the selection and directed evolution of proteins. Ribosome display is typically used for the generation of high-affinity proteins and peptides. This method extends the utility of ribosome display to selecting for stability, defined as the propensity of a molecule to exist in its folded and active state.

Key words: Ribosome display, Stability selection, Directed evolution, Dithiothreitol, Hydrophobic interaction chromatography

1. Introduction

Protein stability can be defined as the propensity of a molecule to exist in its folded and active state. Protein folding is a reversible process as shown in Fig. 1. As a protein folds and unfolds via intermediates, there is the potential for off-pathway folding events that result in aggregates or protein degradation. The process of self-association and aggregation is driven in part by exposed hydrophobic patches (1, 2). Enhanced stability can deliver a number of benefits for example extending in vivo half-life, longer shelf life, reduced propensity to aggregate, and enabling soluble *Escherichia coli* expression without need for refolding from inclusion bodies. However, predicting protein stability and rationally engineering improved stability is challenging (3, 4). In the light of this, an empirical and evolutionary approach to engineer protein stability is recommended. The method described here selects for protein variants that will remain in the folded state and have reduced propensity to misfold and aggregate.

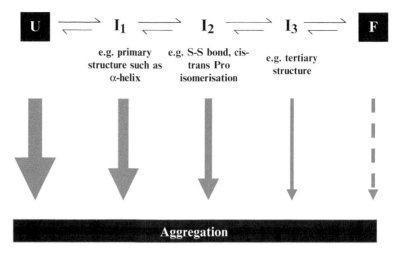

Fig. 1. Schematic illustration of reversible folding and unfolding of proteins and irreversible "off-pathway" aggregation. U unfolded, F folded, I intermediate.

Ribosome display is a powerful in vitro technology for the selection and directed evolution of proteins. The methods for ribosome display affinity selections and generation of cell-free expression systems has been recently reviewed (5, 6). Ribosome display has traditionally been used for generation of high-affinity antibodies and peptides (7, 8). This method extends the utility of ribosome display to enhancing protein stability. It can be applied to antibody single chain variable fragments (scFv) or other single chain proteins as demonstrated below. The principle of ribosome display incorporating stability selection is illustrated in Fig. 2, where the stability selection pressures are applied prior to functional selection for binding. A benefit of this method is that it selects for enhanced stability in parallel with functional activity. The stability selection pressures can be viewed in two parts. Those applied during in vitro translation and those applied post-translation. Many molecules, such as scFv antibodies or four helix bundles e.g. erythropoietin, (Epo) contain disulphide bridges that are crucial for stability and activity. Translation of these proteins in the presence of dithiothreitol (DTT) prevents disulphide bond formation. Mutant molecules can survive the selection process only if they fold correctly in the presence of DTT and retain their target binding activity. This approach has been successfully applied to scFv antibodies and to Epo (9, 10).

A second set of stability selection pressure can be introduced following translation. The ribosome display complex is incubated with hydrophobic interaction chromatography (HIC) matrixes at increased temperature. HIC is a technique for the separation of bio-molecules based on differences in their surface hydrophobicity (11). HIC is based on the hydrophobic attraction between the

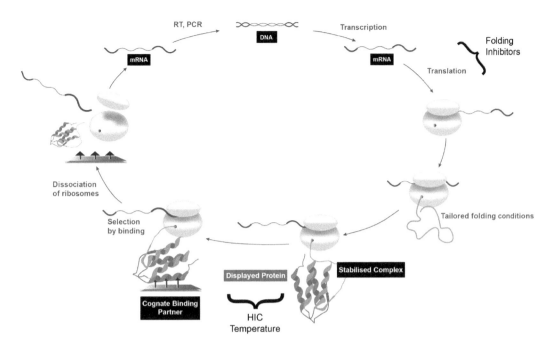

Fig. 2. Illustration of the ribosome display cycle incorporating concurrent stability selection pressures. The first is use of folding inhibitors during translation and then hydrophobic interaction chromatography (HIC) matrix and temperature once the complexes are stabilised.

HIC matrix and the protein molecules. The HIC matrix consists of small non-polar groups (butyl, octyl or phenyl) attached to a hydrophilic polymer backbone (e.g. cross-linked dextran or agarose). Many proteins considered to be hydrophilic, also have sufficient numbers of hydrophobic groups allowing interaction with the HIC matrix. HIC is sensitive enough to interact with nonpolar groups normally buried within the tertiary structure of the protein that are exposed due to incorrect folding. The strength of the interaction is dependent upon the type of matrix, concentration of salt, pH, additives, and temperature. HIC enables the removal of molecules that are misfolded following translation with DTT, are kinetically unstable or prone to aggregation and have exposed hydrophobic patches. Ribosome display is usually carried out at 4°C to ensure the stability of the ribosome–mRNA–polypeptide complex. However, inclusion of a room temperature step for 30 min does not significantly reduce the size and quality of the selection output. The use of HIC to remove unfolded and misfolded peptides has been demonstrated previously (12, 13). However, combining HIC with increased temperature further increases the utility of the method for enhancing stability.

Following selection for stability it is necessary to have appropriate screens for functionality and stability. Potential medium throughput screen for thermodynamic and kinetic stability include the use of sypro orange (14), turbidity (15), or radioimmunoassay (RIA) (9) before use of conventional methods such as dynamic light scattering. If the aim is increased soluble expression, then an expression screen would be most relevant first step.

To illustrate the utility of stability evolution, two examples are provided with this method. Human Epo is a thermodynamically unstable therapeutic protein (12). Improving the stability of Epo and reducing the likelihood of aggregation could have a number of benefits such as reduced cost of goods, improved formulation and route of administration, enhanced pharmacokinetics and pharmacodynamics, and reduced immunogenicity. Random mutagenesis libraries of Epo were stability selected using DTT, HIC, and elevated temperature to destabilise and remove misfolded variants, as well as selected for function by Epo receptor binding. Following screening for improved stability in a radioimmunoassay, improved Epo variants were rapidly identified. To demonstrate improved stability, the wild-type and variant Epo proteins were incubated in PBS at 5 and 45°C for 2 weeks and analysed for the proportions of monomer aggregates and breakdown products. The data for 45°C are presented in Fig. 3 and show that the variant Epo, with four amino acid changes, had significantly improved stability and retained biological activity.

Human granulocyte-colony stimulating factor (G-CSF) is a therapeutic protein that is produced in *E. coli* and then refolded. There are a number of reports of G-CSF variants with improved

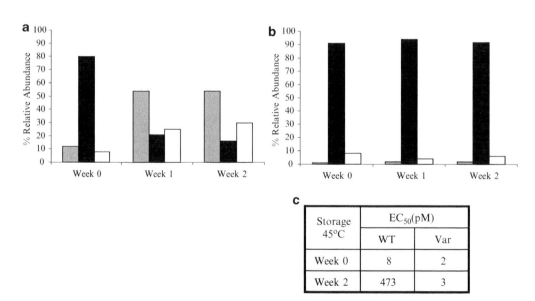

Fig. 3. The proportion of monomer (*black*), aggregates (*grey*) and breakdown products (*white*) determined by SEC-HPLC after 2 weeks of incubation at 45 °C for wild-type erythropoietin (**a**) and stability enhanced erythropoietin (**b**). Biological potency of wild-type and stability enhanced proteins as measured in a cell proliferation assay is also shown (**c**). *SEC-HPLC* size-exclusion high-performance liquid chromatography, *Epo* erythropoietin, *WT* wild type, *Var* stability enhanced variant.

stability, but no reports of successful soluble expression in *E. coli* (13, 14). Following a similar process to that of Epo, and using a high throughput screen for soluble expression, a G-CSF variant with improved soluble expression and retained activity was identified. Wild type G-CSF has a soluble yield of 8 µg/L compared to the variant G-CSF yield of 8 mg/L. The increased soluble expression enables a simplified expression and purification procedure without the need for refolding steps.

2. Materials

2.1. Preparation of the Ribosome Display Construct and Error-Prone Mutagenesis Library Construction

1. DNA encoding the protein of interest in ribosome display format (described in detail in Subheading 3.1, step 1). Store at −20°C.
2. 2× PCR Taq polymerase master mix (or a Taq polymerase with suitable reaction buffer and dNTPs). Store at −20°C.
3. SDCAT specific primer: 5′-AGACCACAACGGTTTCCCTCT AGAAATAATTTTGTTTAACTTTAAGAAGGAGA TATATCCATGG (see Note 1 and Fig. 4).
4. MycRestore primer: 5′-ATTCAGATCCTCTTCTGAGATGAG.
5. T7B primer: 5′-ATACGAAATTAATACGACTCACTATAGGG AGACCACAACGG.
6. T7te primer: 5′-GTAGCACCATTACCATTAGCAAG.
7. T6te primer: 5′-CCGCACACCAGTAAGGTGTGCGGTATC ACCAGTAGCACCATTACCATTAGCAAG.
8. MycgeneIIIshortadapt primer: 5′-ATCTCAGAAGAGGATC TGAATGGTGGCGGCTCCGGTTCCGGTGAT.
9. Nuclease-free water. Store at 4°C.

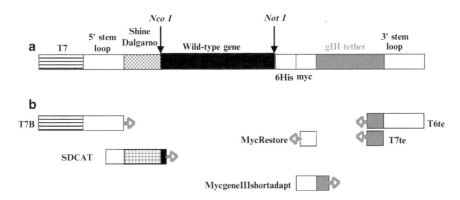

Fig. 4. Schematic illustration of the DNA construct used for *Escherichia coli* ribosome display (**a**) and the locations of primers used during the process (**b**). *T7* T7 RNA polymerase binding site, *Shine Dalgarno* prokaryotic ribosome binding site for translation initiation, *gIII* gene III tether (see Note 6), *6His* 6× histidine tag, *myc* myc tag.

10. 1× TAE gel electrophoresis buffer: 40 mM Tris-base, 1.14% (v/v) glacial acetic acid, 1 mM ethylenediaminetetraacetic acid (EDTA), pH 7.6 (harmful/irritant).

11. 1% (w/v) and 1.5% (w/v) agarose, 1× TAE gel with a DNA staining agent such as 1× SYBR Safe™.

12. DNA gel loading dye.

13. 1-kb Plus DNA ladder marker (Invitrogen) or equivalent. Store at 4°C.

14. Plasmid DNA purification kit (for example mini scale).

15. DNA gel purification kit.

16. 10 mM Tris-HCl, pH 8.0.

17. Diversify™ PCR Random Mutagenesis Kit (Clontech).

2.2. In Vitro Transcription

1. Non-stick, RNAse-free 1.5-ml microcentrifuge tubes (Sarsdedt) (see Note 2).

2. Ribomax Large Scale RNA production system (T7) (Promega).

3. Illustra ProbeQuant G-50 micro column (GE Healthcare).

4. Nuclease-free water. Store at 4°C.

2.3. In Vitro Translation, Stability Selection Using HIC and Functional Selection on Cognate Binding Partner

1. Non-stick, RNAse-free 1.5 and 2.0-ml microcentrifuge tubes (Sarsdedt) (see Note 2).

2. Heparin: 200 mg/ml. Store at 4°C.

3. 10× *E. coli* wash buffer: 0.5 M Tris-acetate pH 7.5, 1.5 M NaCl, 0.5 M magnesium acetate, 1% (v/v) Tween-20. Store at 4°C.

4. Premix X: 250 mM Tris-acetate pH 7.5, 1.75 mM of each amino acid, 10 mM ATP, 2.5 mM GTP, 5 mM cAMP, 150 mM acetylphosphate, 2.5 mg/ml *E. coli* tRNA, 0.1 mg/ml folinic acid, 7.5% PEG 8000. Store at −20°C.

5. Protein disulfide isomerase (PDI): 5 mg/ml in sterile water. Store at −20°C.

6. 2 M potassium glutamate. Store at 4°C.

7. 0.1 M magnesium acetate. Store at 4°C.

8. S30 *E. coli* extract (see Note 3). Store at −70°C.

9. Dithiothreitol (DTT): 500 mM. Store in 100 μl aliquots at −20°C, defrost once and discard any left over.

10. HIC sepharose: Butyl-S Sepharose 4 Fast Flow, Octyl Sepharose 4 Fast Flow, and Phenyl Sepharose 6 Fast Flow (GE Healthcare).

11. 4 M KCl. Prepare by dissolving 149.1 g KCl in 50 ml of milliQ water (or equivalent distilled deionised water) with gentle warming to help dissolve. Filter through a 0.45 μM sterile filter. Store at room temperature.

12. Relevant biotinylated cognate binding protein (see Note 4).
13. 10% (w/v) non-fat dried milk in water, autoclaved. Store at 4°C.
14. Dynabeads M-280 streptavidin beads (Invitrogen). Store at 4°C.
15. EB20: 50 mM Tris-acetate pH 7.5, 150 mM NaCl, 20 mM EDTA. Store at 4°C.
16. *Saccharomyces cerevisiae* RNA (Roche) in nuclease-free water at 10 mg/ml. Store at –20°C.
17. Lysis buffer from the High Pure RNA isolation kit (Roche) or equivalent.

2.4. Purification of Output mRNA and Reverse Transcription into cDNA

1. Non-stick, RNAse-free 1.5-ml microcentrifuge tubes, see Subheading 2.3 item 1.
2. High Pure RNA isolation kit (Roche) or equivalent.
3. Nuclease-free water.
4. 5× First strand buffer (Invitrogen). Store at –20°C.
5. Dithiothreitol (DTT): 100 mM. Store at –20°C.
6. dNTPmix, 25 mM each. Store at –20°C.
7. T7te primer, see Subheading 2.1 item 6.
8. MycRestore primer, see Subheading 2.1 item 4.
9. RNasin: 40 U/μl (Promega). Store at –20°C.
10. Superscript II Reverse Transcriptase: 200 U/μl (Invitrogen). Store at –20°C.

2.5. Relative Quantitation of cDNA Outputs by Real-Time PCR

1. 2× Taqman Universal PCR Master Mix (Applied Biosystems) or equivalent. Store at –20°C for infrequent use or 4°C for regular use.
2. Forward primer: 5′-CTTGATTCTGTCGCTACTGATTA, see Note 5.
3. Reverse primer: 5′-CCATTAGCAAGGCCGGAA, see Note 5.
4. Taqman probe: 5′-FAM-GTCACCAATGAAACCATCGATAGCAGCA-TAMRA, see Note 5.
5. Nuclease-free water.

2.6. PCR Amplification of cDNA for Further Selection or for Sub-Cloning

1. 2× PCR Taq polymerase master mix, see Subheading 2.1 item 2.
2. Dimethylsulfoxide (DMSO).
3. MycRestore primer, see Subheading 2.1 item 4.
4. T7B primer, see Subheading 2.1 item 5.
5. T7te primer, see Subheading 2.1 item 6.
6. T6te primer, see Subheading 2.1 item 7.

7. MycgeneIIIshortadapt primer, see Subheading 2.1 item 8.
8. Nuclease-free water. Store at 4°C.
9. 1× TAE gel electrophoresis buffer, see Subheading 2.1 item 10.
10. 1% (w/v) and 1.5% (w/v) agarose, 1× TAE gels, see Subheading 2.1 item 11.
11. DNA gel loading dye.
12. 1-kb Plus DNA ladder marker (Invitrogen) or equivalent. Store at 4°C.
13. DNA gel purification kit.

3. Methods

The first step in the process is to prepare a linear DNA construct, encoding the protein to be optimised (referred to as the parent protein), in the appropriate format for ribosome display (Subheading 3.1 step 1) and to identify and obtain a suitable cognate binding partner for selection. Then, using the parent protein construct, perform one cycle of ribosome display selection on the cognate binding partner, using a concentration that is around the KD of the interaction, without any stability selection pressure. Begin this with reamplification of the RD construct (Subheading 3.1 step 2) and proceed through transcription (Subheading 3.3), translation (Subheading 3.4 omitting DTT), selection (Subheadings 3.6 and 3.7), and RT-PCR (Subheadings 3.8–3.10 step 5). This step is necessary to confirm that the parent molecule can be displayed functionally on the ribosome, before the more challenging stability selection conditions are employed.

Next, establish the most appropriate stability selection environment by determining how susceptible the parent molecule is to the individual stability selection components. This is important to identify an appropriately stringent selection strategy and is determined empirically as each molecule is unique. Perform selections using the parent molecule and the separate stability selection components. Figure 5 shows a flow diagram illustrating the process of stability and functional selection. The aim is to identify conditions that significantly reduce selection of the parent molecule on its cognate binding partner, i.e. conditions that decrease the positive selection output. This can be assessed in two ways. The first is end-point PCR and DNA electrophoresis to visualise the reduction in the DNA recovered from the selection (Subheading 3.10 steps 1–5). This is straightforward, but there is a risk of over amplifying the products with too many PCR cycles and missing the linear range of the PCR reaction where the DNA levels can be accurately compared. A second method that avoids this problem is real-time

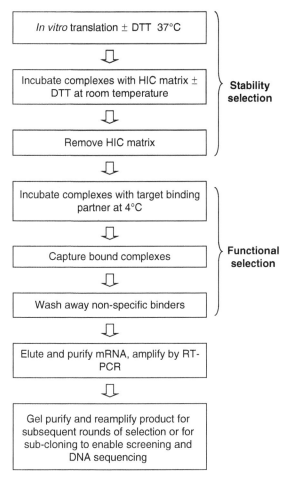

Fig. 5. Flow diagram showing the practical steps of ribosome display selection with the incorporation of stability selection pressures. *HIC* hydrophobic interaction chromatography.

PCR and the calculation of a relative quantitative value (RQV) for each sample (Subheading 3.9). A novice user should attempt a number of trial stability selections with DTT, HIC and temperature to get the timing for preparation of reagents and their use in the protocol.

Once conditions that provide an appropriate stability selection pressure are identified, the next step is to build a library of variants by introducing diversity into the parent molecule. A variety of methods for library building are suitable and essentially any PCR-based method can be used (methods involving cloning may restrict library size due to limits on the transformation efficiency of *E. coli*). Particular regions of the molecule can be targeted for randomisation if there is structural rational; however, in the absence of this, random mutagenesis libraries are recommended and their construction is described here (Subheading 3.2). Additional random mutagenesis can also be performed during the selection cascade, between rounds of selection to maximise the sequence space explored. Once constructed, the library is progressed through rounds of stability selection with increasing stringency. The aim of the stability selection cascade is to incrementally increase the stringency of the stability selection at each round. A selection cascade for a scFv that delivered variants with improved stability is described in ref. 1, 2. Round one should be performed under relatively low stringency conditions e.g. 0.5 mM DTT. For subsequent rounds a range of DTT concentrations can be used and HIC matrices introduced. The selection with the highest stringency that resulted in an enriched DNA product relative to the negative control should be progressed. During the selection cascade, selection outputs should be sub-cloned to allow DNA sequencing and analysis of the variant proteins using appropriate assays to identify improved variants.

3.1. Preparation of Parent Protein in Ribosome Display Format

1. Obtain DNA encoding the protein to be optimised in the correct format for ribosome display by gene synthesis. Figure 4 shows the elements required in the ribosome display construct, as well as the locations of primers used throughout the process. Figure 6 shows the nucleotide sequences that need to be added upstream and downstream of the protein variant coding sequence. Once one construct has been prepared, and inserted

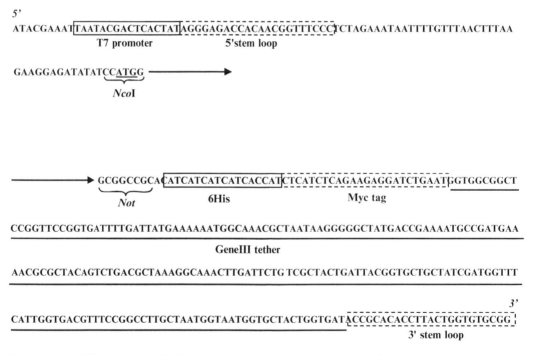

Fig. 6. 5′ and 3′ DNA sequences of the ribosome display construct. The *underlined* ATG in the 5′ region encodes the first methionine of the wild-type molecule as well as being part of the Nco1 cloning site (CC<u>ATG</u>G). The *horizontal arrows* represent the protein or library gene sequence.

into a suitable vector for storage and propagation, this can be used to easily convert other molecules to ribosome display format by sub-cloning alternative inserts between the *Nco*I and *Not*I restriction sites. *Nco*I and *Not*I are also used for sub-cloning selection outputs into other vectors for analysis; therefore, ensure that the DNA sequence contains no internal *Nco*I or *Not*I restriction sites. The DNA sequence should also contain no stop codons. The restriction sites can be modified to suit other required cloning strategies if required, for example the cloning out of selected populations into expression vectors for analysis.

2. Amplify the full-length linear construct with the T7B and T6te primers. Set up the following PCR reaction including a control reaction with no template present: 50 µl of 2× PCR master mix, 2 µl of 10 µM T7B primer, 2 µl of 10 µM T6te primer, 50–200 ng of DNA template and add nuclease-free water to final volume of 100 µl.

3. PCR amplify DNA using the following conditions: 94°C for 3 min, 25 cycles of (94°C for 30 s, 55°C for 30 s, and 72°C for 105 s), 72°C for 5 min.

4. Check a 5 µl sample of the PCR product on a 1% agarose, 1× TAE gel for size and purity. Proceed to in vitro transcription (Subheading 3.3) or store the PCR product at −20°C.

3.2. Error-Prone Mutagenesis Library Construction

Error-prone PCR (EP-PCR) is performed using the Diversify™ PCR Random Mutagenesis Kit. The conditions in the following protocol are for the highest error rate (8.1 nucleotide changes per 1,000 bp); however, the error rate can be adjusted by choosing different manganese sulphate and/or dGTP concentrations. Use the full length construct described in Subheading 3.1 step 1 as the template.

1. Set up the following PCR reaction and include a no template control: 5 µl of 10× Titanium Taq buffer, 4 µl of 8 mM MnSO$_4$, 5 µl of 2 mM dGTP, 1 µl of 50× Diversify dNTP mix, 1 µl of 10 µM MycRestore primer, 1 µl of 10 µM SDCAT specific primer, 1 µl of Titanium Taq DNA polymerase, 10 ng of purified DNA template, and nuclease-free water to final volume of 50 µl (see Note 7).

2. Amplify using the following conditions: 94°C for 3 min, 25 cycles of (94°C for 30 s, 68°C for 120 s).

3. Separate the PCR products by electrophoresis on a 1% agarose, 1× TAE gel. Use a separate gel for each library if preparing more than one. Cut the appropriate band from the gel and purify.

4. Additional EP-PCR may be carried out at this point using 10 ng of purified DNA from the first error prone reaction as

the template for a second EP-PCR reaction by repeating Subheading 3.2 steps 1–3.

5. Set up the following reaction to amplify the gene III tether from the synthesised ribosome display construct by PCR and include a no template control; 50 µl of 2× PCR master mix, 2 µl of 10 µM T7te primer, 2 µl of 10 µM MycgeneIIIshortadapt primer, and 10 ng of vector DNA. Add nuclease-free water to final volume of 100 µl. This reaction can be performed in triplicate to generate a large amount of gene III tether DNA for future use.

6. Amplify DNA using the following conditions: 94°C for 3 min, 25 cycles of (94°C for 30 s, 55°C for 30 s, and 72°C for 105 s), 72°C for 5 min.

7. Separate the PCR products by electrophoresis on a 1.5% agarose, 1× TAE gel. The amplified fragment is 221 bp in length.

8. Cut the appropriate band from the gel and purify into 30–50 µl of nuclease-free water or 10 mM Tris-HCl pH 8.0.

9. Set up the following PCR reaction to assemble the error-prone library with the gene III tether and amplify the full-length construct. Include a no template control and a negative control with gene III tether only. 50 µl 2× PCR master mix, 50–200 ng of gel-purified SDCAT/MycRestore EP-PCR product, and 50–200 ng of gel-purified gene III tether. Add nuclease-free water to final volume of 96 µl (leaving 4 µl to allow for primer additions later).

10. Amplify using the following conditions: 94°C for 3 min, five cycles of (94°C for 30 s, 50°C for 30 s, and 72°C for 105 s). At the beginning of the fifth annealing step at 50°C, pause the PCR block and add 4 µl of a mix of SDCAT primer and T7te primer (at 10 µM each). Resume the programme as follows: three cycles of (94°C for 30 s, 35°C for 30 s, and 72°C for 105 s) and 15 cycles of (94°C for 30 s, 50°C for 30 s, and 72°C for 105 s), 72°C for 5 min.

11. Separate the PCR products by electrophoresis on a 1% agarose, 1× TAE gel. Cut the appropriate band from the gel and purify (see Note 8). Elute in 30–50 µl of nuclease-free water or 10 mM Tris-HCl pH 8.0. Store the purified DNA at −20°C.

12. Check a 5 µl sample of the PCR product on a gel for size and purity and determine the concentration of the recombined product.

13. Amplify the full-length construct as described in Subheading 3.1 steps 2–4. Store the PCR product at −20°C (see Note 9).

14. If the size of the construct is correct and there are no nonspecific products visible (see Note 10), proceed to in vitro transcription (Subheading 3.3) (see Note 11).

3.3. In Vitro Transcription to Generate mRNA

1. Assemble the transcription reaction as follows in a non-stick, RNAse-free microcentrifuge tube in the order listed. 10 µl of 5× Transcription Buffer, 15 µl of 25 mM each rNTP mix (prepared by pooling equal volumes of 100 mM rUTP, rATP, rGTP, and rCTP), 20 µl of linear DNA template (PCR product), and 5 µl of T7 polymerase enzyme mix.

2. Mix well by pipetting or gentle vortexing and pulse-spin (a brief centrifugation) to ensure all liquid is at the bottom of the tube. Incubate at 37°C for 2–3 h.

3. Pulse-spin to collect condensation from the tube lid.

4. Purify the mRNA using a ProbeQuant G50 micro column as follows. Vortex the column to resuspend the matrix, break off the tip and place into a microcentrifuge tube. Loosen the cap ¼ of a turn and centrifuge for 1 min at $735 \times g$ to remove the storage buffer. Discard the flow-through, shake off any residual liquid at the tip of the column and place the column into a fresh RNAse-free microcentrifuge tube. Carefully pipette the 50 µl transcription reaction onto the centre of the surface of the resin. Replace the cap and loosen it ¼ turn, spin 2 min at $735 \times g$ to elute the purified mRNA.

5. Transfer the mRNA immediately to ice.

6. Quantify the mRNA by measuring OD A260 of a sample (for example a 1 in 150 dilution) in a spectrophotometer. The concentration should be higher than 2,000 ng/µl. If the concentration is lower, the transcription should be repeated. If the concentration is persistently low, repeat the PCR amplification of the template using more PCR cycles.

7. Use the mRNA immediately for in vitro translation (Subheading 3.4) or freeze the mRNA and store at −70°C until required.

3.4. In Vitro Translation in the Presence and Absence of DTT

1. Prepare the buffer(s) that will be required to stop the translation reactions and stabilise the ribosome complexes prior to selection. Firstly prepare 4× HB buffer in a 50 ml RNAse-free tube as follows: 5 ml of 10× *E. coli* wash buffer, 11.25 ml of sterile filtered milliQ water (or equivalent), and 625 µl of 200 mg/ml heparin. Then for each selection, prepare 600 µl of 1× HB buffer containing DTT and KCl as required and pre incubate at the appropriate temperature for the selection. For selections that do not incorporate a DTT or HIC stability step, the stop buffer is 1× HB buffer. Refer to Table 1 for an example of stop buffer preparation.

2. A 100 µl in vitro translation reaction is required for each selection condition and control. Typically, include a positive control selection of the library on the cognate binding partner but without stability selection pressure and a negative control

Table 1
A table to illustrate how to tailor the stop buffer to the stability selections

Stability selection conditions required	DTT (mM)	–	5	10
	KCl (M)	–	–	1
	Temperature (°C)	4	RT	RT
		μl per translation		
Stop buffer preparation (final volume 600 μl)	4× HB	150	150	150
	Water	450	444	288
	500 mM DTT	–	6	12
	4 M KCl	–	–	150
	Temperature (°C)	4	RT	RT

Water should be RNAse-free or high-quality laboratory-purified water. *RT* laboratory room temperature

selection of the library without the addition of the cognate binding partner or with the addition of an irrelevant protein. Prepare two separate translation master mixes, one with and one without DTT, for the required number of translations being performed. Combine the following in the order listed below, except for the *E. coli* S30 extract, in an RNAse-free microcentrifuge tube on ice.

μl per 100 μl translation	Without DTT	With 10 mM DTT
Nuclease-free water	17.4	15.4
2 M potassium glutamate	11	11
0.1 M magnesium acetate	7.6	7.6
5 mg/ml PDI	2	2
Premix X	22	22
500 mM DTT	0	2
E. coli S30 extract	40	40

Mix gently by pipetting up and down after addition of premix X. The S30 *E. coli* extract should be thawed and added to the master mix at the latest possible moment (step 4) after preparation of mRNA (step 3).

3. If mRNA has been produced on a previous day and stored at −70°C, thaw it quickly by holding between the fingertips and place immediately on ice when thawed. For each translation reaction, add 10 μl of mRNA at 1 μg/μl (approximately

2×10^{13} molecules) to the bottom of a pre-chilled 1.5-ml RNAse-free microcentrifuge tube on ice. Immediately refreeze any unused mRNA.

4. Complete the translation mastermix by adding 40 µl of S30 *E. coli* extract per translation and mix by gentle pipetting. Do not vortex. Start the translation reaction as soon as possible.

5. Add 100 µl of translation mastermix to the 10 µl of mRNA and mix gently by pipetting. Immediately transfer the tube to a 37°C heat block for exactly 7 min (see Notes 12 and 13).

6. After translation, immediately pipette the reactions into the pre-prepared RNAse-free microcentrifuge tubes containing the 500 µl of the appropriate stop buffer. For HIC selection, proceed to Subheading 3.5.

7. For samples that are not being selected for stability, centrifuge at maximum speed for 5 min at 4°C in a pre-chilled bench top microcentrifuge and replace on ice. Proceed to selection on the cognate binding partner (Subheading 3.6).

3.5. Stability Selection Using HIC

1. Prepare the HIC matrix prior to selection. Resuspend the HIC sepharose as recommended by the supplier. Transfer a volume of the HIC matrix slurry to 1.5-ml RNAse-free microcentrifuge tube and centrifuge $4,000 \times g$ for 2 min to collect the HIC matrix. Remove the liquid phase and repeat until a 1 ml bed volume is achieved. Wash the HIC matrix twice with the appropriate stop buffer.

2. For stability selections with temperature and HIC, add the 500 µl of stabilised mRNA complexes from Subheading 3.4 step 6 to the 1 ml of HIC matrix and incubate at room temperature for 25 min with gentle end-over-end rotation.

3. Centrifuge at $4,000 \times g$ for 2 min at room temperature to collect the HIC matrix.

4. Transfer the 500 µl supernatant to a fresh RNAse-free microcentrifuge tube and centrifuge at maximum speed for 5 min at 4°C in a pre-chilled bench top microcentrifuge and replace on water-ice. Proceed to selection on the cognate binding partner (Subheading 3.6).

3.6. Recovery of Stability Selected Ribosome Complexes on Cognate Binding Partner

1. Prepare de-biotinylated milk for selection (see Note 14). Add 100 µl of streptavidin beads to 1 ml of 10% (w/v) autoclaved non-fat dried milk in water in an RNAse-free, microcentrifuge tube. Incubate with end-over-end mixing for at least 10 min. Collect the beads using a magnetic particle concentrator and transfer the milk to a fresh tube. Add 50 µl of de-biotinylated autoclaved milk to each selection (see Note 15).

2. Add the biotinylated cognate binding partner at the required concentration to the positive selections, but do not exceed

30 μl. An irrelevant protein can be used for negative control selections to test for specificity.

3. Incubate the selections at 4°C for the required time (see Note 16) with gentle end-over-end rotation.

4. Prepare streptavidin beads. Use 1 μl of beads per "nM" of biotinylated cognate binding partner used, down to a minimum volume of 50 μl. Wash the beads four times in 1× HB buffer in RNAse-free microcentrifuge tubes and resuspend in 1× HB to the original volume. Store on ice until required.

5. Capture the bound complexes by addition of washed streptavidin beads to each selection. Incubate for 1–2 min at 4°C.

6. Wash the beads five times with 800 μl of 1× HB buffer at 4°C to remove non-specifically bound complexes. Resuspend the beads in 220 μl of EB20 containing *S. cerevisiae* mRNA (at 10 μg/ml final concentration) at 4°C. Incubate for 10 min at 4°C (see Note 17).

7. Transfer 200 μl of EB20 which now contains eluted mRNA into 400 μl of lysis buffer from the High Pure RNA Isolation kit in a fresh RNAse-free microcentrifuge tube. Vortex immediately to mix well.

3.7. Purification of Output mRNA

Isolate mRNA using the High Pure RNA Isolation kit and a bench top microcentrifuge pre-chilled at 4°C if possible (see Note 18).

1. Transfer samples from Subheading 3.6 step 7 to the upper chamber of the filter tubes, centrifuge for 30 s at ~10,000×g and discard the flow-through.

2. Dilute 10 μl of DNase I in 90 μl DNase incubation buffer for each sample. Pipette 100 μl of DNAse I solution onto each filter. Incubate for 10 min at 15°C.

3. Follow the manufacturer's instructions (sequential washing with 500 μl of wash buffer 1, 500 μl of wash buffer 2, and 200 μl of wash buffer 2) for washing the captured mRNA.

4. Check filter tubes for residual liquid. If present, centrifuge for an additional 15 s at ~10,000×g.

5. Discard the collection tubes and insert each filter tube into a fresh, pre-chilled RNAse-free microcentrifuge tube.

6. For elution of mRNA, add 40 μl of elution buffer (or nuclease-free water) to the centre of each filter and centrifuge at 4°C for 1 min at ~10,000×g.

7. Discard the filter columns and immediately place the eluted mRNA on ice. Proceed directly to reverse transcription (Subheading 3.8).

3.8. Reverse Transcription to Convert Selection Output mRNA into cDNA

Perform reverse transcription (RT) of mRNA using the primer T7te to generate full-length cDNA that is ready for PCR amplification and a subsequent round of selection without any additional processing. Additional RT reactions with a more upstream primer (for example MycRestore) may also be performed in case the full length outputs are poor quality and/or low yield.

1. Prepare a master mix for the required number of RT reactions to be performed (one per selection plus a negative control). Mix on ice in a fresh RNAse-free microcentrifuge tube per reaction: 4 μl of 5× First Strand Buffer, 2 μl of 100 mM DTT, 0.25 μl of 100 μM reverse primer e.g. T7te, 0.5 μl of 25 mM dNTP mix, 0.5 μl of 40 U/μl Rnasin, and 0.5 μl of 200 U/μl Superscript II Reverse Transcriptase.

2. Mix well and aliquot 7.75 μl of the RT master mix per reaction into the bottom of an RNAse-free thin-walled 0.2-ml PCR tube or a microcentrifuge tube.

3. Mix the eluted mRNA from Subheading 3.7 step 7 by pipetting up and down and add 12.25 μl of mRNA to the tube. Mix well but gently by pipetting up and down. Refreeze any unused mRNA as soon as possible and store at −70°C.

4. Incubate RT reactions at 50°C for 30 min.

5. Transfer completed RT reactions to ice or store at −20°C if PCR is to be performed at a later time.

3.9. Real-Time PCR Relative Quantitation of cDNA Outputs

1. Relative quantitation of cDNA yields can be assessed by real-time PCR by adding the following in a PCR plate: 12.5 μl of 2× Taqman Universal PCR master mix, 5 μl of cDNA from Subheading 3.8 step 5 (this may need to be diluted prior to real-time PCR), 2 μl of 10 μM forward primer, 2 μl of 10 μM reverse primer, 1 μl of 5 μM Taqman probe, and 15 μl of nuclease-free water.

2. Mix by pipetting several times, briefly centrifuge to remove air bubbles, and amplify DNA using the following conditions: 50°C for 2 min, 95°C for 10 min, 39 cycles of (95°C for 15 s, plate read, 60°C for 1 min).

3. Compare Ct values (cycle number to threshold) to determine relative cDNA levels in positive versus negative selections (see Note 19).

3.10. PCR Amplification of cDNA to Generate dsDNA for Additional Rounds of Selection or Cloning and Characterisation

1. Prepare a PCR master mix for the required number of reactions to be performed. Mix on ice in a fresh RNAse-free microcentrifuge tube per reaction: 34.5 μl of nuclease-free water, 50 μl of 2× PCR master mix, 0.25 μl of 100 μM reverse (RT) primer e.g. T7te or MycRestore, 0.25 μl of 100 μM T7B, and 5 μl of DMSO.

2. Mix well by pipetting and aliquot 90 µl of master mix into the bottom of a thin-walled 0.2-ml PCR tube. Keep on ice.

3. Add 10 µl of cDNA from Subheading 3.8 step 5 each PCR tubes and mix well. Store any unused cDNA at −20°C.

4. Amplify using the following conditions: 94°C for 3 min, 25–40 cycles (see Note 20) of (94°C for 30 s, 55°C for 30 s, and 72°C for 105 s), 72°C for 5 min.

5. Compare 5 µl of the PCR sample on a 1% (w/v) agarose, 1× TAE electrophoresis gel. There should only be one band per lane corresponding to the appropriate size for the construct and primers used. Non-specific bands of lower molecular weight may be amplified preferentially at the next PCR and eliminate the specific product. Lower molecular weight smears suggest degradation of the specific product. Repeating the RT and PCR with different combinations of primers can often result in these problems being rectified. If problems persist, however, the initial construct may need to be made again and the selections repeated.

6. To process selection outputs for further rounds of selection or for vector sub-cloning, run the remaining PCR reaction on a large 1% agarose, 1× TAE gel. Cut the appropriate band from the gel and purify. Store purified DNA at −20°C. This DNA can be used for sub-cloning to analyse the sequence and function of individual clones in this selection output as desired. To carry out a further selection on this output, and if the RT-PCR was successful with T7te, proceed to the next step. If use of an upstream primer such as MycRestore was required to produce a good quality product, rebuild the full length construct by addition of the gene III tether and amplification with T7B and T6te as described in Subheading 3.2 steps 5–14.

7. Re-amplify the output DNA by PCR as follows: 50 µl of 2× PCR master mix, 2 µl of 10 µM T7B primer, 2 µl of 10 µM reverse primer T6te, 5 µl of gel-purified product from Subheading 3.9 step 6, and 41 µl nuclease-free water.

8. Amplify DNA using the following conditions: 94°C for 3 min, 20 cycles of (94°C for 30 s, 55°C for 30 s, and 72°C for 105 s), 72°C for 5 min.

9. Check a 5 µl PCR sample on a 1% agarose, 1× TAE electrophoresis gel for size and purity using the same criteria as in step 5. Store DNA at −20°C.

10. Begin the next round of selection as described in Subheading 3.4.

4. Notes

1. If the construct is to be prepared by assembly PCR rather than gene synthesis, the SDCAT primer should be extended at the 3′ end to include the first 12–15 nucleotides of the wild-type gene. Ideally the primer should end with one of more G or C. If the ribosome display construct is synthesised directly the additional nucleotides are not required.

2. Non-stick, RNAse-free microcentrifuge tubes are preferred for steps involving pure mRNA, i.e. transcription, cell-free translation, and selection steps, and for elution of purified mRNA. In addition, the lids of these Sarstedt tubes tend to snap less in a microcentrifuge without an internal lid, when eluting from columns with the lids open.

3. S30 *E. coli* extract is a bacterial cell extract containing ribosomes and translation factors that is wisely used for cell-free translation applications. We use an S30 extract prepared in-house from *E. coli* MRE600; however, S30 extract can also be purchased. Preparation of the S30 extract is a reasonably significant undertaking so for occasional users we recommend using a commercially available cell-free translation reagent. Purified recombinant *E. coli* translation systems that are suitable for ribosome display are also available, for example PURExpress (New England Biolabs). We recommend testing the chosen translation reagent for suitability by performing a trial selection of the wild-type protein on its cognate binding partner, and in particular the optimal translation time should be determined empirically. The ribosome display method described here is only applicable to *E. coli* cell-free translation systems since eukaryotic ribosomes require different buffers for optimal performance.

4. The quality of this molecule is critical for the success of an in vitro selection and the molecule should be as pure as possible. To enable soluble selections, the cognate binding partner should be biotinylated for capture using streptavidin (or alternatively tagged for capture, for example Fc tag for protein G capture) and its activity post-modification confirmed. Alternatively, surface or panning selections are possible and are described in the literature (6).

5. This real-time PCR assay amplifies a portion of the gene III tether. For use of other tethers, alternative real-time PCR oligonucleotides should be designed.

6. The gene III tether sequence is derived from the gene III sequence of filamentous phage M13, spanning amino acids 269–336 (SwissProt P69168). The tether is essentially used to

provide an unstructured portion at the C-terminus of the protein, such that the ribosomal tunnel can cover at least 20–30 amino acid residues of the emerging polypeptide without interfering with the folding of the protein of interest. Other non-structured sequences can be employed.

7. When performing mutagenesis, take care to create mutations only in the parent protein sequence and not in the upstream or downstream regions, since mutations here may disrupt display efficiency and downstream processing. To ensure this, use the SDCAT and MycRestore primers; do not attempt to use the T7B and T6te primers.

8. There may be some non-recombined products present in the sample. A careful gel extraction should eliminate these from the desired product prior to amplification of the full-length construct.

9. Successful error-prone library construction should be verified by cloning (e.g. using NcoI and NotI) into a suitable vector and DNA sequencing of a representative number of clones (e.g. 88).

10. There should be one strong band present of the appropriate size. Non-specific bands of lower molecular weight may be amplified preferentially in subsequent rounds and eliminate the desired product.

11. This PCR product should not be purified for transcription.

12. The start of the in vitro translation reactions can be staggered in convenient intervals, e.g. every 15 s, to allow accurate timing of multiple reactions.

13. Refer to manufacturer's instructions if a commercial translation system is being used e.g. ~7 min for a commercial S30 extract, or 20–30 min for a purified system. Purified systems can employ longer translation times because they typically contain fewer nucleases so mRNA is stable for longer. Ideally the optimal translation time should be determined empirically by performing model selections of the parent protein.

14. It is possible that the biotin found naturally in milk could interfere with the selection process when using streptavidin-biotinylated antigen capture. For this reason, biotin is removed from the autoclaved milk using streptavidin beads prior to using the milk in selections.

15. In some cases this blocking step is not necessary and may even be detrimental to selection. In the event of poor results, the effect of blocking should be investigated by performing selections with and without the addition of milk.

16. The time required for functional selection is dependent upon the binding affinity between the wild-type molecule and the

cognate binding partner. For interactions in the nM range this should be achieved within 2 h. The selection can also be run overnight successfully and this can make the work load more manageable for performing complicated or large experiments.

17. Automated washing and elution (e.g. Thermo's Kingfisher ML) is highly recommended to improve the efficiency and reproducibility of the process, as well as being more user-friendly for bead washing at 4°C.

18. The use of a vacuum aspirator is recommended to remove column flow-through wastes as this helps increase speed and consistency and reduces sample cross-contamination.

19. Following the real-time PCR, determine a Ct value (cycle number to threshold) for each selection condition and calculate the relative quantitative value (RQV) as follows: $RQV = 2^{-deltaCt}$, where delta $Ct = Ct$ of a calibrator sample $-$ Ct of a test sample. The calibrator sample should be the negative control.

20. The optimal number of PCR cycles varies depending on the yield and quality of cDNA. Guidance can be obtained from the model selection of the parent clone and by performing real-time PCR before the endpoint PCR. Alternatively, trial PCRs can be performed to determine a suitable number of cycles. In general, 25 or 30 PCR cycles are a good starting point.

References

1. Baneyx F (1999) Recombinant protein expression in *Escherichia coli*. Curr Opin Biotechnol 10, 411–421
2. Chiti F, Stefani M, Taddei N et al (2003) Rationalization of the effects of mutations on peptide and protein aggregation rates. Nature 424, 805–808
3. Marshall SA, Lazar GA, Chirino AJ et al (2003) Rational design and engineering of therapeutic proteins. Drug Discovery Today 8, 212–221
4. Tokurik N and Tawfik DS (2009) Stability effects of mutations and protein evolvability. Curr Opin Struct Bio. 19, 596–604
5. Hanes J, Jermutus L, Plückthun A (2000) Selecting and evolving functional proteins in vitro by ribosome display. In: Abelson JN, and Simon MI (eds), Methods Enzymol., Academic Press, San Diego, p. 404
6. Thom G (2009) Ribosomal Display. In: Therapeutic Monoclonal Antibodies: From Bench to Clinic. An Z (ed) John Wiley & Sons Inc. Hoboken, NJ, USA
7. Zahnd C, Wyler E, Schwenk JM et al (2007) A designed ankyrin repeat protein evolved to picomolar affinity to Her2. J Mol Biol 369, 1015–1028.
8. Thom G, Cockroft AC, AG Buchanan et al (2006) Probing a protein-protein interaction by in vitro evolution. Proc Natl Acad Sci USA 103, 7619–7624
9. Jermutus L, Honeggar A, Schwesinger F et al (2001) Tailoring *in vitro* evolution for protein affinity or stability. Proc Natl Acad Sci USA 98, 75–80
10. Buchanan A (2007) Unpublished US2007/0298430 A1
11. McCue JT (2009) Theory and Use of Hydrophobic Interaction Chromatography in Protein Purification Applications. Methods in Enzymology 463, p405–414
12. Keefe AD and Szostak J (2001) Functional proteins from a random-sequence library. Nature 410, 715–718
13. Matsuura T and Plückthun A (2003) Selection based on the folding properties of proteins with ribosome display. FEBS Letters 539, 24–28
14. Ericsson UB, Hallberg BM, DeTitta GT et al (2006) Thermofluor-based high-throughput

stability optimization of proteins for structural studies. Analytical Biochemisty 357, 289–298

15. Chennamsetty N, Voynov V, Kayser V et al (2009) Design of therapeutic proteins with enhanced stability. Proc Natl Acad Sci. USA. 106, 11937–11942

16. Arakawa T, Philo JS, and Kita Y (2001) Kinetic and thermodynamic analysis of thermal unfolding of recombinant erythropoietin. Biosci Biotechnol Biochem. 65, 1321–1327

17. Bishop B, Koay DC, Sartorelli AC et al (2001) Reengineering granulocyte colony-stimulating factor for enhanced stability. J Biol Chem. 276, 33465–33470

18. Lou P, Hayes RJ, Chan C et al (2002) Development of a cytokine analog with enhanced stability using computational ultra-high throughput screening. Protein Sci. 11, 1218–1226

19. Douthwaite JA, Groves MA, Dufner P et al (2006) An improved method for an efficient and easily accessible eukaryotic ribosome display technology. Protein Eng. Design Selection. 19, 85–90

Chapter 12

Selection of Lead Antibodies from Naive Ribosome Display Antibody Libraries

Peter Ravn

Abstract

A large antibody fragment library (>10^{12}) has been generated in ribosome display format. The library was constructed in a two-step process. First, variable (V) genes were isolated from human B cells from a panel of 14 donors and cloned into designated ribosome display vectors to create a gene bank. Second, RD-VH and RD-VL genes from individual immunoglobulin families were combined in vitro resulting in 112 scFv ribosome display sub-libraries. These were subsequently pooled to form a master library.

This library was used to isolate a panel of antibodies to the IL4 receptor by three rounds of selections on a soluble target.

Key words: Naive ribosome display library, Ribosome display selection, scFv library, Antibody fragment

1. Introduction

Recombinant human antibodies are routinely isolated from phage display libraries (1). These libraries are selected by capture of scFv displaying phage with antigen and depletion of non-binding clones to enrich the phage pool for specific binders. After 2–4 rounds of selection the desired antibodies can be identified by screening individual clones (1).

Libraries of 10^{10}–10^{11} different clones readily yield hundreds of antibodies against individual targets (2) including antibodies with sub-nanomolar affinities (3, 4). It is generally acknowledged that larger libraries will yield antibodies of higher affinity (5). For phage display libraries, the upper limit may have been reached as libraries in excess of 10^{11} become increasingly difficult to produce and select (culture volume in excess of 10 L is needed for phage production from a library of 10^{11} clones and a large selection volume is needed to ensure coverage of the library diversity).

By contrast, ribosome display libraries encoding in excess of 10^{11} different proteins can be generated and selected in a small volume in a process similar to the phage display selection (for review see refs. 6, 7).

Antibody ribosome display libraries can be constructed in vitro to encode in excess of 10^{12} scFv combinations on the genetic level. In vitro transcription of ribosome display libraries is easily achieved with good yield using T7 polymerase systems. The main obstacle in ribosome display selections has been the quality of the in vitro translation of libraries into functional ribosome display complexes of RNA with the ribosome and functional protein.

Recent advances in cell-free protein expression, in particular development of reconstituted systems (8, 9), can be directly applied for ribosome display. The reconstituted systems have been assembled from purified or semi-purified components and have therefore been depleted for RNase and protease activity. Furthermore, the buffer conditions are fully controlled and may assist the folding of the translated protein. All these factors will increase the efficiency of generating complexes with RNA, ribosome and antibody fragment and therefore the functional size of ribosome display libraries. As expected this has an impact on the success of the selection (10, 11).

A plethora of phage display antibody libraries have been generated over the last 20 years, ranging from synthetic to semi-synthetic to fully natural libraries. Natural fully human libraries have been applied successfully in this laboratory (3), and hence this was our choice for a naive ribosome display library. To ensure that all variable-gene families were efficiently cloned and represented in the library, a comprehensive gene bank of immunoglobulin V-genes was constructed from which the ribosome display library was generated.

2. Materials

2.1. Cloning VH and VL for Construction of Gene Bank

1. Human Blood, Peripheral Leukocytes Total RNA (Clontech) at a concentration 1 μg/μl.
   Source: Normal human peripheral leukocytes pooled from 14 male/female Caucasians, ages 19–52. See Note 1 regarding choice of source.
2. Superscript™ III Reverse Transcriptase (200 U/μl, Invitrogen).
3. Primers were synthesised in house. Sequences are summarised in Table 1.
4. RNase Inhibitor 40 U/μl (New England Biolabs).

Table 1
Primer sequences (Note 2)

VH1	VH1a	CCCGGGCCATGGCCCAGGTKCAGCTGGTGCAG
	VH1b	CCCGGGCCATGGCCCAGGTCCAGCTTGTGCAG
	VH1c	CCCGGGCCATGGCCSAGGTCCAGCTGGTACAG
	VH1d	CCCGGGCCATGGCCCARATGCAGCTGGTGCAG
VH2	VH2a	CCCGGGCCATGGCCCAGATCACCTTGAAGGAG
	VH2b	CCCGGGCCATGGCCCAGGTCACCTTGARGGAG
VH3	VH3a	CCCGGGCCATGGCCGARGTGCAGCTGGTGGAG
	VH3b	CCCGGGCCATGGCCCAGGTGCAGCTGGTGGAG
	VH3c	CCCGGGCCATGGCCGAGGTGCAGCTGTTGGAG
VH4	VH4a	CCCGGGCCATGGCCCAGSTGCAGCTGCAGGAG
	VH4b	CCCGGGCCATGGCCCAGGTGCAGCTACAGCAG
	VH5a	CCCGGGCCATGGCCGARGTGCAGCTGGTGCAG
	VH6a	CCCGGGCCATGGCCCAGGTACAGCTGCAGCAG
	VH7a	CCCGGGCCATGGCCCAGGTSCAGCTGGTGCAA
JH pool	JH1	CCCGGGCTCGAGACGGTGACCAGGGTGC
	JH2	CCCGGGCTCGAGACAGTGACCAGGGTGC
	JH3a,b	CCCGGGCTCGAGACGGTGACCATTGTCC
	JH4a,b JH5a,b	CCCGGGCTCGAGACGGTGACCAGGGTTC
	JH4d	CCCGGGCTCGAGACGGTGACCAGGGTCC
	JH6a,b,c	CCCGGGCTCGAGACGGTGACCGTGGTCC
Vκ1	Vκ1a	GGGCCCGTGCACTCRACATCCAGATGACCCAG
	Vκ1b	GGGCCCGTGCACTCGMCATCCAGTTGACCCAG
	Vκ1c	GGGCCCGTGCACTCGCCATCCRGATGACCCAG
	Vκ1d	GGGCCCGTGCACTCGTCATCTGGATGACCCAG
Vκ2	Vκ2a	GGGCCCGTGCACTCGATATTGTGATGACCCAG
	Vκ2b	GGGCCCGTGCACTCGATRTTGTGATGACTCAG
Vκ3	Vκ3a	GGGCCCGTGCACTCGAAATTGTGTTGACRCAG
	Vκ3b	GGGCCCGTGCACTCGAAATAGTGATGACGCAG
	Vκ3c	GGGCCCGTGCACTCGAAATTGTAATGACACAG
	Vκ4a	GGGCCCGTGCACTCGACATCGTGATGACCCAG
	Vκ5a	GGGCCCGTGCACTCGAAACGACACTCACGCAG
Vκ6	Vκ6a	GGGCCCGTGCACTCGAAATTGTGCTGACTCAG
	Vκ6b	GGGCCCGTGCACTCGATGTTGTGATGACACAG
Jκ-pool	Jκ1	GGGCCCGCGGCCGCTTTGATTTCCACCTTGGTCCC
	Jκ2	GGGCCCGCGGCCGCTTTGATCTCCAGCTTGGTCCC
	Jκ3	GGGCCCGCGGCCGCTTTGATATCCACTTTGGTCCC
	Jκ4	GGGCCCGCGGCCGCTTTGATCTCCACCTTGGTCCC
	Jκ5	GGGCCCGCGGCCGCTTTAATCTCCAGTCGTGTCCC
Vλ1	Vλ1a	GGGCCCGTGCACAGTCTGTGCTGACTCAG
	Vλ1b	GGGCCCGTGCACAGTCTGTGYTGACGCAG
	Vλ1c	GGGCCCGTGCACAGTCTGTCGTGACGCAG

(continued)

	Vλ2	GGGCCCGTGCACAGTCTGCCCTGACTCAG
Vλ3	Vλ3a	GGGCCCGTGCACTCTCCTATGWGCTGACTCAG
	Vλ3b	GGGCCCGTGCACTCTCCTATGAGCTGACACAG
	Vλ3c	GGGCCCGTGCACTCTCTTCTGAGCTGACTCAG
	Vλ3d	GGGCCCGTGCACTCTCCTATGAGCTGATGCAG
	Vλ4	GGGCCCGTGCACAGCYTGTGCTGACTCAA
	Vλ5	GGGCCCGTGCACAGSCTGTGCTGACTCAG
	Vλ6	GGGCCCGTGCACTCAATTTTATGCTGACTCAG
	Vλ7	GGGCCCGTGCACAGRCTGTGGTGACTCAG
	Vλ8	GGGCCCGTGCACAGACTGTGGTGACCCAG
	Vλ4/9	GGGCCCGTGCACWGCCTGTGCTGACTCAG
	Vλ10	GGGCCCGTGCACAGGCAGGGCTGACTCAG
Jλ pool	Jλ1	GGGCCCGCGGCCGCTAGGACGGTGACCTTGGTCC
	Jλ2 Jλ3a,b	GGGCCCGCGGCCGCTAGGACGGTCAGCTTGGTCC
	Jλ4	GGGCCCGCGGCCGCTAAAATGATCAGCTGGGTTC
	Jλ5	GGGCCCGCGGCCGCTAGGACGGTCAGCTCGGTCC
	Jλ6a	GGGCCCGCGGCCGCGAGGACGGTCACTTGGTCCA
	Jλ6b	GGGCCCGCGGCCGCGAGGACGGTCACCTTGGTGC
	Jλ7	GGGCCCGCGGCCGCGAGGACGGTCAGCTGGGTGC
	LSEQ	GATTACGCCAAGCTTTGGAGC
	MYC	CTCTTCTGAGATGAGTTTTTG
	VH link	ACCGCCAGAGCCACCTCCGCC
	VL link	GGCGGAGGTGGCTCTGGCGGT
	T6te	CCGCACACCAGTAAGGTGTGCGGTATCACCAGTAGCACCATTACCATTAGCAAG
	T7B	ATACGAAATTAATACGACTCACTATAGGGAGACCACAACGG

5. 2× PCR mix (Abgene).

   Phusion™ High-Fidelity PCR Master Mix with HF Buffer (New England Biolabs).

6. Nuclease-free water (Promega).

7. Restriction enzymes:

   NcoI (10,000 U/ml, New England Biolabs).
   ApaLI (10,000 U/ml, New England Biolabs).
   NotI (10,000 U/ml, New England Biolabs).
   XhoI (20,000 U/ml, New England Biolabs).
   T4 Ligase (400,000 U/ml, New England Biolabs).

8. Qiagen Gel extraction kit/PCR purification kit.

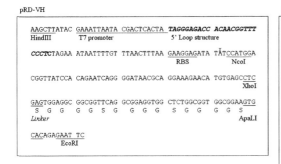
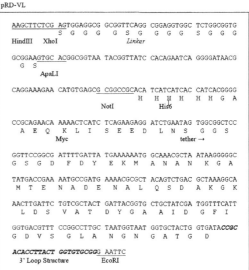

Fig. 1. Ribosome display vectors pRD-VH and pRD-VL. The sequence of the ribosome display elements and the multiple cloning sites in the two vectors.

9. pRD-VH and pRD-VL vectors were constructed in small plasmids with ampicillin resistance by inserting the sequences for ribosome display and V-gene cloning sites between the HindIII and the EcoRI restriction sites. The resulting vectors were 2,014 and 2,218 bp for pRD-VH and pRD-VL, respectively. Figure 1 shows the sequence around the multiple cloning sites.

2.2. Transformation

1. 2×TY is prepared from 16 g Tryptone, 10 g Yeast Extract, and 5 g NaCl in 1 L (see Note 3).
2. 2×TYG is 2×TY supplemented with 1% glucose and chilled on ice.
3. 2×TYAG is 2×TY supplemented with 100 μg/ml Ampicillin and 1% glucose.
4. Milli-Q water is autoclaved and stored 4°C (see Note 3).
5. Bioassay plates (Nunc).
6. QIAGEN Plasmid Plus Midi Kit (Qiagen UK).

2.3. PCR Assembly

See Subheading 2.1 for primers and PCR reagents.

2.4. Transcription

1. RiboMAX Large Scale RNA production system [T7] (Promega).
2. ProbeQuant G50 micro column (Amersham Biosciences).
3. Nuclease-free microfuge tubes.

2.5. Translation

1. In vitro translation mix (see Note 4).
2. 10× ECW: 500 mM Tris-acetate, pH 7.5, 1.5 M NaCl, 500 mM Magnesium acetate, 1% Tween 20. Prepared with

sterile buffers. Mix and filter 0.22 μm. Aliquot and store at 4°C for 1–2 months.

3. 200 mg/ml Heparin sterile-filtered.

4. HB buffer: 5 ml 10× ECW, 45 ml sterile-filtered milliQ water, 625 μl 200 mg/ml heparin. Keep on ice and use the same day of preparation.

5. EB20 buffer: 50 mM Tris-acetate pH 7.5, 150 mM NaCl, 20 mM EDTA (harmful/irritant).

2.6. Selection

1. Dynabeads M280 streptavidin beads (Dynal/Invitrogen).

2. Prepare de-biotinylated, autoclaved milk for selection. Ten percentage non-fat dried milk (e.g. Marvel) in water, autoclaved on liquids cycle. Add 100 μl streptavidin magnetic beads to 1 ml of 10% autoclaved non-fat dried milk (Marvel) in water in an RNAse-free microfuge tube. Incubate with end-over-end mixing for at least 10 min. Collect beads using a magnetic particle concentrator and transfer milk to a fresh tube. Store on ice until required.

3. Biotinylated antigen – commercially or custom made.

4. Magnetic particle collector (Invitrogen) or KingFisher mL (Thermo) in cold room.

2.7. Recovery

1. High Pure RNA Isolation Kit (Roche).

2. 10 μg/ml *Saccharomyces cerevisiae* RNA.

3. Superscript™ III Reverse Transcriptase (200 U/μl, Invitrogen). Kit contains dNTP, DTT, and 5× buffer.

4. 2× PCR mix (Abgene).

2.8. Screening

1. Z-competent™ TG1 cells from Zymo-Research.

2. IPTG, isopropyl β-D-thiogalactoside.

3. StrepMax, Streptavidin-coated plates (Thermo) or Maxisorp (Nunc).

4. Phosphate buffered saline (PBS): prepare 10× stock with 1.37 M NaCl, 27 mM KCl, 100 mM Na_2HPO_4, 18 mM KH_2PO_4 (adjust to pH 7.4 with HCl if necessary) and autoclave before storage at room temperature. PBS is prepared by tenfold dilution of stock in MilliQ-Water.

5. MPBS (Marvel in PBS i.e. 4% MPBS).

6. Penta-His HRP Conjugate Kit (Qiagen).

7. TMB, 3,3′,5, 5′ tetra methyl benzidine – liquid substrate for ELISA (Sigma).

8. 1 M sulphuric acid (harmful/corrosive).

3. Methods

The generation of a naive RD library from total RNA extracts requires a series of steps: First, the variable genes have to be reverse-transcribed from the immunoglobulin RNA and amplified by PCR. Variable genes can be cloned using oligo dT priming (12), however, reverse transcription of the variable genes from pre-prepared total RNA extracts seems to work better using primers closer to the variable gene, as for example a pool of primers annealing to the "J" gene segments (J-primers).

Second, the amplified variable genes should be cloned into designated ribosome display vectors. These vectors encode the functional elements for ribosome display and will ease the final library build. Cloning individual V-gene families separately will generate a gene bank and help ensure that all classes are incorporated in the final library.

Third, variable genes and the associated RD sequences are amplified from the gene bank and VH and VL genes combined in vitro to generate RD-scFv sub-libraries (PCR).

Finally, the library is assembled by pooling the RD-scFv sub-libraries.

Antibodies are conveniently selected from ribosome display libraries by binding to soluble biotinylated antigen, and capture with streptavidin conjugated magnetic beads. Selection outputs can be cloned into scFv expression vectors (phagemid (1) or designated expression vectors (12)) and soluble scFvs expressed to enable screening for binding.

3.1. Cloning VH and VL for Construction of Gene Bank

1. All the variable genes should be transcribed using pools of primers for the VH, VLκ, and VLλ, respectively. Thaw the Human Blood, Peripheral Leukocytes Total RNA on ice, and set up the reverse transcription of immunoglobulin variable genes (VH and VL) accordingly (see Note 5 and 6):

   2.0 μl J-primers (1 μM).

   2.0 μl RNA (1 mg/ml).

   0.5 μl dNTP (25 nM).

   8.5 μl Nuclease-free water.

   Mix and incubate at 65°C for 5 min to destroy any tertiary structures in the RNA. Subsequently cool on ice for a least 1 min. Add the following:

   1 μl DTT (0.1 M).

   1 μl RNasin, 4 μl 5× Buffer.

   1 μl SuperScript III (200 U/μl).

   Mix and incubate at 55°C for 35 min. Subsequently inactivate by incubation at 70°C for 15 min. Apply cDNA directly in

PCR amplification and avoid any short or long term storage at any temperature.

Each reverse transcription is enough for nine individual PCR amplifications.

2. Amplify the V-genes from the single-stranded cDNA by PCR. To ensure that all the variable gene classes are efficiently amplified with minimal bias from the other classes, all the variable genes should be amplified separately using one 5′ primer and the corresponding pool of J primers (i.e. VH1 and JH-pool). PCR reactions can be set up as follows:

50 μl 2× PCR mix.

38 μl Nuclease-free water.

5 μl J-primer mix (10 μM).

5 μl 5′ V primer (10 μM).

2 μl cDNA.

Use a standard PCR program: (1) 96°C 30 s, (2) 96°C 30 s, (3) 50°C 30 s, (4) 72°C 60 s, (5) go to (2), repeat 29 times, (6) 72°C 180 s, and (7) keep at 4°C.

Run 5 μl of the PCR reactions is on a 1% agarose/TAE electrophoresis gel to validate the quality of the PCR product. A distinct band around 350 bp should be visible.

3. Purify the individual PCR products using a standard PCR purification kit. Use two columns per purification and elute the DNA in a total of 150 μl. The purified PCR products can be stored at −20°C for days. The PCR amplifications should be performed multiple times using different vials/lots of RNA (recommend 5–6 times).

4. Thaw the individual PCR purifications. Pool them and set up restriction digest to enable cut and paste cloning of the variable genes into the RD vectors.

Digest the VH PCR products with NcoI and XhoI and the VL PCR products with ApaLI and NotI in double digests [375 μl PCR product, 44 μl 10× (NEBuffer2), 4.4 μl BSA, 5 μl NcoI or ApaLI, and 5 μl XhoI or NotI, respectively (see Note 7)]. Incubate overnight at 37°C.

5. Digest the RD plasmids using the same conditions as above. The final plasmid concentration should be ~50 μg/ml (see Note 8).

6. Separate overnight digests by running on a 1% agarose/TAE electrophoresis gel and excise the variable gene DNA (~350 bp). Purify the DNA using a gel extraction kit using two columns per variable gene purification and one column for each 10 μg plasmid in the plasmid purification. Determine the DNA concentrations (see Note 9).

7. Ligate the purified VH and VL genes into prepared pRD-VH and pRD-VL, respectively. Set up ligations using a 1:3 ratio for vector:insert and a final DNA concentration of 10 μg/ml. Ligation reactions of 100–500 μl (1–5 μg total DNA) are recommended (100 μl example: 650 ng vector, 350 ng insert, 10 μl T4 DNA ligase 10× Buffer, 6 μl ligase, and nuclease-free water to a final volume of 100 μl).
   Incubate ligations at 16°C overnight.

8. Purify ligations using the PCR purification kit (one column per ligation) and elute with 35 μl nuclease-free water.

3.2. Transformation of Variable Gene Libraries for Gene Bank

1. Inoculate 50 ml of 2×TY media with a fresh TG1 colony in a flow-hood. Incubate overnight at 30°C and 300 rpm (see Note 10).

2. In the morning, inoculate six flasks containing 400 ml each of 2×TY with 7.5 ml (per flask) of the overnight culture. Incubate at 30°C and 300 rpm until OD_{600nm} ~0.5–0.6 (this usually takes around 1.5 h).

3. Once optimum OD is reached, chill cells for 45 min on ice in the centrifuge bottles. Subsequently spin at $3,400 \times g$ for 9 min at 2°C.

4. Pour off supernatant. Re-suspend pellet in 25 ml of ice-cold autoclaved Milli-Q water and collect in two bottles. Make up to about 300 ml per pot with Milli-Q water and spin at $3,400 \times g$ for 15 min at 2°C.

5. Pour off supernatant and re-suspend in 25 ml of ice-cold Milli-Q water, then add about 300 ml Milli-Q water per pot and spin as above. Repeat twice.

6. Gently pour off the supernatant. The pellet is very loose at this stage and it is unlikely that all the liquid can be poured off without disturbing the pellet. Re-suspend the pellet in the remaining liquid. This procedure should give about 5 ml of cells. Proceed immediately to the next step as the cells lose their competency very quickly.

7. Electroporate in 2 mm cuvettes (400 μl aliquots) using standard electroporation settings (2.5 kV field strength, 200 Ω resistance, 25 μF capacitance or following manufactures recommendation) (see Note 11).
   Mix 1.6 ml electrocompetent TG1 cells with the cleaned ligation reaction from a sub-library from Subheading 3.1 step 8 and electroporate 4×400 μl (see Note 12).

8. Immediately after electroporating, add 0.8 ml ice-cold 2×TYG to each cuvette and transfer the cells to a 50 ml Falcon tube (one per sub-library!). Rinse each cuvette with a further 0.8 ml 2×TYG and transfer to the Falcon tube (see Note 13).

9. Incubate all samples at 4°C for 45 min to recover.
10. Plate 100 μl samples of a tenfold dilution series of each library on small 2×TYAG plates to determine the number of transformants and plate the rest of the library on two large 2×TYAG bioassay plates in a flow hood (see Note 14).
11. Incubate plates overnight at 30°C.
12. Harvest the libraries by scraping of the bacteria in 10 ml 2×TY supplemented with 25% glycerol.
13. Start cultures for plasmid preparation by inoculating 100 ml 2×TYAG with harvested library glycerol stock. Add stock till the OD_{600nm} is ~0.1–0.2 to ensure that diversity of the library is covered. Incubate at 37°C for 4–5 h.
14. Harvest the bacteria by centrifugation and prepare plasmid preps.

3.3. Assembly of RD scFv Library

1. Amplify RD-VH fragments (5′-RD-VH-linker) from each of the 14 VH libraries and RD-VL fragments (linker-VL-3′-RD) from each of the 13 VLκ and 14 VLλ libraries using the gene bank plasmid preps as templates. To ensure the diversity of the individual variable gene libraries is covered 10 ng plasmid should be used as template (10 ng ~3.5×10^9 copies of plasmid). Use the following PCR reactions:

   50 μl 2× PCR mix.

   36 μl Nuclease-free water.

   10 ng Template.

   2 μl (10 μM) Forward primer.

   2 μl (10 μM) Reverse primer.

   Use T7B and VH-link for VH amplification or T6te and VL-link for VL amplification, respectively. The following PCR program could be used: (1) 96°C 180 s, (2) 96°C 30 s, (3) 50°C 30 s, (4) 72°C 60 s, (5) go to (2) repeat 19 times, (6) 72°C 180 s, and (7) keep at 4°C.

2. Run the PCR products on a 1% agarose/TAE electrophoresis gel and excise the RD-VH and RD-VL fragments; RD-VH (~500 bp) and RD-VL (700 bp). Purify the DNA fragments using a gel-extraction kit and quantify the DNA in the individual preparations (see Note 9).

3. Adjust the concentration of the extracted DNA to 50 μg/ml for all the RD-VH preparations and 70 μg/ml for the RD-VL preparations (see Note 15).

4. For gene classes where multiple primers are used for the construction, pool the RD-VH or RD-VL fragments prior to the assembly PCR (i.e. RD-VH1 contained RD-VH fragments from genebank libraries 1a–1d). Set up the following PCR reaction:

50 μl 2× PCR mix.

35 μl Nuclease-free water.

5 μl RD-VH.

5 μl RD-VL.

Run PCR program: (1) 96°C 30 s, (2) 96°C 30 s, (3) 40°C 30 s, (4) 72°C 60 s, (5) go to (2) repeat five times and (6) pause. Then add:

2.5 μl T7B (10 μM).

2.5 μl T6te (10 μM).

Continue the PCR program: (7) 96°C 30 s, (8) 50°C 30 s, (9) 72°C 60 s, (10) go to (2) repeat 29 times, (11) 72°C 180 s, and (12) keep at 4°C.

Run the PCR products on a 1% agarose/TAE electrophoresis gel. The amplified fragment should be ~1,200 bp in length (see example of library assembly Fig. 2). Excise the RD-scFv constructs and purify these using a gel-extraction kit (see Note 16).

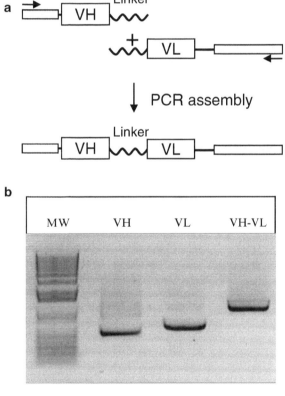

Fig. 2. scFv ribosome display library assembly. (a) Schematic representation of the assembly PCR. RD-VH and RD-VL fragments are assembled by overlapping PCR at the linker region. (b) 1% Agarose/TAE electrophoresis gel of the RD-VH and RD-VL fragments and the assembled RD-scFv (VH-VL).

5. Amplify the assembled mini-libraries using the following PCR reaction:

   50 μl 2× PCR mix.

   1.3 μg Template.

   2 μl T7B (10 μM).

   2 μl T6te (10 μM).

   41 μl Nuclease-free water (adjust volume according to volume of template).

   Run the following PCR program: (1) 96°C 180 s, (2) 96°C 30 s, (3) 50°C 30 s, (4) 72°C 60 s, (5) go to (2) repeat nine times, (6) 72°C 180 s, and (7) keep at 4°C.

   Check 5 μl PCR sample on a 1% agarose/TAE electrophoresis gel for size and purity:
   One strong band should be observed with no smaller bands.

6. Assemble the master library by pooling the sub-libraries. This can be done to favour the sub-libraries of the highest quality or to accommodate all classes at the distribution found in vivo. An example of library assembly is shown in Table 2.

Table 2
Library assembly. The contribution of the individual PBLκ and PBLλ scFv sub-libraries was weighted by the number of germline genes in each variable gene family (i.e. VH1 – VLκ1; 11/51 × 19/40 × 1,000 = 102.5). The PBLκ and PBLλ libraries were subsequently pooled 1:1

PBLκ	VH1	VH2	VH3	VH4	VH5	VH6	VH7
VLκ1	102.5	27.9	204.9	102.5	18.6	9.3	9.3
VLκ2	48.5	13.2	97.1	48.5	8.8	4.4	4.4
VLκ3	37.3	10.3	75.5	37.7	6.9	3.4	3.4
VLκ4	5.4	1.5	10.8	5.4	1.0	0.5	0.5
VLκ5	16.2	4.4	32.4	16.2	2.9	1.5	1.5
VLλ6	5.4	1.5	10.8	5.4	1.0	0.5	0.5
PBLλ	VH1	VH2	VH3	VH4	VH5	VH6	VH7
VLλ1	34.8	9.5	69.6	34.8	6.3	3.2	3.2
VLλ2	34.8	9.5	69.6	34.8	6.3	3.2	3.2
VLλ3	62.6	17.1	125.2	62.6	11.4	5.7	5.7
VLλ4	20.9	5.7	41.7	20.9	3.8	1.9	1.9
VLλ5	20.9	5.7	41.7	20.9	3.8	1.9	1.9
VLλ6	7.0	1.9	13.9	7.0	1.3	0.6	0.6
VLλ7	13.9	3.8	27.8	13.9	2.5	1.3	1.3
VLλ8	7.0	1.9	13.9	7.0	1.3	0.6	0.6
VLλ9	7.0	1.9	13.9	7.0	1.3	0.6	0.6
VLλ10	7.0	1.9	13.9	7.0	1.3	0.6	0.6

3.4. Transcription

1. Assemble the transcription reaction in a RNAse-free microcentrifuge tube as follows:

   10 μl 5× Transcription Buffer.

   15 μl rNTP mix [ATP, CTP, GTP, UTP (25 mM each)].

   20 μl Linear DNA template (>50 ng/μl).

   5 μl T7 enzyme mix (see Note 17).

2. Mix well by pipetting and ensure all contents are at the bottom of the tube. Incubate at 37°C for at least 2 h and no more than 2.5 h.

3. Pulse-spin to collect condensation from the tube lid.

4. Purify mRNA using a G50 micro column (according to manufacturer's instructions): Vortex column to resuspend the matrix. Loosen cap, add to a 2-ml RNase free tube, spin for 1 min, 1,000 rpm to remove storage buffer. Discard flow-through. Add 50 μl transcription reaction. Loosen cap, spin for 2 min at $1,000 \times g$ into fresh 1.5-ml RNAse-free microcentrifuge tube.

5. Place mRNA on ice and quantify by measuring OD A_{260} in a spectrophotometer of 1/150 dilution in duplicate. If the OD A_{260} reading does not lie within the range 0.1–1.0 then repeat the analysis with a more suitable dilution (see Notes 18 and 9).

3.5. Translation

1. For each library, one translation reaction is required (~30 μg mRNA + translation mix) (see Note 19).

2. Allow translation mixes to thaw on ice or on a cool block on ice. Mixes can be encouraged to thaw by holding with fingertips, but *do not allow to warm up*.

3. For each transcription, prepare 2 ml RNAse-free microfuge tube containing chilled HB buffer and store on ice. The HB buffer and the translation mix should bring the final volume to 1,700 μl.

4. If mRNA has been produced previously, thaw it quickly by holding between fingertips and place immediately on ice when thawed.

5. Prepare one translation master mix for all translation reactions according to the manufacture's instruction.

6. Mix the RNA from Subheading 3.4 step 5 by pipetting five times, and add the RNA to the translation mix. Immediately transfer the tube to a 37°C heating block for 30 min.

7. After exactly 30 min, pipette the translation reaction into the pre-chilled 2 ml RNAse-free microfuge tube containing chilled HB buffer to stabilise scFv–ribosome–mRNA complexes. Leave tubes on ice while preparing the antigen for selection (Subheading 3.6 see Note 20).

3.6. RD Selection

1. Chill three RNAse-free selection tubes on ice. One for negative control and two for antigen selection.

2. Mix stopped translation reactions by pipetting gently (5×). Transfer 550 µl stabilised scFv–ribosome–mRNA complexes to each of the three pre-chilled RNAse-free selection tubes (+ + −) per library. Cap the negative selection tube to prevent addition of antigen. Add biotinylated antigen to the positive antigen selections only (do not exceed 30 µl for antigen addition). For information on antigen concentration see Note 21 and for blocking (see Note 22).

3. Incubate selections at 4°C for 2 h or overnight with gentle end-over-end rotation (see Note 23).

4. Prepare streptavidin Magnetic beads. Use 50 µl of beads per selection (see Note 24). Wash beads four times in HB buffer in RNAse-free microfuge tubes and re-suspend in HB to the original volume. Store on ice until required.

5. Capture selected complexes by addition of 50 µl HB-washed streptavidin-coated magnetic beads for each selection (Subheading 3.2 step 4). Incubate for 5 min at 4°C.

6. Wash selections to remove non-specifically bound complexes using a magnetic particle collector or KingFisher mL in cold room. Wash the selections five times with 800 µl ice cold HB buffer (see Note 25).

7. Elute the RNA by transferring the beads to 220 µl EB20 containing *S. cerevisae* mRNA (10 ng/ml final concentration). Mix well and incubate at 4°C for 10 min (see Note 26).

8. Collect the magnetic beads, and transfer 200 µl of the eluted RNA to 400 µl lysis buffer from the High Pure RNA Isolation kit in a fresh RNAse-free microfuge tube. Vortex immediately to mix well. The RNA is stable in this buffer at 4°C for up to 30 min.

3.7. RD Recovery

1. Isolate mRNA using the High Pure RNA Isolation Kit and a bench top microcentrifuge (see Note 27).

2. Transfer samples to the upper chamber of the filter tubes, centrifuge for 30 s at $10,000 \times g$. Discard flow through.

3. Pipette 100 µl DNAse I solution onto each filter and incubate for 10 min at room temperature (see Note 2).

4. Add 500 µl wash buffer I to the upper chambers, centrifuge for 30 s at $10,000 \times g$ and discard the flow-through.

5. Add 500 µl wash buffer II to the upper chambers, centrifuge for 30 s at $10,000 \times g$ and discard the flow-through.

6. Add 200 µl wash buffer II to the upper chambers, centrifuge for 1 min at $13,000 \times g$ and discard the flow-through. Centrifuge for 2 min at $13,000 \times g$ to remove residual wash buffer.

7. Discard the collection tubes and insert filter tubes into fresh, pre-chilled, pre-labelled RNAse-free microfuge tubes.

8. For elution of RNA, add 40 μl elution buffer to the centre of each filter and centrifuge for 1 min at 13,000×*g*.

9. Discard the filter columns, and transfer eluted mRNA immediately onto ice. Proceed straight to the reverse transcription-PCR.

10. Assemble the mastermix for reverse transcription in a RNAse-free microfuge tube on ice as follows:

    4 μl 5× First strand buffer.

    2 μl DTT (100 mM).

    0.25 μl T6te (100 μM).

    0.5 μl dNTP mix (25 mM each).

    0.5 μl RNasin (40 U/μl).

    0.5 μl Superscript II Reverse Transcriptase (200 U/μl).

11. On ice, aliquot 7.75 μl reverse transcription mastermix into the bottom of thin-walled 0.2 ml PCR tubes.

12. Mix eluted mRNA from **step 9** by pipetting up and down 3–5 times and add 12.25 μl to the tube. Mix well but gently by pipetting up and down 3–5 times. Unused mRNA should be snap-frozen on dry ice immediately and stored at −70°C.

13. Incubate reactions in a PCR block at 50°C for 30 min (see Note 29).

14. Transfer completed RT reactions onto ice.

15. PCR amplify the RT products using the following PCR reaction:

    50 μl 2× PCR mix.

    36 μl Nuclease-free water.

    10 μl cDNA.

    2 μl T7B (10 μM).

    2 μl T6te (10 μM).

    The following PCR program can be used: (1) 96°C 30 s, (2) 96°C 30 s, (3) 68°C 30 s, (4) 72°C 60 s, (5) go to (2) repeat 30 times, (6) 72°C 180 s, and (7) keep at 4°C (see Note 30).

16. Compare 5 μl of the PCR samples on a 1% agarose electrophoresis gel.

17. Run the remaining End-Point PCR on large 1% agarose/TAE electrophoresis gel. Excise the ~1,200 bp bands and purify the DNA using a Gel Extraction Kit. Elute the DNA in 40 μl elution buffer. Store purified DNA at −20°C.

18. Re-amplify the output DNA by PCR using the following reaction:

    50 μl 2× PCR mix.

    31 μl Nuclease-free water.

    10 μl Gel extracted PCR product.

    2 μl T7B (10 μM) 2 μl T6te (10 μM).

    The following PCR program could be used: (1) 96°C 30 s, (2) 96°C 30 s, (3) 68°C 30 s, (4) 72°C 60 s, (5) go to (2) repeat 24 times, and (6) 72°C 180 s, (7) keep at 4°C.

19. Check 5 μl PCR sample on a 1% agarose/TAE electrophoresis gel for size and purity (see Note 31).

3.8. Screening

1. Purify the re-amplified selection output using a PCR purification kit and elute the DNA in 50 μl elution buffer.

2. Set up restriction enzyme digests with NcoI and NotI using the re-amplified selection out-put and the scFv-expression plasmid (pCantab6 or pHen2) accordingly:

    1 μl NcoI.

    1 μl NotI.

    0.5 μl BSA (100×).

    2.5 μl NEBuffer 2.

    10 μl Nuclease free water.

    10 μl Purified PCR product or vector.

    Incubate at 37°C for 1 h (see Note 32).

3. Run the digests on a 1% agarose/TAE electrophoresis gel. Identify the scFv genes (750–800 bp) and excise them from the gel. Purify the inserts with a gel-extraction kit and elute in 50 μl elution buffer. Purify the vector by gel-extraction.

4. Measure concentration of DNA (see Note 9).

5. Ligate scFv genes and vectors at a vector:insert ratio of 1:3 and total DNA concentrations of 10 μg/ml in 15 μl final volume reactions (i.e. 100 ng vector, 45 ng insert 1.5 μl T4 DNA ligase 10× buffer, 12.5 μl nuclease free water (adjust to 15 μl final volume), 1 μl T4 DNA ligase). Incubate at RT for 1 h.

6. Transform into Z-competent TG1 cells. Plate 4 dilution steps (tenfold) on 2×TY AG and incubate at 37°C overnight (see Note 33).

7. Pick individual colonies into 100 μl 2×TY 100 μg/ml ampicillin and 1% glucose in 96 cell-well plates and incubate at 250 rpm and 37°C overnight.

8. Use a multi-channel pipette to transfer 10 μl of the overnight cultures to a second 96 cell-well plate containing 140 μl 2×TY

AG (see Note 34). Incubate cultures at 250 rpm and 37°C until the OD_{600nm} is ~0.6 (about 1.5–2 h). (A stock can be made of the first plate, by adding glycerol to a final concentration of 15% and storing the plate at –70°C.)

9. Once OD_{600nm} ~0.6 is reached, add 50 µl 2×TY containing 100 µg/ml ampicillin and 4 mM IPTG (final concentration 1 mM IPTG) (see Note 35). Incubate at 250 rpm and 30°C overnight.

10. Coat a 96 well streptavidin conjugated plate with 50 µl per well of biotinylated antigen (0.1–0.5 µg/ml) in PBS. Incubate for 1 h or overnight at 4°C (see Note 36).

11. Flick out coating solution and add 300 µl per well of 3% Marvel PBS to block and incubate at room temperature for 1 h.

12. Spin the culture plate from step 9 at 4,000×g for 20 min.

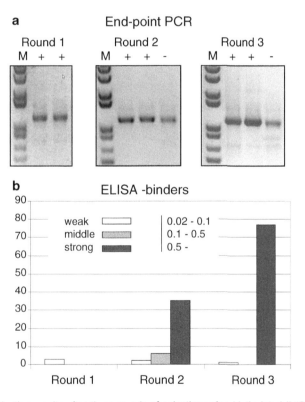

Fig. 3. Selection results after three rounds of selection using biotinylated IL4R. (**a**) 1% agarose/TAE electrophoresis gel of end-point PCR products after rounds 1, 2, and 3. In the first round, all the translated library was selected on antigen (hence no negative selection) and from second round a negative selection control was included. A selection window seems to appear after the third round. (**b**) Selection out-put (end-point PCR) were sub-cloned to pCantab6 and 90 clones from each round was screened for binders in scFv ELISA (coating the plate with 50 ml biotinylated IL4R at 1 µg/ml). The histogram shows the number of antibodies binding to IL4R.

13. Wash the ELISA plate three times with PBS and add 50 μl of 4% MPBS to each well. Then add 50 μl of the supernatant from the well in the culture plate to the wells in the ELISA plate (take care not to transfer any bacteria). Incubate at room temperature for 1 h.
14. Wash three times with PBS.
15. Add 50 μl Penta-His HRP Conjugate (1 in 1,000 dilution in 2% MPBS). Incubate at room temperature for 1 h.
16. Wash three times with PBS.
17. Add 50 μl TMB. Leave at room temperature for 2–15 min. A blue colour should develop.
18. Stop the reaction by adding 50 μl sulphuric acid (1 M). The blue colour should turn yellow. Read the OD at 650 nm and at 450 nm. Subtract OD_{650nm} from OD_{450nm} (see examples of screening results in Fig. 3).

4. Notes

1. Immunoglobulin variable genes (VH and VL) can be amplified from mRNA from various sources. Fresh samples are often of better quality. However, for generation of naive libraries it may be preferred to apply samples from a range of donors. These samples are commercially available from a variety of distributors and may conveniently be used for library generation.
2. Oligos were designed from sequences in V-Base (http://vbase.mrc-cpe.cam.ac.uk) with the appropriate restriction sites for cloning into ribosome display vectors pRD-VH and pRD-VL.
3. All buffers and media are prepared with Milli-Q water. Milli-Q water is water that has a resistivity of 18.2 MΩ-cm and total organic content of less than five parts per billion.
4. A range of in vitro translation mixes are available commercially. They range from more simple S30 *Escherichia coli* extracts and S30 extracts from engineered strains to complex reconstituted in vitro translation mixes where the components are purified individually (i.e. PureSystem, Cosmo Bio and PureExpress®, NEB). The reconstituted systems are recommended as they offer increased control over RNAse activity, buffer conditions, and protease levels, which all affect the performance of the ribosome display selection (10, 11).

    An in-house reconstituted system was used for this work, but the protocols are compatible with the other reconstituted systems. We have validated the protocols with the PureSystem (11) and the PureExpress®.

5. We generally obtained results when total RNA was used in the reverse transcription compared to isolated mRNA. If fresh blood samples are used, it may well be better to use a mRNA extraction kit as an alternative to the extraction of total RNA.

6. Reverse transcription can be performed with a range of reverse transcription polymerases. Superscript III from Invitrogen has worked reliably in our hands.

7. NEBuffer2 is the best choice for the NcoI and XhoI double digest with 100% activity in both. Consecutive digest using NEBuffer3 for NotI and any of NEBuffer1,2,4 for ApaLI is recommended. However, a double digest is possible in NEBuffer2 using high concentrations of NotI and extended digest time.

8. Double digest of plasmids are most efficient at relative low concentrations of plasmid. At 50 µg/ml final concentrations the plasmids are efficiently cut even in double digests, and can easily be purified using a PCR purification kit. Omitting the gel-extraction of the vector is possible if the fragment cut out of the vector is small (<40–50 bp) and will substantially increase the yield.

9. Measure OD_{260nm}, and calculate based on $OD_{260nm} = 1$ for DNA at a concentration of 50 µg/ml and for RNA at a concentration of 37 µg/ml.

10. A variety of protocols for preparing electro-competent cells are available online. In general, three factors are of importance: high quality water; pre-chilling water, buffers and equipment to 2–4°C and Time. The essence of the protocols is to wash the cells *fast* and *efficiently* to deplete conducting materials.

    Alternatively, commercially available sources of electro-competent cells can be used i.e. New England Biolabs, Invitrogen and Lucigen among many others.

11. The electro-competent cells can be pre-tested by electroporating 400 µl cells without DNA. If it arcs (pops) it is an indication that the conductivity of the sample is too high. This can be caused by high salt concentration (insufficient washing) or too high cell concentration. The cells can be diluted by adding additional a 10–25% of their volume of ice-cold Milli-Q water and try again. Always ensure that the cuvette is dry on the outside prior to the transformation.

12. When the cells and the DNA are mixed, the electroporation should be done immediately. There is no advantage of incubating cells and DNA; by contrast, the cells may lose potency. Some electro-competent cells have recommended optimal DNA concentration for transformation. TG1, ER2537 and SS320 will perform better at high DNA concentrations 2–10 µg/ml (final concentration) whereas DH5α had decreased efficiency at higher concentrations.

13. It is essential to recover the cells immediately after transformation. Even short delays can reduce efficiency several fold.

14. The library transformations will generally be in a volume of 8 ml if the transformation is performed as described. Serial tenfold dilutions of the library (50 μl in 450 μl 2×TY) will give the best determination of size. Plate 100 μl of each dilution. Multiplication factors would be 8×10^2 for the first dilution, 8×10^3 for the second dilution, etc. Five dilutions will usually cover the library size.

    Single colonies from the dilution series can be used to validate the quality of the library by either PCR screening or sequencing.

15. The molar concentrations of RD VH at 50 μg/ml and RD VL at 70 μg/ml are roughly the same, giving total numbers of $\sim 4.5 \times 10^{11}$ molecules in the 5 μl. This will easily cover the diversity of the individual variable gene libraries. The diversity of the assembled scFv sub-libraries in excess of 4.5×10^{11} encoded antibodies (assuming the diversities sub-libraries to be combined are large enough that the product of the two is in excess of 4.5×10^{11}).

16. 1.3 μg Gel-extracted DNA is roughly 10^{12} molecules, which should cover the diversity of the assembled libraries (see Note 15).

17. A transcription reaction with 20 μl PCR product at 50 μg/ml has ~1 μg linear library template. 1 μg DNA at 1,200 bp $\sim 7.6 \times 10^{11}$, which will be the theoretical maximal size of the library for ribosome display. If larger libraries are required, the reactions can be scaled up to accommodate the need for more DNA.

18. A transcription reaction containing 1 μg of PCR product should yield >3 μg/μl (total 150 μg) mRNA. If recoveries of mRNA are less than this a comparative transcription should be made against 1 or 2 single clone control PCR products, which should yield mRNA in the region of 3–5 μg/μl. This will determine whether the low yield is down to reagents or PCR product. If the problem is the PCR product then the quality and quantity of this should be re-assessed and if necessary a re-amplification by PCR needs to be carried out.

19. 30 μg RNA at 1,200 bp represents $\sim 4.5 \times 10^{13}$ molecules, which easily covers the diversity at the DNA level (see Note 17). The number of RNA molecules should be higher than the number of ribosomes in the translation mix to preferentially generate monovalent display (one ribosome–protein complex per RNA molecule).

20. The stalled ribosome complexes are relatively stable and can be left on ice for hours. However, it is recommended to proceed directly without any unnecessary pause.

21. In a first round selection it is recommendable to use as much antigen as possible. Practically, this will most often be 2.5–0.2 μM (i.e. 30 μl of an antigen of 1 mg/ml and molecular weight of 20 kDa would result in a final concentration of 2.5 μM). The antigen concentration can be dropped between rounds to increase the selection pressure, however this should not be attempted until a selection window (difference in selection out-put between the positive and negative selections) has been observed. For most selections such a window will appear after the second or third round of selections. If the window does not appear after the third or fourth round, something may have gone wrong. Investigate earlier rounds.

    A "selection" positive control can be included by selecting the library with Protein A conjugated beads, which will pull out scFvs with VH3 genes.

22. Many selections perform better in the absence of milk (especially overnight selections), but this may not be true for all. In the event of a poor selection, consider adding/omitting milk. For blocking, add 50 μl de-biotinylated sterile milk to the selection (see Subheading 2.6 item 2 for details on de-biotinylation of milk).

23. 1–2 h Incubation with antigen is usually sufficient. For convenience, overnight incubation may be preferable.

24. 50 μl Streptavidin magnetic beads (Dynabeads® M-280 Streptavidin) will capture up to 5 μg antibody (~33 pM). The molar capacity will go up for smaller antigens. If selections are performed at high antigen concentrations, more magnetic beads will be needed to retrieve the binders from the selections.

25. Kingfisher is convenient and reliable for the washing. If single tubes are used for washing, it is convenient to set up and pre-chill the required amount of tubes with 800 μl HB buffer.

26. The *S. cerevisae* mRNA is added as "carrier" to increase the efficiency of the RNA purification, and may decoy potential RNases present.

27. A bench top microcentrifuge with the capacity to chill to 4°C may be favourable. Initially, hold the centrifuge at 15°C for the DNase I incubation step and continue at 4°C. Furthermore, a vacuum aspirator for sucking out flow-through is helpful when dealing with many samples. However, neither the cooling centrifuge nor the aspirator is vital for the selection.

28. Prepare the reverse transcription (RT) Mastermix during this time (see Subheading 3.3 step 4).

29. Prepare the PCR Mastermix during this time, *see* Subheading 3.7 step 15.

30. There is a range of new polymerases for PCR amplification that offer increased efficiency for amplification of templates at

very low concentration. As an example, the Phusion (Hot start) polymerase from New England Biolabs gave better yields in some selections with very low recovery in the first selection round.

31. If the size is correct and there are no lower molecular weight products visible, the T7B-T6te-amplified cDNA is ready for another round of selection starting with transcription (Subheading 3.4) or subcloning (Subheading 3.8) for characterisation of individual antibodies from outputs.

32. Any scFv expression vector will be acceptable. Expression using pCantab6 and pHen2 will express the scFvs as his6 and myc tagged proteins, which will enable purification and detection. Furthermore these vectors will enable phage display ELISA if desired.

    Other expression vectors may give better yields and can be applied (12).

    Concentration of the vector in the digest should be <50 μg/ml for optimal cleavage.

33. There are several options in choosing the competent cells. Competent cells can easily be prepared in-house, but can conveniently be purchased. TG1 is regularly used with phage display for the phage particle production, but can be induced to express soluble scFv using the established protocols (similar to the one described in Subheading 3.8). There are several suppliers of competent TG1, i.e. electro-competent TG1 from Lucigen or Z-competent cells from Zymo-research. Alternatively, scFvs can be cloned into a dedicated expression plasmid and expressed in optimised expression cell-lines at substantially higher yields (12).

34. Alternatively, use a 96 well transfer device to transfer a small inoculum (<2 μl) from this plate to a second 96 cell-well plate containing 150 μl 2×TY containing 100 μg/ml ampicillin and 0.1% glucose per well. This works equally well, but cultures may require 4–5 h incubation at 37°C to reach $OD_{600nm} \sim 0.6$ compared to 1.5–2 h.

35. The OD will vary from well to well. $OD_{600nm} \sim 0.6$ is optimal for induction, and should be a general guideline. When the cultures in the wells start looking cloudy they are generally ready for induction.

36. The streptavidin conjugated plates are conveniently used because the coating is fast, and the antigen presentation will resemble the selection. Alternatively, non-biotinylated antigen can be coated directly on plates (i.e. Nunc Maxisorp plates) at a concentration of 1–10 μg/ml in either PBS or a 50 mM $NaCO_3$ pH 9.6 and incubation overnight at 4°C.

Acknowledgements

Julie Douthwaite and Susan Kunze heroically established the reconstituted translation system used for this work, furthermore Susan Kunze is gratefully acknowledged for her work on the generation of the ribosome display library and selection.

References

1. McCafferty, J., Griffiths, A. D., Winter, G., and Chiswell, D. J. (1990) Phage antibodies: filamentous phage displaying antibody variable domains, *Nature 348*, 552–554.
2. Edwards, B. M., Barash, S. C., Main, S. H., Choi, G. H., Minter, R., Ullrich, S., Williams, E., Du Fou, L., Wilton, J., Albert, V. R., Ruben, S. M., and Vaughan, T. J. (2003) The remarkable flexibility of the human antibody repertoire; isolation of over one thousand different antibodies to a single protein, BLyS, *Journal of molecular biology 334*, 103–118.
3. Lloyd, C., Lowe, D., Edwards, B., Welsh, F., Dilks, T., Hardman, C., and Vaughan, T. (2009) Modelling the human immune response: performance of a 1011 human antibody repertoire against a broad panel of therapeutically relevant antigens, *Protein Eng Des Sel 22*, 159–168.
4. Vaughan, T. J., Williams, A. J., Pritchard, K., Osbourn, J. K., Pope, A. R., Earnshaw, J. C., McCafferty, J., Hodits, R. A., Wilton, J., and Johnson, K. S. (1996) Human antibodies with sub-nanomolar affinities isolated from a large non-immunized phage display library, *Nature biotechnology 14*, 309–314.
5. Perelson, A. S., and Oster, G. F. (1979) Theoretical studies of clonal selection: minimal antibody repertoire size and reliability of self-non-self discrimination, *Journal of theoretical biology 81*, 645–670.
6. Groves, M. A., and Osbourn, J. K. (2005) Applications of ribosome display to antibody drug discovery, *Expert opinion on biological therapy 5*, 125–135.
7. Schaffitzel, C., Hanes, J., Jermutus, L., and Pluckthun, A. (1999) Ribosome display: an in vitro method for selection and evolution of antibodies from libraries, *Journal of immunological methods 231*, 119–135.
8. Shimizu, Y., Inoue, A., Tomari, Y., Suzuki, T., Yokogawa, T., Nishikawa, K., and Ueda, T. (2001) Cell-free translation reconstituted with purified components, *Nature biotechnology 19*, 751–755.
9. Ying, B. W., Taguchi, H., Ueda, H., and Ueda, T. (2004) Chaperone-assisted folding of a single-chain antibody in a reconstituted translation system, *Biochemical and biophysical research communications 320*, 1359–1364.
10. Ohashi, H., Shimizu, Y., Ying, B. W., and Ueda, T. (2007) Efficient protein selection based on ribosome display system with purified components, *Biochemical and biophysical research communications 352*, 270–276.
11. Villemagne, D., Jackson, R., and Douthwaite, J. A. (2006) Highly efficient ribosome display selection by use of purified components for in vitro translation, *Journal of immunological methods 313*, 140–148.
12. Ravn, P., Danielczyk, A., Jensen, K. B., Kristensen, P., Christensen, P. A., Larsen, M., Karsten, U., and Goletz, S. (2004) Multivalent scFv display of phagemid repertoires for the selection of carbohydrate-specific antibodies and its application to the Thomsen-Friedenreich antigen, *Journal of molecular biology 343*, 985–996.

Chapter 13

Evolution of Disulfide-Rich Peptide Aptamers Using cDNA Display

Yuki Mochizuki and Naoto Nemoto

Abstract

Protein scaffolds containing some disulfide bonds (e.g., Knottin, Kunitz domain, etc.) are promising candidates for molecular recognition. cDNA display has been developed to screen functional disulfide-rich peptide aptamers from a vast library by promoting disulfide bond shuffling after the synthesis of peptides in a cell-free translation system. Here we present a detailed protocol for the selection of disulfide-rich peptide aptamers against interleukin 6 receptor (IL-6R) from a 35-amino acid peptide library containing 32 amino acids in the random region, which is linked to its genotype by cDNA display.

Key words: cDNA display, mRNA display, cDNA–protein fusion, In vitro selection, Peptide aptamer, Disulfide-rich peptide, Protein scaffold, Protein engineering, Directed evolution

1. Introduction

In vitro display technologies have become more and more important for drug discovery. Especially, in vitro virus (1) and mRNA display (2) are promising and powerful techniques to explore functional peptides and proteins from a huge library (>10^{12}). The versatile puromycin linker, which we developed for cDNA display (3), can be useful and convenient for preparing the puromycin-attached mRNA template of in vitro virus and mRNA display and for stabilization of mRNA–protein fusion as described in the previous chapter (see Chapter 8). In addition, cDNA display has the advantage of rapidly purifying the mRNA/cDNA–protein fusion molecules, which facilitates buffer exchange and the succeeding modification reaction by a simple manipulation of mRNA/cDNA–protein fusions (cDNA display molecules) immobilized on

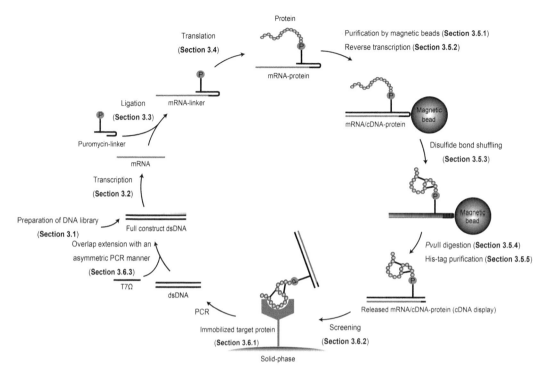

Fig. 1. Schematic diagram of the screening cycle of disulfide-rich peptide aptamers using cDNA display. The initial DNA library is prepared first (Subheading 3.1). The DNA library is transcribed into mRNA (Subheading 3.2). An mRNA is ligated with a puromycin linker (Subheading 3.3). This ligated product is translated in vitro and an mRNA–protein fusion molecule is formed (Subheading 3.4). This mRNA–protein fusion is purified on magnetic beads (Subheading 3.5.1) and is reverse transcribed to form an mRNA/cDNA–protein fusion (Subheading 3.5.2). The disulfide bonds of mRNA/cDNA–protein fusion are shuffled by the oxidative conditions (Subheading 3.5.3). The disulfide bond-shuffled mRNA/cDNA–protein fusion is released from the beads by *pvu*II digestion (Subheading 3.5.4) and purified by His-tag purification (Subheading 3.5.5). In the screening process (Subheading 3.6.2), the purified mRNA/cDNA–protein fusion binds to the target molecule. The screened molecule is amplified by PCR, and full construct double stranded (ds) DNA is prepared by overlap extension for the next screening cycle or analysis.

streptavidin-magnetic beads through the puromycin linker (3). This allows the disulfide-rich libraries to be easily synthesized on the beads (see Fig. 1).

The following section describes a basic protocol of in vitro selection against soluble extracellular domain of interleukin 6 receptor (sIL-6R) with a 35-residue peptide library containing a 32-residue random region. A novice user may find this protocol difficult. We therefore recommend that a beginner of cDNA display perform the simple model experiment of in vitro selection with cDNA display several times (see Chapter 8) and become familiar with each process of the selection cycle in cDNA display before attempting disulfide-rich peptide aptamer selections. The puromycin linker for cDNA display is indispensable for this in vitro selection. We also recommend that a user prepare 1 nmol of puromycin linker molecules according to the procedure in Chapter 8 before performing the in vitro selection against a new target molecule.

2. Materials

2.1. Preparation of Full DNA Constructs of Library

1. Oligodeoxynucleotides.

    T7Ω: 5′-GATCCCGCGAAATTAATACGACTCACTATA GGGGAAGTATTTTACAACAATTACCAACAACAACAA CAAACAACAACAACATTACATTTTACATTCTACAA CTACAAGCCACCATG-3′ (117 mer) (see Note 1).

    RND35aa: 5′-CAACTACAAGCCACCatgGGATGTxyzxyz xyzxyzxyzxyzxyzxyzxyzxyzxyzxyzxyzxyzxyzxyzxyzx yzxyzxyzxyzxyzxyzxyzxyzxyzxyzxyzxyzxyzxyzxyzG GGGGAGGCAGCCATC-3′ (145 mer) where is x; T 13%, C 20%, A 35%, G 32%, y; T 24%, C 22%, A 30%, G 24%, z; T 37%, C 37%, A 0%, G 26% (see Note 2) (4).

    Newleft: 5′-GATCCCGCGAAATTAATACGACTCACTATA GGG-3′ (33 mer).

    LHR: 5′-TTTCCCCGCCGCCCCCCGTCCT-3′ (22 mer).

    LHR-His: 5′-TTTCCCCGCCGCCCCCCGTCCTGCTTCC GCCGTGATGATGATGATGAT GGCTGCCTCCCCC-3′ (61 mer).

2. Klenow fragment (2140, Takara and various suppliers). Store at −20°C.
3. 10× Klenow fragment reaction buffer: 100 mM Tris–HCl (pH 7.5), 70 mM, $MgCl_2$, 1 mM DTT.
4. Taq DNA polymerase (various suppliers) or other thermostable polymerase. Store at −20°C.
5. 10× PCR buffer (various suppliers, supplied with DNA polymerase in general).
6. 2.5 mM each dNTP Mix (various suppliers).
7. Nuclease-free water.
8. Thermal cycler.
9. SYBR® Gold nucleic acid gel stain (S11494, Invitrogen): other DNA staining reagents can be used.
10. Fluorescence scanner.
11. SPIN-X (8160, COSTA).

2.2. In Vitro Transcription

1. RiboMAX™ Large-Scale RNA Production System-T7 (P1300, Promega). Store at −20°C.

    5× T7 transcription buffer.

    25 mM each rNTP Mix (see Note 3).

    Enzyme Mix (T7).

    RQ1 RNase-Free DNase.

    Nuclease-free water.

2. Full construct library DNA (see Note 4).
3. Heat block or Thermal cycler.
4. RNeasy Mini kit (74104, Qiagen).

2.3. Ligation of mRNA to Puromycin Linker

1. T4 RNA Ligase (Takara or other supplier).
2. 10× T4 RNA Ligase buffer: 500 mM Tris–HCl (pH 7.5), 100 mM $MgCl_2$, 100 mM DTT, 10 mM ATP.
3. Polynucleotide kinase (M0201S, New England Biolabs). Store at −20°C.
4. Puromycin linker. Prepare according to the protocol in Chapter 8. Store at −20°C (see Note 5)
5. mRNA template (library). Store at −80°C (see Note 6).
6. Nuclease-free water.
7. Thermal cycler.
8. RNeasy Mini kit (74104, Qiagen).
9. SYBR® Gold nucleic acid gel stain (S11494, Invitrogen); other DNA-staining reagents can be used.
10. Fluorescence scanner (e.g., Pharos FX from Bio-Rad, Typhoon from GE Healthcare).

2.4. In Vitro Translation (Synthesis of mRNA–Protein Fusion)

1. Rectic Lysate IVT™ kit (1200, Ambion). Store at −80°C.
   Reticulocyte lysate.
   20× Translation mix (−Met).
   20× Translation mix (−Leu).
   Nuclease-free water.
2. mRNA-Puromycin linker conjugate (ligated product). Store at −20°C.
3. 3 M KCl prepared with nuclease-free water. Store at 4°C.
4. 1 M $MgCl_2$ prepared with nuclease-free water. Store at 4°C.
5. Heat block.

2.5. Synthesis of mRNA/cDNA–Protein Fusion

2.5.1. Purification of mRNA–Protein Fusion from the Lysate with Streptavidin (SA) Magnetic Beads

1. SA magnetic beads: MAGNOTEX-SA (9088, Takara) (see Note 7).
2. Solution A: diethylpyrocarbonate (DEPC)-treated 100 mM NaOH, DEPC-treated 50 mM NaCl.
3. Solution B: DEPC-treated 100 mM NaCl.
4. 2× Binding buffer: 20 mM Tris–HCl (pH 8.0), 2 mM EDTA, 2 M NaCl, 0.2% Triton X-100. This buffer is supplied with MAGNOTEX-SA.

                        5. Ribosome releasing buffer: 50 mM Tris–HCl (pH 8.0), 20 mM EDTA prepared with nuclease-free water.
                        6. Magnetic separator (such as DynaMag™ series from Invitrogen).

2.5.2. Reverse Transcription

1. SuperScript III™ Reverse Transcriptase (18080–093, Invitrogen). Store at −20°C.
   SuperScript III Reverse Transcriptase (RT) (200 U/μl).
   5× First-Strand buffer: 250 mM Tris–HCl (pH 8.3), 375 mM KCl, 15 mM $MgCl_2$.
   100 mM DTT.
2. 2.5 mM each dNTP Mix (various suppliers).
3. Nuclease-free water.
4. Magnetic separator.
5. Temperature-controllable tube rotator (such as SNP-24B, NISSIN) (see Note 8).

2.5.3. Disulfide Bond Shuffling

1. Protein disulfide-isomerase (PDI) (7318, Takara). Store at −20°C (see Note 9).
2. 5× Oxidative folding buffer: 1 M phosphate buffer (pH 7.5), 50 mM reduced glutathione, 5 mM oxidized glutathione. Store at −20°C.

2.5.4. Restriction Enzyme Digestion (Release of mRNA/cDNA–Protein Fusion from SA Magnetic Beads)

1. *Pvu*II restriction enzyme (1076A, Takara).
2. 10× *Pvu*II buffer (−DTT): 100 mM Tris–HCl (pH 7.5), 100 mM $MgCl_2$, 500 mM NaCl (see Note 10).
3. 0.1% BSA: supplied with *Pvu*II restriction enzyme from Takara.
4. Nuclease-free water.
5. Magnetic separator.
6. Temperature-controllable tube rotator.

2.5.5. His-tag Purification of mRNA/cDNA–Protein Fusion

1. Ni-NTA magnetic agarose beads (36111, Qiagen).
2. Ni-NTA binding buffer: 50 mM NaH_2PO_4 (pH 8.0), 300 mM NaCl, 10 mM imidazole, 0.05% Tween 20.
3. Ni-NTA wash buffer: 50 mM NaH_2PO_4 (pH 8.0), 300 mM NaCl, 20 mM imidazole, 0.05% Tween 20.
4. Ni-NTA elution buffer: 50 mM NaH_2PO_4 (pH 8.0), 300 mM NaCl, 250 mM imidazole, 0.05% Tween 20.
5. Magnetic separator.
6. Temperature-controllable tube rotator.

2.6. Selection of Peptide Aptamers Against IL-6R from RND35aa Library

2.6.1. Immobilization of IL-6R on NHS-Activated Sepharose Beads for Selection

1. Recombinant Human sIL-6 Receptor (200-06R, PEPRO TECH, INC.). Store at −20°C.
2. NHS-activated Sepharose 4 Fast Flow (17-0906-01, GE Healthcare).
3. 1 mM HCl. Store at 4°C.
4. 200 mM phosphate buffer (pH 7.0). Store at 4°C.
5. 20 mM phosphate buffer (pH 7.5–8.0) or HEPES buffer (pH 7.5). Store at 4°C.
6. 10× Phosphate buffered saline (PBS) buffer: 1.37 M NaCl, 27 mM KCl, 100 mM Na_2HPO_4, 18 mM KH_2PO_4, pH 7.4; adjust pH with 6N HCl. Store at 4°C.
7. 1 M Tris–HCl, pH 7.5.
8. pH indicator paper.
9. NAP5 column (17-0853-02, GE Healthcare).

2.6.2. Selection of Aptamers Against IL-6R

1. IL-6R-immobilized beads (prepared at Subheading 3.6.1).
2. Purified cDNA display library.
3. 2× Binding buffer: same as written in Subheading 2.5.
4. 10× Selection buffer: 500 mM Tris–HCl (pH 7.6), 10 mM EDTA, 5 M NaCl, 1.0% Tween 20. Prepare with nuclease-free water and store at 4°C.
5. 1 M DTT.
6. Magnetic separator.
7. Temperature-controllable tube rotator (see Note 10).
8. Quick-Precip™ Plus Solution (70437, Edge Biosystems) for ethanol precipitation.
9. Taq DNA polymerase (various suppliers) or other thermostable polymerase. Store at −20°C.
10. 10× PCR buffer (various suppliers, supplied with DNA polymerase in general).
11. 2.5 mM each dNTP Mix (various suppliers).
12. Thermal cycler.
13. QIAquick PCR Purification kit (28106, QIAGEN).

3. Methods

A variety of libraries can be used for in vitro selection by cDNA display. Here, we present a 35-residue peptide library (RND35aa) containing 32 random residues and three constant residues [Met (start codon), Gly and Cys], which generated peptide aptamers against IL-6R. The sequence space is huge (i.e., the number of combinatorial sequences is around 4×10^{41}). On the other hand, the practical screening size per selection cycle is generally up to 10^{14} molecules, which is a very small part of this. In this study, we planned to select disulfide-rich peptide aptamers. Thus, we introduced a Cys residue at the N-terminal of the library sequence, which would be expected to bridge with Cys residues in the random region. Because we can search for only a part of the sequence space of the full library, many other peptide aptamers with different sequences may be obtained against the same target molecule when a new selection is performed.

3.1. Preparation of Full DNA Constructs of Library

The full DNA construct of the library consists of T7Ω-kozak-atg-GC-$(X)_{32}$-GGGS-His×6-GGS-LHR where T7Ω is the promoter and 5′ untranslated region (5′ UTR), "atg" is the translation initiation codon, GC (G = glycine; C = cysteine) is the constant region of the library, $(X)_{32}$ is the 32 random residues region, GGGS and GGS (G = glycine; S = serine) are the spacer regions, His×6 is the His-tag, and LHR is the sequence required for hybridization of mRNA with the puromycin linker.

1. Prepare the Klenow fragment reaction solution as follows: 5 μl of 10× Klenow fragment reaction buffer, 4 μl of 2.5 mM dNTP mix, 100 pmol of "LHR-His" fragment (61 mer), equal mol of "RND32aa" (145 mer) fragment and nuclease-free water up to 48 μl. To anneal fragments, heat to 60°C and then reduce to 20°C for 20 min. Add 2 μl of Klenow fragment enzyme (5 U/μl) and perform the extension reaction at 37°C for 1 h to generate the double-stranded (ds) DNA (190 mer) (see Fig. 2).

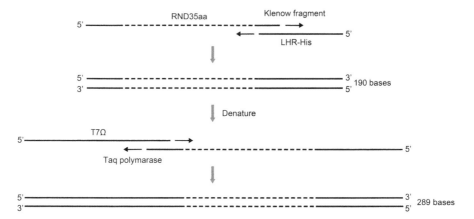

Fig. 2. Schematic diagram of the preparation of the full construct library DNA (Subheading 3.1). The LHR-His is annealed to the RND35aa, which contains a random region (represented by a *broken line*) and is extended with the Klenow fragment. After denaturing the extended dsDNA, the T7Ω hybridizes with the single stranded DNA and is extended by Taq polymerase in order to join a promoter region.

2. Prepare a 1.5 or 2 mm-thick 7.5% polyacrylamide gel containing 8 M urea. And the dsDNA is analyzed by polyacrylamide gel electrophoresis (PAGE) on the above gel.

3. Excise the expected band (190 mer) containing the synthesized dsDNA from the gel and purify the DNA from the excised gel with an appropriate gel purification kit.

4. To synthesize full construct library, perform overlap extension by asymmetric PCR manner on purified template from step 3 with the "T7Ω" fragment. Prepare the PCR reaction solution as follows: mix 50 μl of 10× Taq buffer, 40 μl of 2.5 mM dNTP mix, 2.5 μl of Taq polymerase (5 U/μl), add 50 pmol of above purified template and ten times pmol of "T7Ω" fragment, and nuclease-free water up to 500 μl and divide into five aliquots. The thermal cyclic reaction program is as follows: 95°C for 2 min, repeat 18 cycles of three steps, 95°C for 25 s, 69°C for 20 s, 72°C for 30 s, followed by 72°C for 2 min using a thermal cycler.

5. Purify the product with the QIAquick PCR Purification kit and perform ethanol precipitation with Quick-Precip™ Plus Solution.

6. Prepare a 1.5 or 2-mm-thick 7.5% polyacrylamide gel containing 8 M urea. And the purified product is analyzed by PAGE on the above gel.

7. Excise the expected band (289 mer) containing the synthesized full construct library DNA from the gel and crush in a microcentifuge tube.

8. Add 400 μl of nuclease-free water and vortex for 10 min. Transfer the slurry gel to a Spin-X filter, centrifuge at $10,000 \times g$ for 15 min at 25°C. Perform ethanol precipitation with Quick-Precip™ Plus Solution. Add 10 μl of nuclease-free water.

3.2. In Vitro Transcription

1. Prepare the transcription reaction: mix 20 μl of 5× T7 Transcription buffer, 30 μl of 25 mM each rNTP Mix, the full construct library DNA (from Subheading 3.1) to the final concentration of 100 nM, and add nuclease-free water up to 90 μl. Heat the reaction solution at 65°C for 5–10 min, then cool on ice. After it has cooled, add 10 μl of Enzyme Mix (T7) and incubate at 37°C for 2–4 h (see Note 7).

2. Add 1 μl of RQ1 RNase-free DNase/μg of template DNA into the transcription reaction solution, incubate at 37°C for 15 min.

3. Purify the transcript with the RNeasy Mini kit according to the attached protocol. The concentration should be around 20 pmol/μl and the ratio of 260/280 nm should be above 1.8. The mRNA can be stored at –20°C.

3.3. Ligation of mRNA to Puromycin linker

1. Synthesize the puromycin linker (see Chapter 8).
2. Determine the volume of the translation required to carry three pool equivalents through the translation and purification steps to just before the affinity selection step. Assume an overall yield of 1% through all of the steps from translation up until the affinity selection step.
3. Calculate the volume of the ligation reaction based on the number of pmol required for translation, assuming a conservative yield of 80% of the input mRNA.
4. For preparing around 50 pmol of the ligated product, add 3 μl of 10× T4 RNA ligase buffer to the mixture of 100 pmol of puromycin linker and 60 pmol of mRNA and add up to 28 μl of nuclease-free water. Incubate at 94°C for 1 min and then reduce to 25°C for over 30 min using a thermal cycler.
5. Add 1 μl of T4 polynucleotide kinase (10 U/μl) and 1 μl of T4 RNA ligase (30 U/μl) then mix gently (do not use a vortex). Incubate at 25°C for 60–120 min.
6. Purify the ligated product with the RNeasy Mini Kit. The ligated product should be analyzed on 8 M urea denaturing 8% PAGE imaging the fluorescent of FITC derived on the puromycin linker using a fluorescence scanner and subsequently stained with the SYBR® Gold and visualized by the fluorescence scanner.

3.4. In Vitro Translation (Synthesis of mRNA–Protein Fusion)

When you perform the following procedure for the first time, you should confirm mRNA–protein fusion on 8 M urea denaturing 6% SDS–PAGE after translation with 0.5–5 pmol of the ligated product in a 25-μl scale.

1. Prepare a translation reaction equal to the volume calculated in step 3 of Subheading 3.3; each cold 25 μl reaction contains 0.5–5 pmol of the ligated product from Subheading 3.3, 0.625 μl of 20× Translation Mix (-Met), 0.625 μl of 20× Translation Mix (-Leu), 17 μl of reticulocyte lysate and add up to 25 μl of nuclease-free water. Incubate at 30°C for 20 min using a heat block (see Note 8).
2. Add 3 μl of 1 M $MgCl_2$ (final concentration; 79 mM) and 10 μl of 3 M KCl (final concentration; 789 mM). Incubate at 37°C for 90 min using the heat block.

3.5. Synthesis of mRNA/cDNA–Protein Fusion

3.5.1. Purification of mRNA–Protein Fusion from the Lysate with Streptavidin Magnetic Beads

1. Take 20 μl of Streptavidin (SA) magnetic beads for 10 pmol of the ligated product and wash the beads twice with 20 μl of solution A and once with 20 μl of solution B (see Note 9).
2. Mix the translation reaction with an equal volume of 2× binding buffer, suspend the bead with the mixture in the microcentrifuge tube and incubate using a rotator at room temperature for 20 min (see Note 10).

3. Place the microcentrifuge tube on a magnetic separator for 1 min and discard the supernatant, wash the beads three times with 200 μl of 1× binding buffer (see Note 11).

4. Add 20 μl of 1× ribosome releasing buffer (same volume as the initial volume of the magnetic beads taken from the vial) and incubate using the rotator at room temperature for 10–15 min (see Notes 10 and 12).

5. Place the microcentrifuge tube on the magnetic separator for 1 min and discard the supernatant, wash the beads twice with 1× First-Strand buffer and discard the supernatant for following "Reverse Transcription".

3.5.2. Reverse Transcription

1. Suspend the above beads (Subheading 3.5.1) with 15 μl of nuclease-free water, 6 μl of 5× First-Strand buffer, 6 μl of 2.5 mM each dNTP mix, 1 μl of SuperScript III Reverse Transcriptase (200 U/μl) and 1.5 μl of 100 mM DTT. Incubate at 45°C for 30 min using the rotator (see Note 10).

2. Place the microcentrifuge tube on the magnetic separator for 1 min and discard the supernatant, wash the beads twice with 1× oxidative folding buffer and discard the supernatant for following "Disulfide Bond Shuffling".

3.5.3. Disulfide Bond Shuffling

1. Suspend the above beads (Subheading 3.5.2) with 1× oxidative folding buffer (equal volume with initial volume of magnetic beads taken from the vial) and add PDI at an eqimolar ratio with the input cDNA–protein fusion. Incubate using the rotator at room temperature for 1 h (see Notes 10 and 13).

2. Place the microcentrifuge tube on the magnetic separator for 1 min and discard the supernatant, wash the beads with 1× *Pvu*II buffer (−DTT) and discard the supernatant for following "Restriction Enzyme Digestion" (see Note 14).

3.5.4. Restriction Enzyme Digestion (Release of mRNA/cDNA–Protein Fusion from SA Magnetic Beads)

1. Suspend the above beads (Subheading 3.5.3) with 15 μl of nuclease-free water, 2 μl of 1× *Pvu*II buffer (−DTT), 2 μl of 0.1% BSA and 1 μl of *Pvu*II (10 U/μl). Incubate using the rotator at 37°C for 1 h (see Note 10).

2. Place the microcentrifuge tube on the magnetic separator for 1 min and collect the supernatant.

3.5.5. His-tag Purification of mRNA/cDNA–Protein Fusion

1. Put 20 μl of the Ni-NTA magnetic agarose beads into a microcentrifuge tube. Place the microcentrifuge tube on the magnetic separator for 1 min and discard the supernatant, wash the beads three or four times with Ni-NTA binding buffer. Suspend the beads with the supernatant (from Subheading 3.5.4) and incubate using the rotator at room temperature for 1 h.

2. Place the microcentrifuge tube on the magnetic separator for 1 min and discard the supernatant, wash the beads twice with Ni-NTA wash buffer.

3. Elute the cDNA display molecules from the beads with 10, 5, and 5 μl of Ni-NTA elution buffer. Collect all eluates (total volume is 20 μl) (see Note 15).

3.6. Selection of Peptide Aptamers Against IL-6R from RND35aa Library

3.6.1. Immobilization of IL-6R on NHS-Activated Sepharose Beads for Selection

You should keep the materials cold during the following experiment.

1. Prepare 80 μg of IL-6R and 400 μl of NHS-activated Sepharose 4 Fast Flow (see Note 16). Wash the beads twice with 500 μl of 1 mM HCl (see Note 17). Suspend the beads with 100 μl of 200 mM phosphate buffer (pH 7.0). Measure the pH of the supernatant with a pH indicator paper, immediately. It should be around 7–7.5 (see Note 18).

2. Immediately add 40 μl 1× PBS buffer of IL-6R (2 mg/ml). Tap/mix thoroughly. Incubate using the rotator at room temperature for 10 min. Periodically, tap/mix thoroughly (see Note 19). Add 200 μl of 20 mM phosphate buffer (pH 7.8) or HEPES buffer (pH 7.5). Seal and rotate using the rotator at 4°C overnight.

3. Next day, add 20 μl of 1 M Tris–HCl (pH 7.5) and rotate using the rotator at 4°C for 2 h to stop the reaction.

4. Spin and store the supernatant to determine the coupling efficiency. Wash the beads three times with 500 μl of 1× PBS buffer.

5. Make up to 400 μl or the desired volume with 1× PBS buffer and store at 4°C.

For the determining the coupling efficiency.

1. Desalt the supernatant with NAP5 column or another procedure.

2. Measure the amount of protein by any protein assay (e.g. Bradford Protein Assay).

3.6.2. Selection of Aptamers Against IL-6R

You should prepare the non-coated sepharose beads according to Note 19 before the following procedure.

1. Take 20–25 μl of non-coated sepharose beads and wash with 1× selection buffer. Incubate cDNA display molecules (from Subheading 3.5.5) with the beads using the rotator at room temperature for 20–30 min. Spin briefly and collect the supernatant.

2. Take around 20 μl of lL-6R-immobilized beads and wash the beads two or three times with the 1× selection buffer. Suspend

the beads with the supernatant from step 1 (cDNA display molecules) and five times the volume of IL-6R-immobilized beads (=5×20=100 μl) of 1× selection buffer. Incubate using the rotator at room temperature for 30 min (see Note 20).

3. Wash the beads three times with 500 μl of 1× selection buffer. Incubate with 100 μl of 1× selection buffer using the rotator at room temperature for 10 min.

4. Repeat step 3 from one to three times.

5. Elute the cDNA display molecules from the beads with elution buffer (90 μl of selection buffer, 10 μl of 1 M DTT). Incubate using the rotator at room temperature for 10 min. Spin briefly and collect the supernatant. Repeat twice.

6. Perform ethanol precipitation using Quick-Precip™ Plus Solution.

7. Perform PCR with the above precipitant using primers (ΩRT-L: 5'-CAACAACATTACATTTTACATTCTACAACTAC AAGCCACC-3', LHR-HGGS-R: 5'-TTTCCCCGCCGCCC CCCGTCCTGCTTCCGCCGTGATGAT-3'). PCR program is as follows: 95°C for 2 min, repeat 30 cycles of three steps, 95°C for 25 s, 69°C for 20 s, 72°C for 30 s, followed by 72°C for 2 min using a thermal cycler (see Note 21).

8. Analyze and check the PCR product by 8 M urea denaturing PAGE.

9. Purify the PCR product with the QIAquick PCR Purification kit (see Note 22).

10. Join the purified PCR product (selected sequences) with the "T7Ω" fragment for the succeeding next screening cycle according to the Subheading 3.1 steps 4–8 (see Note 23).

11. Take aliquot (1/10 volume) of the solution containing the full construct dsDNA and dilute it with nuclease-free water for direct PCR sequencing and storage.

12. The reminder (9/10 volume) can be used for the transcription process in the next screening cycle.

4. Notes

1. The long oligonucleotide (>80 mer) should be prepared by polyacrylamide gel electrophoresis (PAGE) purification grade.

2. NNN codons (where = A/T/G/C is equimolar) contain a higher frequency of stop codons. This is fatal for the template containing the long random region (>30 amino acids = 90 mer).

3. This reagent can be prepared by mixing an equal volume of 100 mM rNTP (rATP, rGTP, rCTP and rUTP) supplied in the RiboMAX™ Large-Scale RNA Production Systems.

4. The DNA template should be purified from PAGE gel.

5. The concentration of linker is conveniently around 20 pmol/μl in the working solution for the following procedure.

6. It is convenient for the working solution of the mRNA template to prepare in 10–20 pmol/μl.

7. The reaction can be scaled down as desired. The amount of mRNA will depend on the incubation time; however, there is a possibility of mRNA degradation if incubated for long time.

8. Wheat germ extract can be also used instead of rabbit reticulocyte lysate for the cell-free translation system. However, the reaction time should be restricted to around 10 min because wheat germ extracts contain a considerably high amount of nucleases in comparison with rabbit reticulocyte lysate. The yield with wheat germ in 5 min is as same as one with rabbit reticulocyte lysate in 20 min.

9. The binding capacity of beads depends on the length or bulkiness of molecules binding on the beads. Confirm the amount of beads to collect the target molecules completely before the following procedures.

10. Keep the beads in suspension during the incubation because it is very important to avoid the aggregation of beads and non-specific adsorption to the cap of the microcentifuge tube.

11. The volume of the binding buffer for washing depends on the amount of beads. Repeat washing until the supernatant becomes clear.

12. After the translation reaction, some ribosomes (polysome) bind to an mRNA, these ribosomes inhibit the reverse transcription. Thus, it is necessary to remove all ribosomes from an mRNA to finally synthesize a complete cDNA.

13. The alternative method is air oxidation. Prepare the 1× PBS buffer instead of the oxidative folding buffer. Incubate the beads of Subheading 3.5.2 with 1× PBS buffer at 4°C overnight using the rotator.

14. The *Pvu*II buffer supplied with the *Pvu*II enzyme generally contains DTT. Thus, you should prepare the *Pvu*II buffer (−DTT) yourself to keep the disulfide bonds of the random peptide library.

15. The volume of the elution buffer depends on choice but should be small to allow reduction in imidazole concentration by dilution for a downstream process such as selection.

16. You can see the beads easily by using a clear 1.5 ml microcentrifuge tube.
17. Wash the beads stringently and remove the residual HCl solution completely.
18. Execute this step as fast as possible, because a N-hydroxysulfosuccinimide (NHS) ester group may hydrolyze quickly after adding the phosphate buffer (pH 7.0).
19. For preparation of the control (non-coated beads), add only 40 μl of 50 mM Tris–HCl (pH 7.5) instead of IL-6R.
20. Various volumes of IL-6R-immobilized beads may be used (e.g., 50 μl for the first round, 20 μl for the second to fourth round and 10 μl for the fifth round).
21. An error-prone PCR can be performed in this step for an evolution experiment.
22. We recommend that part (1 μl) of this product should be stored for the sequencing analysis.
23. The reaction volume can be scaled down as desired.

Acknowledgments

The contents of this work comprise much of the collected wisdom of a number of colleagues. The authors especially thank Dr. Junichi Yamaguchi and Dr. Mohammed Naimuddin for substantial contribution in this project. We also would like to thank Prof. Yuzuru Husimi for helpful comments on the manuscript.

References

1. Nemoto, N., Miyamoto-Sato, E., Husimi, Y. and Yanagawa, H. (1997) In vitro virus: Bonding of mRNA bearing puromycin at the 3′-terminal end to the C-terminal end of its encoded protein on the ribosome *in vitro*. *FEBS lett.* **414**, 405–408.
2. Roberts, R.W. and Szostak, J.W. (1997) RNA-peptide fusions for the in vitro selection of peptides and proteins, *Proc. Natl. Acad. Sci. USA* **94**, 12297–12302.
3. Yamaguchi, J., Naimuddin, M., Biyani, M., Sasaki, T., Machida, M., Kubo, T., Funatsu, T., Husimi, Y., Nemoto, N. (2009) cDNA display: a novel screening method for functional disulfide-rich peptides by solid-phase synthesis and stabilization of mRNA–protein fusions. *Nucleic Acids Res.* **37**, e108.
4. LaBean, T.H. and Kauffman, S.A. (1993) Design of synthetic gene libraries encoding random sequence proteins with desired ensemble characteristics. *Protein Sci.* **2**, 1249–1254.

Chapter 14

Peptide Screening Using PURE Ribosome Display

Hiroyuki Ohashi, Takashi Kanamori, Eriko Osada,
Bintang K. Akbar, and Takuya Ueda

Abstract

To demonstrate directed protein evolution or selection of functional polypeptides, ribosome display is one of the most ideal technologies of evolutionary engineering. Intrinsic components, such as nucleases in the cell extract-based cell-free protein synthesis systems, reduce the stability of the messenger RNA–ribosome–polypeptide ternary complex, thereby preventing the attainment of reliable results. To overcome this problem, we have developed an effective and highly controllable ribosome display system using the protein synthesizing using recombinant elements (PURE) system. Since the activities of nucleases and other inhibitory factors are very low in the PURE system, the ternary complex is highly stable and the selected mRNA can be reliably recovered. Using this system, we were able to select peptides that specifically bind to monoclonal antibodies from random peptide libraries. The advantages of the modified PURE system for ribosome display strongly substantiate its usability.

Key words: PURE system, Ribosome display, Cell-free protein synthesis system, In vitro selection, Epitope mapping, Peptide screening, Protein engineering

1. Introduction

Ribosome display is one of the most ideal technologies to select functional proteins and peptides from large libraries. It is based on the formation of a messenger RNA (mRNA)–ribosome–nascent polypeptide ternary complex in a cell-free protein synthesis system (1, 2), and thus provides a physical linkage between phenotype and genotype. The sequence information for the polypeptide of interest can be selected by affinity purification of this complex. Screening of single-chain Fv (scFv) with high affinity to the antigens is one of the several applications of ribosome display technology (3, 4). There are some advantages for ribosome display methods, which are performed entirely in vitro over selection technologies involving a step in cells, such as phage display. The biggest advantage is that the diversity of the library is not limited by the transformation efficiency of cells. This technology was originally developed based on a cell-free protein synthesis system using *Escherichia coli* (*E. coli*) S30 extract (1, 5). Since *E. coli* S30 extract contains endogenous components, such as proteases and nucleases, the stability of the mRNA–ribosome–polypeptide ternary complex is reduced due to the degradation of mRNA and nascent polypeptide. Moreover, incomplete information on components within a cell extract increases the difficulty to optimize the system for ribosome display.

Here, we have established a highly controllable cell-free translation system called the protein synthesizing using recombinant elements (PURE) system (6–9). The PURE system contains only the factors, such as ribosome and translation factors, considered to be sufficient for protein synthesis in *E. coli* (Fig. 1). Since the construction of the PURE system is performed using only the purified factors and enzymes, it contains no nucleases and proteases. We have developed a new strategy that makes it possible to

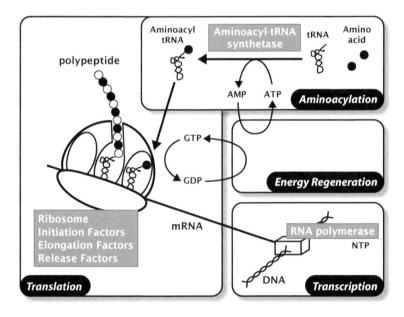

Fig. 1. Outline of protein synthesis reaction in the PURE system. A minimal set of components essential for protein synthesis was purified and reconstituted into a biologically functional system – the PURE system as illustrated. To complete translation of an open reading frame (ORF) encoded in the mRNA sequence, three reaction steps are required on the ribosome: initiation, elongation, and termination. In addition to this main translation reaction, three other reactions are necessary to facilitate protein synthesis: transcription to synthesize mRNA, aminoacylation of tRNAs, and energy source regeneration. All of the protein factors responsible for the protein expression are purified in a histidine-tagged form without loss of activity. The PURE system is reconstituted by also including buffer, tRNA mixtures, and substrates, such as the 20 amino acids and 4 nucleoside triphosphates.

prepare the ternary complex, which is more stable than the complex obtained by the cell extract-based ribosome display, and termed it as "Pure Ribosome Display (PRD)" (10, 11). We have refined the original PURE system for efficient selection of functional polypeptides. Using the developed system, a 12,000-fold enrichment of single-chain Fv (scFv) was achieved in a single round of selection (10).

Furthermore, we demonstrated epitope mapping of antibodies with this system. In only two selection rounds, we were able to select peptides that specifically bind to monoclonal antibodies from random peptide libraries (Fig. 3) (12). This method can be adapted to the selection of peptides binding to other target proteins by substituting these into the protocol.

2. Materials

2.1. Preparation of DNA and mRNA for In Vitro Translation

1. Template DNA; an example of a template DNA construct for PRD is shown in Fig. 2 (see Note 1).
2. mRNA synthesis kit; CUGA7 in vitro Transcription Kit (Wako, Japan).
3. RNA purification kit; RNeasy MinElute Cleanup Kit (Qiagen).

2.2. In Vitro Transcription and Translation

1. Standard PURE system reagent mix for PRD: 0.33–0.5 μM purified *E. coli* ribosome (10), 1.21 μM initiation factor (IF) 1, 0.41 μM IF2, 0.49 μM IF3, 0.64 μM elongation factor (EF)-G, 2.31 μM EF-Tu, 1.64 μM EF-Ts, 0.25 μM release factor (RF) 1, 0.24 μM RF2, 0.17 μM RF3, 0.48 μM ribosome recycle factor, 179 nM alanyl-tRNA synthetase (AlaRS), 31 nM ArgRS, 209 nM AsnRS, 60 nM AspRS, 24 nM CysRS, 60 nM GlnRS, 235 nM GluRS, 43 nM GlyRS, 9 nM HisRS, 379 nM IleRS, 41 nM LeuRS, 56 nM LysRS, 14 nM MetRS, 67 nM

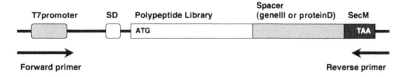

Fig. 2. The DNA construct used for the PURE ribosome display. A T7 promoter (T7pro) and a Shine–Dalgarno (SD) sequence are located upstream of the gene encoding for the displayed polypeptide. It is followed by a Gly/Ser-rich spacer sequence from gene III (G/S spacer) or protein D, and SecM elongation arrest sequence (SecM). The initiation and the termination codon are inserted at the 5'- and 3'-terminus of the gene, respectively. The primers used for reverse transcription-PCR (RT-PCR) after in vitro selection are indicated at the *bottom*.

PheRS, 80 nM ProRS, 19 nM SerRS, 42 nM ThrRS, 14 nM TrpRS, 6 nM TyrRS, 17 nM ValRS, 10 μg/ml T7 RNA polymerase, 585 nM methionyl-tRNA transformylase, 4.0 μg/ml creatine kinase, 3.0 μg/ml myokinase, 1.1 μg/ml nucleoside-diphosphate kinase, 1.0 μg/ml pyrophosphatase, 0.3 mM 20 amino acids each, 56 Abs_{260}/ml *E. coli* tRNA mix (Roche), 50 mM Hepes–KOH, pH 7.6, 100 mM potassium glutamate, 13 mM magnesium acetate, 2 mM spermidine, 1 mM dithiothreitol, 2 mM adenosine-5′-triphosphate (ATP), 2 mM guanosine-5′-triphosphate (GTP), 1 mM cytidine-5′-triphosphate (CTP), 1 mM uridine-5′-triphosphate (UTP), 20 mM creatine phosphate, 10 μg/ml 10-formyl-5, 6, 7, 8-tetrahydrofolic acid. All translation factors and enzymes were prepared from overexpressing *E. coli* cells in a histidine-tagged form (6, 7, see Note 2).

2. Stop buffer: 50 mM Tris–HCl, pH 7.5, 150 mM NaCl, 10 mM magnesium acetate, 0.1% (v/v) Tween 20, 2.5 mg/ml sodium heparin.

2.3. Preparation of Biotinylated Target Protein

1. Target protein: e.g., anti-FLAG M2 antibody (Sigma).
2. Ez-Link Sulfo-NHS-LC-LC-Biotin (Pierce).
3. Phosphate-buffered saline (PBS), pH 7.4.
4. Protein Assay (Bio-Rad).
5. BSA (10 mg/ml): BSA powder is dissolved in water.
6. Dynabeads M-280 Streptavidin (Invitrogen).
7. PBS-T: PBS with 0.1% (v/v) Tween 20.

2.4. In Vitro Selection

1. Wash buffer: 50 mM Tris–HCl, pH 7.5, 150 mM NaCl, 50 mM magnesium acetate, 0.1% (v/v) Tween 20.
2. Elution buffer: 50 mM Tris–HCl, pH 7.5, 150 mM NaCl, 50 mM ethylenediamine tetraacetic acid (EDTA), 10 μg/ml total RNA from *Saccharomyces cerevisiae* (Sigma).
3. RNA purification kit: RNeasy MinElute Cleanup Kit (Qiagen).

2.5. Reverse Transcription-PCR and Sequencing of Selected Clones

1. Reverse transcriptase: SuperScript III Reverse Transcriptase (Invitrogen).
2. DNA polymerase: KOD-Plus-(Toyobo, Japan, see Note 3).
3. PCR product purification kit: MinElute PCR Purification Kit (Qiagen).
4. TA cloning kit: TOPO-TA cloning kit (Invitrogen).

3. Methods

3.1. In Vitro Transcription and Translation

1. Prepare the PURE system reagent without three release factors and ribosome-recycling factor in a 1.5-ml tube, and add water to make a final volume of 29 μl (see Note 2).
2. Add 1 μl of 0.6 pmol/μl DNA solution to the PURE system reagent mixture (see Note 2).
3. Incubate the reaction mixture at 37°C for 20 min to form the mRNA–ribosome–polypeptide ternary complex (see Note 4).
4. To stop the translation reaction, place the tube on ice and add 470 μl of stop buffer (see Note 5).
5. Centrifuge the reaction tube at $14,000 \times g$ for 10 min to remove insoluble components.
6. Transfer the supernatant to a new 1.5-ml tube.

3.2. Preparation of Immunogloblin-Immobilized Beads

1. Prepare 1 mg of the target protein in 0.1 ml of PBS.
2. Prepare 10 mM Ez-Link Sulfo-NHS-LC-LC-Biotin by dissolving 1 mg of reagent powder in 150 μl of water (see Note 6).
3. Mix the protein solution with 100 μl of 10 mM Ez-Link Sulfo-NHS-LC-LC-Biotin and incubate it at room temperature for 30 min.
4. Remove excess nonreacted and hydrolyzed biotin reagent by dialysis against PBS.
5. Determine the concentration of biotinylated protein with Protein Assay (Bio-Rad) using BSA as a standard.
6. Mix 100 μl of Dynabeads M-280 Streptavidin and 10 μg of biotinylated protein.
7. Incubate at room temperature for 30 min.
8. Remove the supernatant and wash the beads three times with PBST.

3.3. In Vitro Selection

1. Add 50 μl of magnetic beads without ligands to the supernatant from 3 to 1. Keep it undisturbed for 30 min at room temperature to eliminate any nonspecific ternary complexes.
2. Remove the magnetic beads.
3. Add 50 μl of the immunogloblin-immobilized beads (2).
4. Incubate the mixture at room temperature for 30 min (see Note 7).
5. Recover the beads and wash them five to ten times with 1 ml of wash buffer (see Note 8).

6. Isolate the mRNA from the beads by incubating it in 60 μl of elution buffer at room temperature for 30 min or by competitive eluting of it with the target protein.

7. Purify the eluted mRNA using an RNA purification kit according to supplier's instruction.

3.4. Reverse Transcription-PCR and Sequencing of Selected Library Clones

1. Using purified mRNA as a template RNA, perform RT and PCR by using primers shown in Fig. 2.

2. Perform RT by SuperScript III Reverse transcriptase (Invitrogen) according to supplier's instruction.

3. Mix the PCR reaction mixture as follows:

10× reaction mix buffer	20.0 μl
2 mM dNTPs mix	20.0 μl
25 mM MgSO$_4$	16 μl
Forward primer (10 μM)	8.0 μl
Reverse primer (10 μM)	8.0 μl
KOD-Plus-	4.0 μl
cDNA solution	20.0 μl
RNase-free water	To 200.0 μl

4. Perform PCR as follows:

94°C for 2 min	1 cycle
94°C for 15 s and 68°C for 45 s	20–30 cycles
68°C for 5 min	1 cycle

5. Subject the PCR reaction mixture to agarose gel electrophoresis to analyze the amplified products.

6. Purify the amplified DNA by MinElute Gel purification kit and subject it to an additional selection round, if necessary.

7. Perform TA cloning by TOPO-TA cloning kit according to the manufacturer's specification for cloning from the selected library. The sequences of the bound clones to FLAG-M2 antibody are shown in Fig. 3.

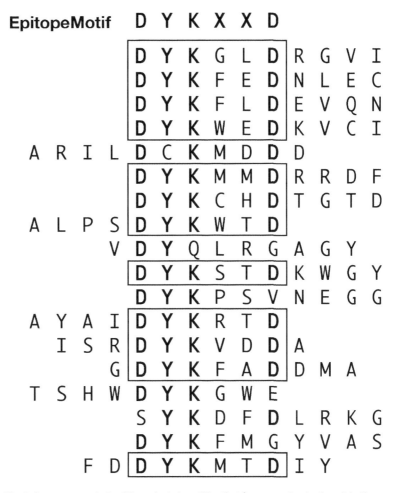

Fig. 3. Sequence analysis of the selected peptides that form a randomized peptide library. Using this technology, we selected peptides that specifically bind to anti-FLAG M2 antibody from a random peptide library. When selection was performed against the anti-FLAG M2 antibody, selected peptides contained previously characterized consensus epitope (DYKXXD). Isolated clones after the second round of selection were sequenced. Sequences with FLAG epitope motif are boxed. Reprinted from Osada et al. (12).

4. Notes

1. The template DNA requires a T7 promoter sequence and ribosome-binding site (Shine–Dalgarno sequence) upstream of the open reading frame (ORF). ORF must initiate with an initiation codon (ATG). To prevent steric hindrance between the displayed polypeptide and the ribosome, a spacer sequence, such as gene III, a filamentous phage, or protein D, a phage Lambda capsid protein, is introduced downstream of the polypeptide of interest. Furthermore, the SecM elongation arrest

sequence is positioned downstream of the spacer sequence in order to stabilize the mRNA–ribosome–polypeptide ternary complex. This sequence has been shown to interact with the ribosomal polypeptide tunnel (13). To form the mRNA–ribosome–polypeptide ternary complex using the PURE system reagent without release factors, the ORF requires a stop codon (TAG, TGA, or TAA) at the 3′ terminus (10, 12). The PURE system is now commercially available from Wako Pure Chemical Industries Ltd. (Japan) and NEW ENGLAND BioLabs inc. as PURESYSTEM kits and PURExpress In Vitro Protein Synthesis kits, respectively. If the kits contain release factors, omit the stop codon. The DNA encodes a random peptide library which can be prepared with PCR.

2. PURE system reagent containing 1–3 µM ribosome is generally used for a maximum yield of protein synthesis. In contrast, we have demonstrated that maximum mRNA recovery was achieved with 1 pmol of mRNA and 10 pmol of ribosome (0.33 µM) in 30 µl of reaction (10). Release factors are omitted from the reagent for formation and stabilization of the mRNA–ribosome–polypeptide ternary complex. The reagent composition can be easily modified according to the displayed polypeptide. For example, addition of molecular chaperones, such as hsp70 and hsp60, can increase solubility of the displayed polypeptide. When a polypeptide with disulfide bonds is displayed and selected, addition of protein disulfide isomerase with 1 mM oxidized glutathione and 0.1 mM reduced glutathione in place of dithiothreitol can improve correct folding of displayed polypeptide. One can also add DNA as the template instead of mRNA.

3. It is recommended that a high-fidelity DNA polymerase be used to predict a correct epitope sequence from the selected peptides and to minimize the effects of errors with a polymerase after selection.

4. For the formation of mRNA–ribosome–polypeptide complex, a 20-min incubation is sufficient. Longer incubation may destabilize the mRNA–ribosome–polypeptide ternary complex.

5. Unrelated RNA (e.g., *S. cerevisiae* RNA) can also be added to stop buffer to suppress nonspecific binding of mRNA to the target and magnet beads and to inhibit degradation of mRNA.

6. Because the NHS-ester moiety readily hydrolyzes, Ez-Link Sulfo-NHS-LC-LC-Biotin is dissolved in water just before use.

7. Because the mRNA–ribosome–polypeptide ternary complex is unstable in the cell extract system, the reaction mixture must be placed at a temperature of 4°C after translation reaction. In contrast, the mRNA–ribosome–polypeptide complex is highly

stable in the PURE system; hence, it is possible to perform selection step at room temperature.

8. Selection efficiency is also dependent on the ratio of the target to the displayed polypeptide and the stringency of washing step.

References

1. Hanes, J. and Pluckthun, A. (1998) In vitro selection and evolution of functional proteins by using ribosome display. Proc. Natl. Acad. Sci. USA 94, 4937–4942.
2. Hanes, J., Jermutus, L. and Pluckthun, A. (2000) Selecting and evolving functional proteins in vitro by ribosome display. Methods Enzymol. 328, 404–430.
3. Hanes, J., Jermutus, L., Weber-Bornhauser, S., Bosshard, H.R. and Pluckthun, A. (1998) Ribosome display efficiently selects and evolves high-affinity antibodies in vitro from immune libraries. Proc. Natl. Acad. Sci. USA 95, 14130–14135.
4. Binz, H. K., Amstutz, P., Kohl, A., Stumpp, M. T., Briand, C., Forrer, P., Grutter, M. G. and Pluckthun, A. (2004) High-affinity binders selected from designed ankyrin repeat protein libraries. Nat. Biotechnol. 22, 575–582.
5. Zahnd, C, Amstutz, P, Pluckthun, A. (2007) Ribosome display: selecting and evolving proteins in vitro that specifically bind to a target. Nat. Methods 4, 269–279.
6. Shimizu, Y., Inoue, A., Tomari, Y., Suzuki, T., Yokogawa, T., Nishikawa, K. and Ueda, T. (2001) Cell-free translation reconstituted with purified components. Nat. Biotechnol. 19, 751–755.
7. Shimizu, Y., Kanamori, T. and Ueda, T. (2005) Protein synthesis by pure translation systems. Methods 36, 299–304.
8. Shimizu, Y. and Ueda, T. (2010) PURE technology. Methods Mol. Biol. 607, 11–21
9. Ohashi, H., Kanamori, T., Shimizu, Y. and Ueda, T. (2010) A highly controllable reconstituted cell-free system – a breakthrough in protein synthesis research. Curr. Pharm. Biotechnol. 11, 267–271.
10. Ohashi, H., Shimizu, Y., Ying B. W. and Ueda T. (2007) Efficient protein selection based on ribosome display system with purified components. Biochem. Biophys. Res. Commun. 352, 270–276.
11. Ueda, T., Kanamori, T. and Ohashi, H. (2010) Ribosome display with the PURE technology. Methods Mol. Biol. 607, 219–225.
12. Osada, E., Shimizu, Y., Akbar, B. K., Kanamori, T. and Ueda, T. (2009) Epitope mapping using ribosome display in a reconstituted cell-free protein synthesis system. J. Biochem. 145, 693–700.
13. Nakatogawa, H. and Ito, K. (2002) The ribosomal exit tunnel functions as a discriminating gate. Cell 108, 629–636.

Chapter 15

Rapid Selection of High-Affinity Binders Using Ribosome Display

Birgit Dreier and Andreas Plückthun

Abstract

Ribosome display has proven to be a powerful in vitro selection and evolution method for generating high-affinity binders from libraries of folded proteins. It has been successfully applied to single-chain Fv fragments of antibodies and alternative scaffolds, such as *D*esigned Ankyrin Repeat *P*roteins (DARPins). High-affinity binders with new target specificity can be obtained from highly diverse DARPin libraries in only a few selection rounds. In this protocol, the selection from the library and the process of affinity maturation and off-rate selection are explained in detail.

Key words: Ribosome display, In vitro selection, In vitro translation, *D*esigned Ankyrin Repeat *P*roteins, Affinity maturation

1. Introduction

Ribosome display is a potent in vitro method to select and evolve proteins or peptides from a naïve library with very high diversity to bind to any chosen target of interest (1–4). The background and mechanism are summarized and discussed in the accompanying chapter (5). We report here the most recent version of the standard protocols for selection from a complex library and affinity maturation using off-rate selection.

The protocol detailed here starts from a library of about 10^{12} DNA molecules in the form of a PCR fragment. Whether this number corresponds to the functional library size, i.e., whether all these 10^{12} molecules are different and at least potentially functional, depends both on the *design quality* of the input library template and the *amount* of the input template used from which this PCR fragment is generated.

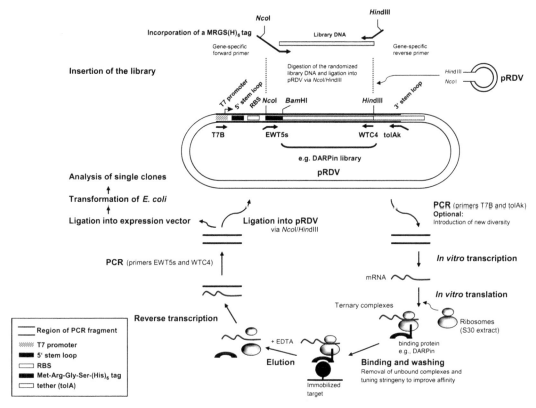

Fig. 1. Scheme of the ribosome display cycle, illustrated for selection of high-affinity DARPins. In ribosome display, all steps of the selection are performed in vitro. The cycle begins with a DNA library (*top*) in the form of a PCR fragment encoding a library of the protein of interest. This cassette is ligated into a vector in vitro which provides a promoter and ribosome-binding site (RBS). The ORF of interest (*light grey*) is fused to an additional protein region (the "*spacer*" or "*tether*," *checkered white*). This *tether or spacer*, here used as an unstructured region from the *E. coli* TolA protein, has the sole function of allowing the protein domain of interest to emerge from the ribosomal tunnel. A PCR is then carried out from the promoter to the middle of the *tether*. Importantly, the PCR fragment does not encode a stop codon at the end. Each member of the library pool is then transcribed from double-stranded DNA into mRNA and is subsequently translated by the ribosomes present in the S30 extract, leading to ternary complexes consisting of ribosomes, mRNA, and the DARPin encoded by that particular mRNA. Since there is no stop codon on the mRNA, the protein is not released from the ribosome. It is believed to be still covalently attached to the tRNA within the ribosome, with the tether in the tunnel, and the domain of interest outside and already folded. Selection can be achieved by binding the protein–ribosome–mRNA complexes to the desired immobilized target, followed by removal of unbound or nonspecifically bound protein by stringent washing. Affinity can be increased by addition of an excess of nonlabeled target (off-rate selection) (see Subheading 3.12.2). Particular selectivity in binding can be achieved by adding an unwanted target as a competitor. Selection for other properties, such as stability, requires other selection pressures at this step (see Note 25). Binders can be easily recovered by destruction of the protein–ribosome–mRNA complex using EDTA and recovery of the genetic information of the binders by RT-PCR using the inner primers WTC4 (annealing to the sequence encoding the C-terminus of the DARPin sequence, which can be replaced by a primer specific for other library folds) and EWT5s (pRDV-specific primer overlapping with the RBS and beginning of the Met-Arg-Gly-Ser-(His)$_6$ tag). The inner primer set is used to amplify the selected clones, which often is not possible with the outer primer set due to incomplete synthesis or degradation of the mRNA. For further selection rounds, the PCR product pool is subcloned into pRDV via the restriction endonucleases, *Nco*I and *Hin*dIII, followed by a second PCR with the outer primers, T7B and tolAk. T7B introduces the T7 promoter sequence and part of the stabilizing 5′ stem loop sequences that are part of the pRDV vector. The tolAk primer binds in the sequence of the tolA spacer region and introduces a stabilizing 3′ stem loop. If further diversity is required, an error-prone PCR can be included at this step. The amplified PCR product then serves as template for in vitro transcription, initiating the next round of selection. At the end of the selection rounds (typically, 2–5), the resulting PCR product pool can be directly subcloned via the restriction endonucleases, *Bam*HI and *Hin*dIII, into an expression vector in order to screen for binders.

A key feature of ribosome display, in contradistinction to most other selection technologies (5), is that it incorporates PCR into the procedure and thus allows a convenient incorporation of a diversification ("randomization") step. Thereby, ribosome display allows refinement and affinity maturation not only of preexisting binders (6–9), but also of the whole pool during selection from a complex library, if desired. This is one of the major advantages of ribosome display over other selection strategies. The diversity of the library members can be easily manipulated at any selection step by introduction of additional mutations, e.g., by using DNA shuffling (10) and/or error-prone PCR (11). This additional randomization step can readily be integrated in the protocol (see Subheading 3.12.1). In combination with off-rate selection (see Subheading 3.12.2) where binders to the biotinylated target protein are competed with a molar excess of nonbiotinylated target protein, many initial leads were improved for affinity in the range of low nM to low pM (6–9, 12). The theoretical considerations for designing efficient off-rate selection experiments were recently summarized (13).

We have previously applied the ribosome display protocol to antibody scFv fragments as libraries of folded proteins (4, 14, 15). Here, we give a protocol for the in vitro selection of protein scaffolds with more favorable biophysical properties than antibody fragments. The general workflow is outlined in Fig. 1 (a similar protocol had been published previously (16)). One of the most promising scaffolds are the *Designed Ankyrin Repeat Proteins* (DARPins) which are devoid of disulfide bonds, highly soluble, and highly stable, and therefore achieve high expression levels in *E. coli* (17). They also fold well in the in vitro translation inherent in ribosome display and are thus readily enriched for binding specificity. Using ribosome display, DARPins have been evolved to bind various targets with affinities all the way down to the picomolar range (8, 9, 18–23).

2. Materials

2.1. General

1. 96-well Maxisorp plates or strips (Nunc).
2. Adhesive plate sealers (Thermo Scientific).
3. Sterile, RNase-free ART filter tips (Molecular Bio Products).
4. Sterile, RNase-free HydroLogix 1.5- and 2.0-mL tubes (Molecular Bio Products).
5. Roche High pure RNA isolation kit (Roche).
6. illustra MicroSpin™ G-50 Columns (GE Healthcare).
7. NucleoSpin® Extract II DNA purification kit (Macherey-Nagel).

2.2. Selection (see Note 1)

1. Tris-buffered saline (TBS): 50 mM Tris, 150 mM NaCl; adjust pH to 7.5 with HCl at 4°C; filter through 0.22-μm pores.
2. TBST: TBS containing 0.05% Tween-20.
3. Stock solutions for wash buffer (WB) and elution buffer (EB):
   (a) 2 M Tris–acetate; adjust pH to 7.5 at 4°C with acetic acid.
   (b) 5 M NaCl.
   (c) 2 M magnesium acetate.
   (d) 250 mM EDTA; adjust pH to 8.0 by NaOH addition.
   Sterile filter all solutions. For alternative buffer composition, see Note 2.
4. WB/Tween-20 (WBT):
   50 mM Tris–acetate, pH 7.5 at 4°C, 150 mM NaCl, 50 mM magnesium acetate, 0.05% Tween-20; adjust pH to 7.5 with acetic acid at 4°C; filter through 0.22-μm pores.
5. EB: 50 mM Tris–acetate, pH 7.5 at 4°C, 150 mM NaCl, 25 mM EDTA; adjust pH to 7.5 with acetic acid at 4°C; filter through 0.22-μm pores.
6. *Saccharomyces cerevisiae* RNA (Fluka): Dissolve to 25 μg/μL in H_2O; aliquot and store at −20°C.
7. 10% BSA in H_2O; filter through 0.22-μm pores and store at −20°C.
8. Neutravidin and/or streptavidin (Pierce): 1.2 mg/mL (20 μM) in TBS; store at −20°C.
9. Streptavidin-coated MyOne T1 magnetic beads (Invitrogen).
10. Reagents for biotinylation of the target: Either for chemical biotinylation use a NHS-biotin reagent (e.g., from Pierce EZ-link™ SulfoNHS-LC-biotin) or for enzymatic biotinylation use an AviTag together with the *E. coli* biotinylation enzyme BirA (24) (reagents from Avidity).

2.3. Cleanup of mRNA After In Vitro Transcription

1. 6 M LiCl; filter through 0.22-μm pores.
2. 3 M sodium acetate; filter through 0.22-μm pores.
3. 70% EtOH diluted with H_2O and 100% EtOH; filter through 0.22-μm pores.
4. illustra MicroSpin™ G-50 Columns (GE Healthcare).
5. DNAseI (10 U/μL; Roche).

2.4. Reverse Transcription, PCR, and Cloning

1. Primer dissolved to 100 μM in H_2O; aliquot and store at −20°C.
   EWT5s: 5′-TTCCTCCATGGGTATGAGAGGATCG-3′.
   WTC4: 5′-TTTGGGAAGCTTTTGCAGGATTTCAGC-3′.

T7B: 5′-ATACGAAATTAATACGACTCACTATAGGGAGACCACAACGG-3′.

tolAk: 5′-CCGCACACCAGTAAGGTGTGCGGTTTCAGTTGCCGCTTTCTTTCT-3′.

2. AffinityScript™ Multiple Temperature Reverse Transcriptase (50 U/μL; Stratagene) and 10× buffer; see Note 3.
3. 100 mM DTT in H2O; aliquot and store at −20°C.
4. RNasin® Ribonuclease Inhibitor (20–40 U/μL; Promega).
5. Vent$_R$® DNA Polymerase (2 U/μL; New England Biolabs) and 10× Thermopol buffer; see Note 4.
6. Platinum® Taq DNA Polymerase (5 U/μL; Invitrogen) and 10× polymerase buffer.
7. dNTPs: 5 mM each (Eurogentec); aliquot and store at −20°C.
8. Nucleotide analogs dPTP and 8-oxo-dGTP (Jena Biosciences) at 100 μM in H_2O.
9. Dimethyl sulfoxide (DMSO; Fluka, 41640).
10. Restriction endonucleases: *Bam*HI (20 U/μL), *Hin*dIII (20 U/μL), *Nco*I (10 U/μL), and 10× buffer all from New England Biolabs.
11. T4 DNA ligase (5 U/μL; Fermentas) and 10× ligase buffer.
12. Ribosome display vector pRDV (GenBank accession code AY327136; please note the revised sequence) (18).

2.5. In Vitro Transcription

1. T7 RNA polymerase (20 U/μL; Fermentas), see Note 5.
2. RNasin® Ribonuclease Inhibitor (20–40 U/μL; Promega).
3. 100 mM DTT in H_2O; aliquot and store at −20°C.
4. T7 RNA polymerase buffer (5×): 1 M HEPES, 150 mM magnesium acetate, 10 mM spermidine, 200 mM DTT; adjust pH to 7.6 with KOH; aliquot and store at −20°C.
5. 50 mM NTP mix: 50 mM adenosine 5′-triphosphate (ATP; Sigma–Aldrich), 50 mM uridine 5′-triphosphate (UTP; Sigma–Aldrich), 50 mM guanosine 5′-triphosphate (GTP; Sigma–Aldrich), 50 mM cytidine 5′-triphosphate (CTP; Sigma–Aldrich) in H_2O; aliquot and store at −20°C.

2.6. In Vitro Translation

1. Protein disulfide isomerase (PDI; Sigma–Aldrich): 22 μM in H_2O; aliquot and store at −80°C.
2. Heparin (Sigma–Aldrich): 200 mg/mL heparin in H_2O; do not filter; aliquot and store at −20°C.
3. L-Methionine (Sigma–Aldrich): 200 mM L-methionine in H_2O; do not filter; aliquot and store at −20°C.
4. STOP mix: 1 mL WBT buffer/0.5% BSA plus 12.5 μL heparin.

2.7. S30 Extract

1. *E. coli* strain MRE600 (25) lacking ribonuclease I activity.
2. Incomplete rich medium: 5.6 g KH_2PO_4, 28.9 g K_2HPO_4, 10 g yeast extract, 15 mg thiamine for 1 L medium. Autoclave and add 50 mL 40% glucose (w/v) and 10 mL 0.1 M magnesium acetate, both sterile filtered.
3. S30 buffer: 10 mM Tris–acetate (pH 7.5 at 4°C), 14 mM magnesium acetate, 60 mM potassium acetate. Chill to 4°C before use.
4. Preincubation mix (must be prepared directly before use): 3.75 mL 2 mM Tris–acetate (pH 7.5 at 4°C), 71 µL 3 M magnesium acetate, 75 µL amino acid mix (10 mM of each of the 20 amino acids; Fluka), 300 µL 0.2 M ATP, 50 units pyruvate kinase (Fluka), 0.2 g phosphoenolpyruvate trisodium salt (Fluka); add to 10 mL H_2O.
5. Dialysis tubing with a MW cutoff of 6,000–8,000 Da (Spectrum Laboratories).

2.8. PremixZ

1. Set up premixA (the final concentration is fivefold lower in the final volume of the in vitro translation reaction, see Subheading 3.4): 250 mM Tris–acetate (from a 2 M stock solution, pH 7.5 at 4°C), 18 µM anti-ssrA oligonucleotide (5′-TTAAGCTGCTAAAGCGTAGTTTTCGTCGTTT GCGACTA-3′) from a 200 µM stock solution, 1.75 mM of each amino acid except for methionine, 10 mM ATP from a 1 M stock solution, 2.5 mM GTP from a 0.2 M stock solution, 5 mM cAMP (Sigma–Aldrich) from a 0.4 M stock solution, 150 mM acetyl phosphate (Sigma–Aldrich) from a 2 M stock solution, 2.5 mg/mL *E. coli* tRNA from strain MRE600 (Roche) from a 25 mg/mL stock solution, 0.1 mg/mL folinic acid (Sigma–Aldrich from a 10 mg/mL stock solution).
2. Set up an in vitro translation reaction (see Subheading 3.4), and use the above premixA, but titrate the optimal concentration of the following components for the final premixZ composition to achieve optimal performance of each newly generated S30 extract. Optimize the final concentrations for the in vitro translation in the order shown below.

    Magnesium acetate (MgAc) usually in the range of 7–15 mM from a 0.2 M stock solution, potassium glutamate (KGlu) usually in the range of 180–220 mM from a 2 M stock solution, and PEG-8000 usually in the range of 2–15% (w/v) from a 40% stock solution. Adjust the premixA with the optimal composition of MgAc, KGlu, and PEG to obtain the premixZ (we are usually using concentrations of 21.4 mM MgAc, 481 mM KGlu, and 7% PEG-8000 in the premixZ). Aliquot the premixZ and flash freeze in liquid nitrogen. Long-time storage should be at –80°C, but the premixZ is stable for several months at –20°C and can be frozen several times. If not noted otherwise, reagents are purchased from Sigma–Aldrich.

2.9. β-Lactamase Assay

Used to test the activity of the S30 extract and to optimize the premixZ.

1. Prepare β-lactamase mRNA from the pRDV template DNA encoding the double Cys→Ala mutant of β-lactamase (26) using PCR with the T7B and tolAk primers (Fig. 1) (see Subheading 3.2), followed by in vitro transcription and purification of mRNA (see Subheadings 3.2 and 3.3).

2. Set up in vitro translation reactions containing 2 μg RNA, 0.5 μL 200 mM methionine, 10 μL S30 extract, and 8.2 μL premixZ and add to 22 μL H$_2$O. For optimization of the activity of the S30 extract, use premixA and adjust the concentration of magnesium acetate, potassium acetate, and PEG-8000.

3. Incubate at 37°C for 10 min.

4. Add 88 μL STOP mix.

5. Use 5 μL of stopped in vitro translation for the activity assay with the chromogenic substrate nitrocefin (Glaxo Research, obtained from Oxoid Ltd.) (27).

6. Dilute nitrocefin 1:20 in β-lactamase buffer (100 mM sodium phosphate buffer, pH 7.0) from a stock solution (1 mg nitrocefin dissolved in 500 μL DMSO and stored at −20°C). For one reaction, use 20 μL diluted nitrocefin together with 5 μL translation plus 175 μL β-lactamase buffer in a 200 μL reaction.

7. Measure OD$_{486nm}$ immediately. Follow the kinetics for approximately 12 min measuring at least once every minute.

2.10. DARPin Expression and Binding Analysis of Single Clones

1. *E. coli* strain XL-1 blue (Stratagene).
2. Expression plasmid pDST67 (22, 28), which is a derivative of pQE30 (QIAGEN).
3. 2×TY media: 5 g NaCl, 16 g tryptone, 10 g yeast extract per liter. Adjust pH to 7.2 with NaOH.
4. 96-well deep well plates (ABgene).
5. TBS: 50 mM Tris, 150 mM NaCl; adjust pH to 7.4 with HCl.
6. TBST: TBS containing 0.05% Tween-20.
7. 10% BSA in H$_2$O.
8. Mouse-anti-RGS(His)$_4$ antibody (QIAGEN).
9. Goat-anti-mouse IgG coupled to alkaline phosphatase (Sigma–Aldrich).
10. pNPP substrate (*p*-nitrophenyl phosphate disodium salt; Fluka): 1 M stock in pNPP buffer (50 mM NaHCO$_3$, 50 mM MgCl$_2$); aliquot and store at −20°C.
11. B-PER II detergent solution (Pierce).

3. Methods

3.1. Insertion of the Library

The ribosome display vector pRDV is used to ligate the library of interest using gene-specific primers and insertion via the restriction endonuclease sites *Bam*HI and *Hin*dIII as indicated in Fig. 1 (18, 29) or *Nco*I and *Hin*dIII. The general elements that need to be present on a template used for ribosome display are the T7 RNA polymerase promoter sequence to initiate efficient transcription and a ribosome binding site (RBS) for docking of the ribosome to initiate translation. The PCR fragment (between the primers T7B and tolAk, Fig. 1) that serves as the template for transcription ends without a stop codon in the ORF. At both the 5' and 3' ends of the mRNA, stabilizing stem loops are incorporated to protect the mRNA from exonuclease degradation (1, 30). The absence of a stop codon in the resulting mRNA prevents termination of translation. The fact that the library is fused in frame to a spacer (or tether) sequence (e.g., derived from the *E. coli tolA* gene) allows the nascent protein chain to exit the ribosome and fold outside of the ribosomal tunnel. The original pRDV contains an N-terminal FLAG tag instead of an N-terminal MRGS(His)$_6$ tag as shown here for the case of the DARPin libraries (18). Both tag variants lead to efficient initiation of in vitro translation and yield ternary complexes in good yields.

3.2. Transcription of PCR Products

1. To obtain a length-defined fragment of DNA as template for in vitro transcription, use the outer primers T7B and tolAk in the following PCR reaction to introduce the T7 RNA polymerase promoter sequence, RBS, the stabilizing 5' and 3' stem loops and the tolA spacer sequence:

5.0 μL	10× Thermopol buffer
2.0 μL	dNTPs (final concentration 200 μM each)
2.0 μL	DMSO (final concentration 5%)
0.5 μL	T7B primer (final concentration 1 μM)
0.5 μL	tolAk primer (final concentration 1 μM)
5.0 μL	Library DNA [either of the initial library or of the amplified DNA after selection which has been ligated to pRDV (see Note 6)]
0.5 μL	Vent DNA polymerase (2 U/μL)
Add to 50 μL with H$_2$O	

2. Perform a hot start to increase specificity and use the following cycling parameters (see Note 7): 3 min at 95°C, 25 cycles: 30 s at 95°C, 30 s at 55°C, 45 s at 72°C, final extension 5 min at 72°C.

3. Verify the product on an agarose gel.
4. For in vitro transcription, set up the following reaction on ice:

20.0 µL	5x T7 polymerase buffer
14.0 µL	NTPs (final concentration 7 mM each)
4.0 µL	T7 RNA polymerase (20 U/µL)
2.0 µL	RNasin (40 U/µL)
22.5 µL	PCR product without further purification
Add to 100 µL with H$_2$O	

5. Incubate the transcription for 2–3 h at 37°C (see Note 8).

3.3. Cleanup of Template mRNA for In Vitro Translation

1. In order to remove all impurities from the reaction, the RNA needs to be purified. This can be performed in two ways (see Note 9).

3.3.1. LiCl precipitation

1. A LiCl precipitation can be performed to purify the RNA product. For this purpose, add 100 µL ice-cold H$_2$O and 200 µL ice-cold 6 M LiCl to the 100 µL translation reaction and vortex.
2. Incubate on ice for 30 min, and then centrifuge at 20,000×g at 4°C for 30 min.
3. Discard the supernatant and wash the pellet with 500 µL ice-cold 70% EtOH ensuring that the pellet is not disturbed.
4. Remove supernatant and dry pellet in a Speedvac apparatus.
5. Completely dissolve the pellet in 200 µL ice-cold H$_2$O and centrifuge at 20,000×g at 4°C for 5 min to remove remaining precipitates.
6. Transfer 180 µL supernatant to a new tube without disturbing the pellet. Add 20 µL 3 M NaOAc and 500 µL ice-cold 100% EtOH, vortex.
7. Incubate at −20°C for at least 30 min. Vortex and centrifuge at 20,000×g at 4°C for 30 min and discard supernatant.
8. Wash the pellet with 500 µL ice-cold 70% EtOH, dry the pellet in a Speedvac apparatus, and resuspend pellet in 30 µL H$_2$O.

3.3.2. Gel filtration

1. For purification of the RNA, small gel filtration columns (e.g., illustra MicroSpin™ G-50 Columns) can be used.
2. Vortex the column to resuspend the material and break off the bottom of the column.

3. Place the column into a 1.5-mL tube and centrifuge at $735 \times g$ for 1 min to pack the column material.

4. Place the column into a collection tube, apply 50 µL sample from the transcription reaction, and centrifuge at $735 \times g$ for 1 min.

5. Optional: DNAse I treatment before loading the column (see Note 10): Take 43 µL of the transcription reaction and add 2 µL of DNAse I solution (10 U/µL) plus 5 µL 10× dilution buffer supplied with the enzyme. Incubate for 10–15 min at room temperature, and then apply the sample to the column.

6. Aliquot RNA and immediately flash freeze in liquid nitrogen. Store at −80°C.

7. Determine the RNA concentration of a 1:100 dilution by OD_{260nm}. If the transcription worked well, a yield of 3–8 µg/µL for RNA after LiCl/EtOH precipitation (total yield from a 100 µL reaction: 90–240 µg) or 1–3 µg/µL from the illustra MicroSpin™ G-50 Columns (total yield from a 50 µL reaction: 50–150 µg) should be obtained.

3.4. In Vitro Translation

1. For one in vitro translation reaction, set up the following mix on ice:

2.0 µL	200 mM methionine
41 µL	premixZ with optimized composition
x µL	In vitro-transcribed RNA (total 10 µg; see Note 11)
50 µL	S30 extract and add to 110 µL with H_2O (for preparation of the S30 extract, see Subheading 3.11)
Add 0.625 µL PDI if your library scaffold requires the formation of disulfide bonds	

2. Mix carefully by pipetting up and down and incubate the reaction at 37°C for 10 min, the time found optimal for DARPins. The incubation time and temperature must be optimized for each library based on different constructs.

3. Stop the reaction by addition of 440 µL ice-cold STOP mix.

4. Mix by pipetting up and down and centrifuge at $20,000 \times g$ at 4°C for 5 min. Transfer 500 µL supernatant to a fresh tube and use 100 µL per well when performing selection in plates or 250 µL per tube when performing selections in solution for either the target-containing or control reaction (see Subheading 3.5).

3.5. Selection (see Note 12)

3.5.1. Target Protein Preparation

Express and purify the target by methods of your choice, but the target for selection must be of excellent purity and homogeneity. To immobilize the target for capturing the ternary complexes, it is recommended to biotinylate the target. This is the method of immobilization found to be most robust by far to stringent washing, including washing with detergents. The advantage of immobilizing biotinylated targets is that it is very general, and works equally well for proteins, peptides, oligonucleotides, and small molecules. Furthermore, by avoiding any direct binding to plastic surfaces, the structure of the target is maintained. Finally, the nonbiotinylated version of the target is a convenient competitor in off-rate selections and in the specificity screening of single clones. Biotinylation can be achieved in two ways (see Note 13).

1. Fuse the target to an AviTag and biotinylate it in vivo or in vitro using the *E. coli* biotinylation enzyme BirA (24) following the guidelines posted on the Avidity Web page (http://www.avidity.com).

2. Alternatively, biotinylate surface lysine amino acid residues using NHS-biotin reagents from Pierce following the manufacturer's instructions.

3.5.2. Selection in Plates

1. Coat wells of a 96 well Maxisorp plate with 100 μL of a 66 nM neutravidin or streptavidin solution in TBS and close with an adhesive plate sealer (see Notes 14 and 15). Store overnight at 4°C or for 1 h at room temperature. Invert plate and shake out solution, dry on paper towels, and wash wells three times with 300 μL TBS.

2. Block the wells with 300 μL 0.5% BSA in TBST per well, and seal and incubate on an orbital shaker for 1 h at room temperature. Shake out blocking solution and dry on paper towels.

3. Immobilize 100 μL biotinylated target at a concentration of 100–200 nM (see Note 16) in TBST/0.5% BSA and TBST/0.5% BSA only for control wells. Seal and incubate on an orbital shaker at 4°C for 1 h. Wash plate three times with 300 μL ice-cold TBST and once with 300 μL ice-cold WBT. Remove WBT only when the stopped translation reaction can be added to the wells (see Subheading 3.4).

4. Add the stopped in vitro translation, seal the plate, and incubate the binding reaction at 4°C for 1 h. Wash the wells with 300 μL ice-cold WBT containing 0.1% BSA for eight to ten times. Use two fast washes removing the buffer immediately, followed by incubations starting at 5 min and extending to 15 min in later rounds. In these longer incubations, binders with fast off-rates dissociate and subsequently are washed away.

5. For elution of the RNA, add 100 μL EB containing EDTA to release the mRNA from the captured protein–mRNA–ribosome complexes and freshly add *S. cerevisiae* RNA (final concentration

50 μg/mL) to block the surface of the tubes and perhaps to act as competing substrate for any residual RNases. Incubate at 4°C for 10 min and add to 400 μL lysis buffer of the High Pure RNA purification kit on ice. Repeat the elution step and collect the second elution in the same tube. After vortexing, the RNA is stable and can be processed at room temperature until elution from the column (see Subheading 3.6).

3.5.3. Selection in Solution

1. Starting from the stopped and centrifuged in vitro translation reaction (see Subheading 3.4), divide the reaction into two aliquots of 250 μL and add 250 μL of STOP mix. Add each of the 500 μL of the diluted stopped translation reactions to 40 μL of streptavidin-coated magnetic beads that were washed two times with 500 μL TBS and blocked with 500 μL TBST/0.5% BSA for 1 h in a 2-mL tube as preselection step (see Note 15). Rotate head over end at 4°C for 1 h.

2. Transfer the supernatant to a blocked 2-mL tube and add to 100–200 nM of biotinylated target (omit target in the control reaction) and incubate rotating at 4°C for 1 h (see Note 17).

3. Transfer the supernatant to a blocked tube containing 40 μL of streptavidin-coated magnetic beads and capture the ternary complexes rotating at 4°C for 30 min. Wash with 500 μL ice-cold WBT containing 0.1% BSA as indicated above (see Subheading 3.5.2, step 4). Separate captured complexes using a magnetic separator between each washing step.

4. Proceed with the elution and purification of RNA as described for the selection on plates (see Subheading 3.5.2, step 5).

3.6. Recovery of Eluted RNA

1. Apply the lysis buffer/eluate mixture from Subheading 3.5.2, step 5, on the column of the High Pure RNA isolation kit (see Note 18; *Optional*: As a positive control, also purify 2 μL of the input RNA from the in vitro transcription diluted in 200 μL EB) and centrifuge at $8,000 \times g$ for 1 min. Discard the flow-through.

2. Add 100 μL diluted DNAse I solution (1.8 U/μL) directly onto the column filter and incubate at room temperature for 15 min (see Note 19). Add 500 μL wash buffer 1 and centrifuge at $8,000 \times g$ for 1 min. Discard flow-through

3. Wash with 500 μL wash buffer 2, centrifuge, and discard flow-through.

4. Add 100 μL wash buffer 2 and centrifuge at $13,000 \times g$ for 2 min to remove any residual EtOH.

5. Elute with 50 μL elution buffer and incubate for 2 min before centrifugation at $8,000 \times g$ for 1 min into a fresh 1.5-mL RNAse-free tube.

6. Freeze the remaining sample of eluted RNA in liquid nitrogen and store at −80°C (see Notes 11 and 20).

3.7. Reverse Transcription of DARPin-Encoding mRNA

1. Transfer two times 12.5 μL of eluted RNA to fresh 1.5-mL tubes (see Note 21).
2. Denature the eluted RNA at 70°C for 10 min and chill on ice.
3. Set up the following RT mix (total of 7.75 μL) per RT reaction on ice:

0.25 μL	WTC4 primer (final concentration 1.25 μM)
0.5 μL	dNTPs (final concentration 125 μM of each nucleotide)
0.5 μL	RNasin (40 U/μL)
0.5 μL	AffinityScript™ Multiple Temperature Reverse Transcriptase (50 U/μL)
2.0 μL	10× AffinityScript buffer
2.0 μL	DTT (final concentration 10 mM)
2.0 μL	H$_2$O

4. Distribute 7.75 μL RT mix per RT reaction to the 12.25 μL samples of denatured RNA.
5. Incubate at 50°C for 1 h.
6. Use 2–5 μL as template for PCR using the inner primers, WTC4 and EWT5s (see Subheading 3.8).
7. Freeze the rest of the cDNA in liquid nitrogen and store at −20°C.

3.8. Amplification of cDNA Coding for DARPins

The standard protocol for Vent DNA polymerase (NEB) is shown below. If another DNA polymerase or primers are used, the reaction conditions might have to be adapted.

1. Set up the following reaction mix per sample:

2–5 μL	cDNA from Subheading 3.7, step 6
5.0 μL	10× Thermopol buffer (NEB)
2.0 μL	dNTPs (final concentration 200 μM of each nucleotide)
2.5 μL	DMSO (final concentration 5%)
0.5 μL	WTC4 primer (final concentration 1 μM)
0.5 μL	EWT5s primer (final concentration 1 μM)
0.5 μL	Vent DNA Polymerase (2 U/μL)
Add to 50 μL with H$_2$O (see Note 22)	

2. Perform a hot-start PCR to reduce unspecific amplification. Use the following cycling parameters: 3 min at 95°C, 25 cycles:

30 s at 95°C, 30 s at 55°C, 45 s at 72°C, final extension 5 min at 72°C (see Note 23).

3. Verify the product on an agarose gel (see Note 24).

3.9. Incorporation of Promoter Elements, RBS, tolA Spacer, and RNA-Stabilizing Stem Loops

1. Purify PCR products from Subheading 3.8, step 3, either by excision of the according bands from the agarose gel and subsequent purification or in later rounds, when only one single band is observed, by direct purification over commercially available columns, e. g., of the NucleoSpin extract II kit. Elute in a small volume of 20 µL.

2. Digest ≥150 ng of the PCR product with the corresponding restriction enzymes, e.g., NcoI and HindIII for DARPin selections, in a final volume of 30 µL at 37°C for 2 h (see Note 25).

3. Purify digested PCR product using the NucleoSpin® Extract II DNA purification kit. Elute in 15 µL elution buffer supplied with the kit.

4. Ligate the PCR fragments into the ligation-ready pRDV plasmid using 100 ng of digested pRDV and the digested PCR product with a molar ratio of vector to insert of 1:5–7 in a final volume of 10 µL. Add 1 U of T4 DNA ligase and 1 µL ligase buffer. Incubate for 30–60 min at room temperature. Use this ligation as PCR template with the T7B and tolAk primers (see Subheading 3.2) or perform an error-prone PCR to increase diversity (see Subheading 3.12.1).

3.10. Initial Analysis of Selected Individual Library Members in a 96-Well Format (28)

1. After RT-PCR (see Subheadings 3.7 and 3.8), prepare the DARPin pool after enrichment has been observed for subcloning into a prokaryotic expression plasmid using the endonucleases BamHI and HindIII. Enrichment is indicated by a much stronger PCR band recovered from a well with immobilized target than from a control well without immobilized target.

2. Ligate the PCR fragment into pDST67 (22, 28) as fusion with the sequence coding for an N-terminal MRGS(H)$_6$ tag for purification.

3. After transformation of E. coli XL1-Blue, pick single clones and inoculate in deep 96-well plates in 1 mL 2×TY/1% glucose/amp (100 µg/mL) and grow overnight at 37°C while shaking at 540 rpm on an orbital shaker.

4. Transfer 100 µL of each culture to 900 µL fresh media and grow for 1 h at 37°C.

5. Induce with 0.5 mM IPTG (add 100 µL media containing 5.5 mM IPTG) and grow for an additional 3–5 h at 37°C.

6. Harvest cells by centrifugation at 400×g for 10 min, and discard supernatant.

7. Resuspend pellet in 50 µL B-PER II detergent and lyse cells for 15–30 min on an orbital shaker.

8. Add 1 mL TBST/0.1% BSA and centrifuge to remove debris.

9. Use 10 to 100 μL of the crude extract for ELISA (see step 13). If high-affinity binders are expected, e.g. after affinity maturation, using 10 μL of a 1:100 predilution of the extract with TBST/0.1% BSA can give you a better indication of the affinity of the binders.

10. For ELISA, coat wells with 100 μL of 66 nM neutravidin in TBS for 1 h at room temperature or overnight at 4°C. Wash two times with 300 μL TBS. Dry plate on paper towels after each step.

11. Block with 300 μL TBST/0.5% BSA for 1 h at room temperature.

12. Invert plate and shake out liquid and immediately add 100 μL of the biotinylated target (10–100 nM) in TBST/0.1% BSA. Incubate for 1 h at 4°C or room temperature on an orbital shaker (see Note 26). Wash three times with 300 μL TBST.

13. Add 100 μL DARPin extract from step 9. Incubate for 1 h at 4°C or room temperature on an orbital shaker. Wash three times with 300 μL TBST.

14. Add 100 μL mouse-anti-RGS(His)$_4$ antibody in a 1:5000 dilution. Incubate for 1 h at 4°C or room temperature on an orbital shaker. Wash three times with 300 μL TBST.

15. Add 100 μL goat-anti-mouse antibody coupled to alkaline phosphatase in a 1:20,000 dilution. Incubate for 1 h at 4°C or room temperature on an orbital shaker. Wash three times with 300 μL TBST.

16. Add 100 μL pNPP substrate solution. Incubate until color development and determine OD_{405nm}.

3.11. Preparation of S30 Extract (31–33)

1. Grow a 100-mL culture of *E. coli* MRE600 in incomplete rich medium overnight at 37°C.

2. Transfer 10 mL of the overnight culture in 1 L of fresh media in a 5-L baffled shaker flask and grow until OD_{600nm} of 1.0–1.2 at 37°C while shaking. This procedure can be scaled up to your needs and 1-L culture usually yields 10–15 mL of S30 extract. The S30 extract is stable for years when stored at −80°C.

3. Chill cultures for 10 min on an ice water bath with gentle shaking.

4. Centrifuge cells at 3,500×*g* at 4°C for 15 min and discard supernatant.

5. Wash the pellet three times with 50 mL of ice-cold S30 buffer per 1-L culture. It is best to resuspend cells with plating beads or on a magnetic stirrer using a sterile magnetic stir bar.

6. Freeze the cell pellet in liquid nitrogen and store for a maximum of 2 days at −80°C or continue immediately.

7. Resuspend the cell pellet (use 50 mL ice-cold S30 buffer per 1 L of culture), centrifuge at $4,000 \times g$. Discard supernatant and resuspend pellet in 4 mL S30 buffer per g wet cells (typically, 1 L of culture yields 1.5–2.0 g cell pellet).

8. Lyse the cells by one single passage through a French press applying 1,000 psi or an EmulsiFlex at ~17,000 psi.

9. Centrifuge cells at 20,000 rpm (SS-34 in a Sorvall centrifuge) at 4°C for 30 min. Transfer supernatant to clean centrifuge bottle(s) and repeat this step.

10. Add 1 mL of preincubation mix to each 6.5 mL of cleared supernatant (usually, 1-L culture yields 8–10 mL of S30 extract) and slowly shake at 25°C for 1 h. In this time, all endogenous mRNAs are translated and cellular nucleases degrade mRNA and DNA (34).

11. Dialyze the S30 extract (MW cutoff of 6,000–8,000 Da) against a 50-fold volume of S30 buffer at 4°C three times for 4 h.

12. Centrifuge S30 extract at $6,000 \times g$ in a tabletop centrifuge at 4°C for 10 min. If the library members and target are devoid of disulfide bonds, 1 mM DTT can be added to the extract. Aliquot at 4°C in suitable volumes (e.g., 55 µL is sufficient for one in vitro translation reaction, 110 µL for two) since it should not be refrozen to guarantee best activity. Flash freeze in liquid nitrogen and store at –80°C.

3.12. Affinity Maturation

To increase the affinities of the library members, it is best to select for those having the lowest dissociation rate constant (off rate) from the target (7–9, 35–37). This off-rate selection can be applied for the improvement of known binders (after mutagenizing the gene for defined binders and thus creating a new library), but also during the initial selection from the original library. In this off-rate selection step, an excess of nonbiotinylated target is added after the binding reaction to the biotinylated target has already been equilibrated for >1 h. Any fast dissociating binder is immediately occupied by nonbiotinylated target and thereby prevented from being captured with biotinylated target on streptavidin or neutravidin. Conversely, any high-affinity binder with a slow off rate retains its biotinylated target and thus can be captured. The optimal duration of competitor incubation and the excess concentrations depend on the expected off rates. Considerations as to which parameters to choose have been recently discussed (13). As a general guideline, we recommend to perform the affinity maturation over several selection blocks (usually, three blocks seem sufficient) each containing a round of randomization (see Subheading 3.12.1), a round of off-rate selection (see Subheading 3.12.2), and a low-stringency round (see Subheading 3.5.3) for recovery of rare tight binders from a high background of unselected library members. For the off-rate selections in block 1, we recommend to start with a modest stringency that can be increased in block 2 and 3, for example: for the first off-rate selection, use a 2-h incubation with a 10- to 100-fold excess of competitor, later use a 1,000- to 10,000-fold excess of competitor. Then, proceed with washing and elution of the bound ternary complexes as above.

3.12.1. Introduction of Additional Diversity Applying Error-Prone PCR

1. Set up PCR reactions on template DNA from Subheading 3.9, step 4, introducing different mutational rates using various concentrations of the nucleotide analogs dPTP and 8-oxo-dGTP in the range of 1 to 20 µM (see Note 27):

1 µL	pRDV_DARPin template (10 ng/µL)
4 µL	dNTPs each (final concentration 250 µM)
1–20 µM	dPTP and 8-oxo-dGTP, each
0.5 µL	T7B primer (final concentration 1 µM)
0.5 µL	tolAk primer (final concentration 1 µM)
5 µL	10× polymerase buffer
3 µL	$MgCl_2$ (final concentration 1.5 mM)
0.2 µL	Platinum® Taq DNA polymerase
Add to 50 µL with H_2O	

2. Apply the following cycling parameters (must be adapted according to primers and template): 3 min at 95°C, 25 cycles: 30 s at 95°C, 30 s at 50°C, 1 min at 72°C, final extension 5 min at 72°C.

3. Verify the product on an agarose gel.

4. Mix PCR products in equimolar amounts to serve as template for the in vitro transcription (see Subheading 3.2, step 4).

3.12.2. Competition with Non-labeled Target (Off-Rate Selection)

This protocol describes a selection strategy to enrich binders with a slow off rate.

Ribosomal complexes are incubated with low amounts of biotinylated target (in solution or immobilized) before adding unbiotinylated target in large excess as competitor. Considerations for selection conditions have recently been published (see ref. 13, Note 28).

1. For prepanning to remove all sticky ribosomal complexes, e.g., containing misfolded DARPins after randomization, add two times 500 µL diluted and stopped translation mix (*from* Subheading 3.4, step 4; see Subheading 3.5.3, step 1) to 20–50 µL of BSA-blocked streptavidin magnetic beads in a blocked, RNAse-free, 2.0-mL tube. Remember to set up two reactions, one containing the target and one not containing the target as negative control.

2. Incubate at 4°C for 30–60 min with head-over-end rotation.
3. After separation of the magnetic beads on a magnetic stand, remove the supernatant carefully and transfer the translation mix to a BSA-blocked 2.0-mL tube. Add biotinylated target in the range of 0.1–10 nM to the selection reaction (see Note 28) and buffer only to the tube containing the control.
4. Allow for equilibration of the DARPin/target complexes at 4°C with head-over-end rotation for 1–14 h.
5. For competition of the complexed DARPins, add a large excess of nonbiotinylated target. The ratio varies dependent on the expected affinity of the binders in the pool (see ref. 13, Note 28).
6. Incubate at 4°C with head-over-end rotation for 1–14 h.
7. Add the binding reactions to 20–50 µL fresh streptavidin magnetic beads previously blocked with BSA in a blocked 2.0-mL tube and capture the complexes remaining on the biotinylated target during a 30-min incubation rotating at 4°C.
8. Wash with 500 µL ice-cold WBT containing 0.1% BSA in each step as indicated (see Subheading 3.5). Separate captured complexes using a magnetic separator between each washing step.
9. Proceed with the elution and purification of RNA as described (see Subheading 3.5.2, step 5).

4. Notes

1. Use RNAse-free water, chemicals, and consumables. Most commercially available water is RNase free or can be generated using a membrane microfiltration system, e.g., MilliQ from QIAGEN, to produce ultrapure water. Alternatively, you can use 0.1% diethylpyrocarbonate (DEPC) which reacts with histidine residues, but also other nucleophilic groups, and therefore inactivates RNases, but for the same reason cannot be used for, e.g., Tris buffers. Chemicals used for RNA should be kept separate from the common chemical shelf and handled only with gloves and a flamed spatula to avoid RNase contamination. Purchase only RNase-free plastic consumables. If necessary, you can bake glass bottles and pipettes at 180°C for 6 h.
2. The buffer composition may be adjusted to the requirements of the library and target, but it is important that the wash buffer contains 50 mM Mg^{2+} to stabilize the ribosome. It is recommended to test buffer conditions with a known binder to ensure stability of the nascent chain complex.
3. Different reverse transcriptases (AffinityScript™ Multiple Temperature Reverse Transcriptase, SuperScript™ II

(Invitrogen, No. 18064-022), ThermoScript™ (Invitrogen, No. 12236-014) and QuantiTect (QIAGEN, No. 205310)) were tested for efficiency on DARPin sequences. With exception of QuantiTect, the yield obtained was comparably high with all other reverse transcriptases.

4. Previously Phusion™ High-Fidelity DNA Polymerase (New England Biolabs) has also been used (31, 35). Different DNA polymerases were tested, e.g., Vent$_R$® DNA Polymerase (New England Biolabs), Herculase® II Fusion DNA Polymerase (Stratagene), and Expand High Fidelity PCR System (Roche Diagnostics). The DNA polymerase mix from the Expand Hi Fidelity PCR System gave the lowest yield of PCR product while Herculase II gave the highest amount of side products. Therefore, we now routinely use Vent$_R$® DNA Polymerase for amplification of DARPin sequences. Since the yield was highest with the Herculase II, it might be a good alternative to increase the yield of PCR product or for amplification of other library scaffolds.

5. Use the homemade RNA polymerase buffer (see Subheading 2.5) as indicated for maximum yield of RNA. Commercial buffers have not worked very well at this step when the PCR product is directly used without further purification.

6. In round one, ensure that the number of molecules actually exceeds the library size, but no more members than ribosomes present in the translation reaction can be displayed. Nonetheless, under standard conditions as described here, ribosomes should be in excess. In a newly constructed library, the diversity cannot be higher than the number of DNA molecules used in this step. What limits the functional library in ribosome display is also discussed elsewhere (5). In later rounds, an enrichment is obtained, and it is generally sufficient to use ~50 ng of pRDV_DARPin template.

7. The PCR products can be used without additional purification. We highly recommend to use nonpurified PCR product at this step, since purified PCR product generally yields a greatly reduced amount of mRNA.

8. Optionally, the transcribed RNA can be analyzed on a denaturing formaldehyde agarose gel following standard procedures (38). The mRNA product should give a sharp band. A smear or no product indicates RNase contamination, which needs to be eliminated and the step repeated. If the band is sharp but the yield is lower than expected, obtain more starting DNA template by not purifying the PCR product that is used as template, as the quality is usually sufficient even without purification (see Note 7), and do use the homemade RNA polymerase

buffer (see Subheading 2.4) for better transcription yield. If the products are not of the expected size, optimize the PCR conditions depending on your template and primers.

9. In our experience, both protocols are yielding high-quality RNA as template for the in vitro translation, but the quality might be still higher using LiCl precipitation (see Subheading 3.3.1). Considerations on which protocol to use might also be the final concentration that is usually obtained (using gel filtration the sample is usually more dilute) or the time it takes (LiCl precipitation is performed over a time frame of 3 h while the purification using gel filtration (see Subheading 3.3.2) can be performed in 10 min).

10. In some cases, the template DNA itself might bind unspecifically to the target, e.g., if the target is highly positively charged, and then it is recommended to remove this contamination by DNAse I treatment before the actual selection. Always freeze small aliquots of DNAse I and store at −20°C. Do not refreeze or vortex solutions containing DNAse I because the enzyme is very sensitive to denaturation.

11. Always freeze RNA immediately after use; only thaw when needed to avoid degradation.

12. For the selection, some general considerations need to be pointed out. Always use the same target preparation through all of the selection and screening rounds, ensure its quality, and account for its stability over the duration of the experiment. If the target denatures, epitopes present in the native protein will vanish, and such binders will be lost. Account for high diversity, especially in the first round, by using sufficient amounts of the starting library. Start selections with a higher number of DNA template molecules than the diversity of the library. Be aware that no matter how large the library is the limitation of molecules that can be displayed depends on the number of ribosomes present in the translation reaction. Fortunately, under the conditions used here, more ribosomes than input DNA molecules are used (5). To extract all putative binders in the library, use a larger surface area to immobilize the target in the first round. The first round should, in general, not be highly selective; it is more important to capture the full diversity of binders, as a binder lost at this stage can never be recovered. In general, it is recommended to perform the selection in duplicates to monitor the selection quality. It is recommended to switch between neutravidin (a chemically modified derivative of avidin) and streptavidin or even switch between selections on immobilized target and target in solution during the selection process to focus selection on binding to the target, rather than on streptavidin/neutravidin or any other surface features. If high-affinity binders in the pM range are needed

(see Subheading 3.12), include the introduction of additional random mutations using error-prone PCR (see Subheading 3.12.1) and increase stringency by applying off-rate selections (see Subheading 3.12.2, Note 27).

Some applications, e.g., for therapy, require high stability of the therapeutic agent (39). Selection for high stability can also be achieved with ribosome display. This is best achieved by first making the whole population unable to fold, by deliberately introducing a reversible destabilization, e.g. by mutating a critical residue, then selecting for compensating mutations, and finally removing the destabilization again. For example, most antibody domains require disulfides for stability, which form only under oxidizing conditions. A destabilization of the antibody fold and increase in aggregation are usually observed when the disulfides are removed (40, 41). Using a reducing environment during the selection, scFv antibody fragments could be evolved that were able to fold under reducing conditions, correlating with conditions in the cytosol, and they showed higher stability than the starting molecule in the absence of the disulfide bonds (35), but also after the disulfide bonds were allowed to form again. Antibody fragments with these improved biophysical property can be used in biomedical applications with disulfides formed, but they also make an intracellular application (as "intrabodies") (42) more feasible. In addition, rational design of the antibody framework (43) could contribute to the development of stability-improved, antibody-based therapeutics.

13. Using the AviTag has the advantage that all biotinylated proteins are labeled uniformly and remote from epitopes which might interfere with later use and are labeled only once, leading to a more homogenous target preparation. Avoid the presence of a Met-Arg-Gly-Ser-$(His)_6$ tag ("RGS-His-tag") on the biotinylated target; rather, use a $(His)_6$ tag for purification, since the detection of DARPins bound to the target is performed using an anti-RGS$(His)_4$ antibody (see Subheading 3.10). Make sure that your target sample is devoid of free biotin. Biotin removal requires an extensive dialysis, for example four times against a 100-fold volume buffer for 4 h each. Nonbiotinylated target can be removed using a monomeric avidin column following the manufacturer's instructions (Pierce, No. 53146).

14. Use one well as nontarget control and two wells with immobilized target in later rounds as mutual controls for enrichment. When starting from the libraries in round one, it is recommended to use a larger surface, e.g., four wells with immobilized target.

15. To remove unspecifically binding ribosomal complexes, it is recommended to use a preselection on BSA-blocked wells coated only with neutravidin or streptavidin, but omitting the

target protein, except for round one, where this "prepanning" should not be performed. For prepanning, the preparation of additional wells and incubation of the ternary complexes from the in vitro transcription for 30–60 min is necessary before transferring the solution to the target-coated or control wells.

16. The amount of target can be reduced in later rounds to 1–20 nM.

17. The amount of target can be reduced to 100 pM, e.g., when performing an off-rate selection, and thus a high amount of competitor can be added. At lower target concentrations, the unspecific binding might prevail over target binding, however, and thus specificity of binding must be carefully controlled.

18. RNA isolation can also be performed with the RNeasy mini kit (QIAGEN) with comparable yield of resulting PCR product.

19. This step is highly recommended to avoid amplification of nonselected template DNA that has been carried over through all steps of the selection procedure. See also Note 10 for handling of DNAse I.

20. The RNA should stable for years at –80°C, but we recommend to immediately proceed with cDNA synthesis and PCR amplification for best recovery of sequences of putative binders.

21. Use one sample without addition of reverse transcriptase as control. The result of the following PCR is a measure for the quality of the selection regarding DNA carryover from the input DNA and putative overcycling (see Notes 19 and 23).

22. Always use one reaction containing no template, but all other components. Appearance of a band in this reaction indicates a contamination in one of the selection/amplification reagents. In our experience, the main candidate is the water used. Replace all the reagents immediately to prevent carryover of DNA of unwanted, unselected clones. To minimize expenses, it is recommended to store aliquots of all the reagents before starting selection.

23. Depending on the round of selection, more or fewer cycles could be advantageous. In the first round, 32–40 cycles are recommended to obtain sufficient product, since only a few clones have the desired properties. After more rounds of selection, specific binders are being enriched; therefore, the output of eluted RNA molecules increases. By lowering the cycle numbers in round 2 to between 28 and 35 and in all following rounds to 25, unspecific amplification can be reduced to a minimum. In addition, note that when the selection pressure increases, for example, after off-rate selection, the yield of PCR product might decrease. In this case, use more cycles.

24. If the quality and amount (<10 ng/μL) of the PCR product was not satisfactory, repeat the PCR. Never reamplify the PCR

product because this might lead to unspecific amplification of unwanted by-products.

25. In parallel, digest the ribosome display vector (pRDV) with the same restriction enzymes, e.g., *Nco*I and *Hin*dIII, for DARPin selections. Purify the plasmid backbone using extraction from a preparative agarose gel. It is recommended to use a larger preparation to last for several selection rounds and/or multiple target selections: Test the quality of the digested plasmid by ligation and transformation and/or PCR on the ligation mix to evaluate the level of religation and therefore quality of the ligation-ready plasmid.

26. The incubation temperature depends on the stability of the target and library scaffold.

27. Add different concentrations of the nucleotide analogs, for example 0, 3, and 10 μM. Up to 20 μM can be used, but the amount of product is greatly reduced at this concentration. The mutational load per kb under the conditions described is 1.5 mutations with 1 μM nucleotide analogs and 3.2 mutations with 3 μM nucleotide analogs. These numbers refer to fresh nucleotides and can vary if the nucleotides are no longer incorporated well, e.g., by hydrolysis of the triphosphate. The use of a low to medium mutational load per selection, but repeating over several rounds, might be beneficial over a high mutational load which might result in a high number of misfolded library members in the pool.

28. Perform one cycle of nonstringent selection, including an error-prone PCR followed by a round of off-rate selection without error-prone PCR. The rationale is that error-prone PCR generates many nonfunctional molecules. First, *all* functional molecules should be recovered by a nonstringent selection, and then from this pool of functional (randomized) molecules the best ones should be recovered. Use these to perform a stringent round using off-rate selection. Here are some general considerations (13): Subsequent selection rounds with modest selection pressure are preferred over high-stringency selection rounds because this leads to higher diversity. Start at 10- to 100-fold excess competitor; increase to 100- to 10,000-fold in later rounds, if feasible. In any case, the highest possible ratio of unbiotinylated:biotinylated target should be used to maximize the selection outcome as there is a greater margin for error in selection time in the presence of competitor (as the optimal selection time is generally unknown). Too long an incubation time eliminates the kinetic selection pressure because the system is at or near equilibrium. An incubation for 2 h is usually a good starting point, but can be increased to 14 h. This longer incubation time should be favored if the amount of target is limited and thus used at very low concentrations (100 pM) and the

highest possible unbiotinylated:biotinylated target ratio is used (e.g., a 1,000- to 10,000-fold access of competitor).

These two rounds should be followed again by a nonstringent round without any additional selection pressure simply to amplify the rare molecules. Perform this cycle of error-prone PCR, off-rate selection, and nonstringent round two to three times before analyzing single clones (see Subheading 3.10). For stringent selections, use 0.1–10 nM (depends on the availability of biotinylated target and expected affinity of the clones), and for nonstringent selections use 100 nM biotinylated target.

Acknowledgments

We thank many former and current members of the Plückthun laboratory, mentioned in the references, for establishing and continuously optimizing the ribosome display protocol.

Support

Work on ribosome display was supported by the Swiss National Science Foundation.

References

1. Hanes, J. & Plückthun, A. (1997) *In vitro* selection and evolution of functional proteins by using ribosome display. *Proc. Natl. Acad. Sci. U S A* **94**, 4937–4942.
2. Hanes, J., Jermutus, L., Weber-Bornhauser, S., Bosshard, H. R. & Plückthun, A. (1998) Ribosome display efficiently selects and evolves high-affinity antibodies *in vitro* from immune libraries. *Proc. Natl. Acad. Sci. U S A* **95**, 14130–14135.
3. Mattheakis, L. C., Bhatt, R. R. & Dower, W. J. (1994) An in vitro polysome display system for identifying ligands from very large peptide libraries. *Proc. Natl. Acad. Sci. U. S. A.* **91**, 9022–9026.
4. Hanes, J., Jermutus, L. & Plückthun, A. (2000) Selecting and evolving functional proteins *in vitro* by ribosome display. *Methods Enzymol.* **328**, 404–430.
5. Plückthun, A. (2012) Ribosome Display: a perspective. *Methods Mol. Biol.* **805**, 3–28.
6. Hanes, J., Schaffitzel, C., Knappik, A. & Plückthun, A. (2000) Picomolar affinity antibodies from a fully synthetic naive library selected and evolved by ribosome display. *Nat. Biotechnol.* **18**, 1287–1292.
7. Luginbühl, B., Kanyo, Z., Jones, R. M., Fletterick, R. J., Prusiner, S. B., Cohen, F. E., Williamson, R. A., Burton, D. R. & Plückthun, A. (2006) Directed evolution of an anti-prion protein scFv fragment to an affinity of 1 pM and its structural interpretation. *J. Mol. Biol.* **363**, 75–97.
8. Zahnd, C., Wyler, E., Schwenk, J. M., Steiner, D., Lawrence, M. C., McKern, N. M., Pecorari, F., Ward, C. W., Joos, T. O. & Plückthun, A. (2007) A designed ankyrin repeat protein evolved to picomolar affinity to Her2. *J. Mol. Biol.* **369**, 1015–1028.
9. Dreier, B., Mikheeva, G., Belousova, N., Parizek, P., Boczek, E., Jelesarov, I., Forrer, P., Plückthun, A. & Krasnykh, V. (2011) Her2-specific multivalent adapters confer designed tropism to adenovirus for gene targeting. *J. Mol. Biol.* **405**, 410–426.
10. Stemmer, W. P. (1994) Rapid evolution of a protein in vitro by DNA shuffling. *Nature* **370**, 389–391.

11. Zaccolo, M., Williams, D. M., Brown, D. M. & Gherardi, E. (1996) An approach to random mutagenesis of DNA using mixtures of triphosphate derivatives of nucleoside analogues. *J. Mol. Biol.* **255**, 589–603.

12. Zahnd, C., Spinelli, S., Luginbühl, B., Amstutz, P., Cambillau, C. & Plückthun, A. (2004) Directed in vitro evolution and crystallographic analysis of a peptide binding scFv antibody with low picomolar affinity. *J. Biol. Chem.* **279**, 18870–18877.

13. Zahnd, C., Sarkar, C. A. & Plückthun, A. (2010) Computational analysis of off-rate selection experiments to optimize affinity maturation by directed evolution. *Protein Eng. Des. Sel.* **23**, 175–184.

14. Schaffitzel, C., Zahnd, C., Amstutz, P., Luginbühl, B. & Plückthun, A. (2001). In vitro selection and evolution of protein-ligand interactions by ribosome display. In *Protein-Protein Interactions, A Molecular Cloning Manual* (Golemis, E., ed.), pp. 535–567. Cold Spring Harbor Laboratory Press, New York.

15. Schaffitzel, C., Zahnd, C., Amstutz, P., Luginbühl, B. & Plückthun, A. (2005). In vitro selection and evolution of protein-ligand interactions by ribosome display. In *Protein-protein interactions: A molecular cloning manual* 2nd edit. (Golemis, E. & Adams, P., eds.), pp. 517–548. Cold Spring Harbor Laboratory Press, Cold Spring Harbor, NY.

16. Dreier, B. & Plückthun, A. (2010) Ribosome Display, a technology for selecting and evolving proteins from large libraries. *Methods Mol. Biol.* **687**, 283–306.

17. Binz, H. K., Stumpp, M. T., Forrer, P., Amstutz, P. & Plückthun, A. (2003) Designing repeat proteins: well-expressed, soluble and stable proteins from combinatorial libraries of consensus ankyrin repeat proteins. *J. Mol. Biol.* **332**, 489–503.

18. Binz, H. K., Amstutz, P., Kohl, A., Stumpp, M. T., Briand, C., Forrer, P., Grütter, M. G. & Plückthun, A. (2004) High-affinity binders selected from designed ankyrin repeat protein libraries. *Nat. Biotechnol.* **22**, 575–582.

19. Amstutz, P., Binz, H. K., Parizek, P., Stumpp, M. T., Kohl, A., Grütter, M. G., Forrer, P. & Plückthun, A. (2005) Intracellular kinase inhibitors selected from combinatorial libraries of designed ankyrin repeat proteins. *J. Biol. Chem.* **280**, 24715–24722.

20. Zahnd, C., Pécorari, F., Straumann, N., Wyler, E. & Plückthun, A. (2006) Selection and characterization of Her2 binding-designed ankyrin repeat proteins. *J. Biol. Chem.* **281**, 35167–35175.

21. Schweizer, A., Roschitzki-Voser, H., Amstutz, P., Briand, C., Gulotti-Georgieva, M., Prenosil, E., Binz, H. K., Capitani, G., Baici, A., Plückthun, A. & Grütter, M. G. (2007) Inhibition of caspase-2 by a designed ankyrin repeat protein: specificity, structure, and inhibition mechanism. *Structure* **15**, 625–636.

22. Huber, T., Steiner, D., Röthlisberger, D. & Plückthun, A. (2007) In vitro selection and characterization of DARPins and Fab fragments for the co-crystallization of membrane proteins: The Na(+)-citrate symporter CitS as an example. *J. Struct. Biol.* **159**, 206–221.

23. Veesler, D., Dreier, B., Blangy, S., Lichière, J., Tremblay, D., Moineau, S., Spinelli, S., Tegoni, M., Plückthun, A., Campanacci, V. & Cambillau, C. (2009) Crystal structure of a DARPin neutralizing inhibitor of lactococcal phage TP901-1: comparison of DARPin and camelid VHH binding mode *J. Biol. Chem.* **384**, 30718–30726.

24. Schatz, P. J. (1993) Use of peptide libraries to map the substrate specificity of a peptide-modifying enzyme: a 13 residue consensus peptide specifies biotinylation in Escherichia coli. *Biotechnology (N. Y.)* **11**, 1138–1143.

25. Wade, H. E. & Robinson, H. K. (1966) Magnesium ion-independent ribonucleic acid depolymerases in bacteria. *Biochem. J.* **101**, 467–479.

26. Laminet, A. A. & Plückthun, A. (1989) The precursor of β-lactamase: Purification, properties and folding kinetics. *EMBO J.* **8**, 1469–1477.

27. O'Callaghan, C. H., Morris, A., Kirby, S. M. & Shingler, A. H. (1972) Novel method for detection of beta-lactamases by using a chromogenic cephalosporin substrate. *Antimicrob. Agents Chemother.* **1**, 283–288.

28. Steiner, D., Forrer, P. & Plückthun, A. (2008) Efficient selection of DARPins with sub-nanomolar affinities using SRP phage display. *J. Mol. Biol.* **382**, 1211–1227.

29. Zahnd, C., Amstutz, P. & Plückthun, A. (2007) Ribosome display: selecting and evolving proteins in vitro that specifically bind to a target. *Nat. Methods* **4**, 269–279.

30. Hajnsdorf, E., Braun, F., Haugel-Nielsen, J., Le Derout, J. & Regnier, P. (1996) Multiple degradation pathways of the rpsO mRNA of Escherichia coli. RNase E interacts with the 5′ and 3′ extremities of the primary transcript. *Biochimie* **78**, 416–424.

31. Amstutz, P., Binz, H. K., Zahnd, C. & Plückthun, A. (2006). Ribosome Display: In Vitro Selection of Protein-Protein Interactions. In *Cell Biology – A Laboratory Handbook* 3rd edit. (Celis, J., ed.), Vol. 1, pp. 497–509. 4 vols. Elsevier Academic Press.

32. Chen, H. Z. & Zubay, G. (1983) Prokaryotic coupled transcription-translation. *Methods Enzymol.* **101**, 674–690.
33. Pratt, J. M. (1984). Coupled transcription-translation in prokaryotic cell-free systems. In *Current Protocols* (Hemes, B. D. & Higgins, S. J., eds.), pp. 179–209. IRL Press, Oxford.
34. Kushner, S. R. (2002) mRNA decay in Escherichia coli comes of age. *J. Bacteriol.* **184**, 4658–4665; discussion 4657.
35. Jermutus, L., Honegger, A., Schwesinger, F., Hanes, J. & Plückthun, A. (2001) Tailoring *in vitro* evolution for protein affinity or stability. *Proc. Natl. Acad. Sci. U.S.A.* **98**, 75–80.
36. Hawkins, R. E., Russell, S. J. & Winter, G. (1992) Selection of phage antibodies by binding affinity. Mimicking affinity maturation. *J. Mol. Biol.* **226**, 889–896.
37. Yang, W. P., Green, K., Pinz-Sweeney, S., Briones, A. T., Burton, D. R. & Barbas, C. F., 3rd. (1995) CDR walking mutagenesis for the affinity maturation of a potent human anti-HIV-1 antibody into the picomolar range. *J. Mol. Biol.* **254**, 392–403.
38. Sambrook, J. & Russel, D. W. (2001). *Molecular cloning: A laboratory handbook*. 3rd edit, Cold Spring Harbor Laboratory Press, Cold Spring Harbor, NY.
39. Willuda, J., Honegger, A., Waibel, R., Schubiger, P. A., Stahel, R., Zangemeister-Wittke, U. & Plückthun, A. (1999) High thermal stability is essential for tumor targeting of antibody fragments: Engineering of a humanized anti-epithelial glycoprotein-2 (epithelial cell adhesion molecule) single-chain Fv fragment. *Cancer Res.* **59**, 5758–5767.
40. Wörn, A. & Plückthun, A. (2001) Stability engineering of antibody single-chain F_v fragments. *J. Mol. Biol.* **305**, 989–1010.
41. Proba, K., Wörn, A., Honegger, A. & Plückthun, A. (1998) Antibody scFv fragments without disulfide bonds made by molecular evolution. *J. Mol. Biol.* **275**, 245–253.
42. Chames, P. & Baty, D. (2000) Antibody engineering and its applications in tumor targeting and intracellular immunization. *FEMS Microbiol. Lett.* **189**, 1–8.
43. Honegger, A., Malebranche, A. D., Röthlisberger, D. & Plückthun, A. (2009) The influence of the framework core residues on the biophysical properties of immunoglobulin heavy chain variable domains. *Protein Eng. Des. Sel.* **22**, 121–134.

Chapter 16

mRNA Display-Based Selections Using Synthetic Peptide and Natural Protein Libraries

Steve W. Cotten, Jianwei Zou, Rong Wang,
Bao-cheng Huang, and Rihe Liu

Abstract

mRNA display is a powerful in vitro selection technique that can be applied toward the identification of peptides or proteins with desired properties. The physical conjugation between a protein and its own RNA presents unique challenges in manipulating the displayed proteins in an RNase-free environment. This protocol outlines the generation of synthetic peptide and natural proteome libraries as well as the steps required for generation of mRNA–protein fusion libraries, in vitro selection, and regeneration of the selected sequences. The selection procedures for the identification of Ca^{2+}-dependent, calmodulin-binding proteins from synthetic peptide and natural proteome libraries are presented.

Key words: mRNA display, Genotype–phenotype conjugation, In vitro selection, Synthetic combinatorial peptide library, Natural proteome library, Conditional protein–protein interaction

1. Introduction

mRNA display is a genotype–phenotype conjugation method that allows the amplification-based, iterative rounds of in vitro selection to be applied to peptides and proteins (1–4). mRNA display can be used to display both long natural proteome and short synthetic peptide libraries with high diversity. Compared to prior peptide or protein selection methods, mRNA display has several major advantages. First, the genotype is covalently linked to and is always present with the phenotype. This stable linkage makes it possible to use any desired conditions during the selection process and titrate the stringency of the selection. Second, the complexity of the peptide or protein library can be close to that of the RNA or DNA pools. Peptide or protein libraries containing as many as 10^{12}–10^{14} unique sequences can be readily generated and selected, a few orders of magnitude higher than that can be achieved using phage display and other selection platforms. Therefore, both the likelihood of isolating rare sequences and the diversity of the sequences isolated in a given selection are significantly increased. In principle, mRNA display can be used for any in vitro selection that aims at identifying peptide or protein sequences with desired properties. The various applications of mRNA display can be classified based on the type of library constructed for the selection. The first type of libraries that can be used for mRNA display-based selection are synthetic combinatorial peptide libraries. These libraries can be designed from structured protein scaffolds containing totally or partially randomized amino acids on surface loops or from unstructured peptides consisting of randomized residues (5–12). The second type of libraries are natural proteome libraries derived from the mRNAs of any organism, tissue, or treatment (i.e., drug/environmental insult) of interest (13–21). To date, mRNA display has been successfully applied in the identification of drug-binding targets, mapping of the protein–protein interactions and DNA–protein interaction networks, elucidation of the enzyme–substrate interactions, and improvement of the binding affinities of existing affinity molecules (13–21).

Much insight into the function of a protein can be gained by studying its interaction with other proteins. Often, such protein–protein interactions only occur under specific conditions. One effective strategy to get a thorough understanding of a protein of interest is to map its conditional protein–protein interactions on a proteome-wide scale. mRNA display can provide a global picture of conditional protein–protein interactions and allows for simultaneous search of the sequence space to understand the nature and specificity of the target protein with its natural or synthetic interacting partners.

Functional selections using mRNA display approach can be challenging due to the necessity of manipulating from nanomolar to low micromolar amounts of radiolabeled proteins in an RNase-free environment. The success of a selection from a highly diversified synthetic peptide or natural proteome library displayed on its own mRNA relies on a selection scheme that allows for specific enrichment of sequences with desired properties while minimizing the isolation of nonspecific sequences. In general, immobilization of a target of interest followed by competitive elution of bound molecules using an excess of unmodified target is an effective approach for specific enrichment. If the protein interaction is conditional, binding and elution steps for the selection can be designed such that the presence or absence of small molecules or conditions, such as Ca^{2+}, cofactors, light, temperature, or ionic strength, dictate when molecules are released from the target. These "binary" binding events used for mRNA display selection are very effective at rapidly enriching target sequences. Selections that use binding events, where conditions that achieve specificity are not yet known, often take many more cycles of selection and can be challenging to enrich the desired functional sequences from nonspecific ones (see Note 1).

We describe here the procedures of using mRNA display to perform two selections against the same target from both a natural proteome and a synthetic combinatorial peptide library, so the interacting partners from both natural proteome and synthetic sequence space can be examined. Specifically, we use the identification of Ca^{2+}-dependent binding synthetic peptides or natural proteins against calmodulin, the major transducer of Ca^{2+} signals in eukaryotes, as an example. Since the desired binding requires Ca^{2+} ion, EGTA can be used to effectively remove Ca^{2+} from solution, resulting in a conformational change in calmodulin and specific release of the bound mRNA–protein fusion molecules under very mild conditions. The selection scheme for this conditional protein–protein interaction is illustrated in Fig. 1. For other protein–protein interactions, gentle and competitive elution should be applied whenever possible.

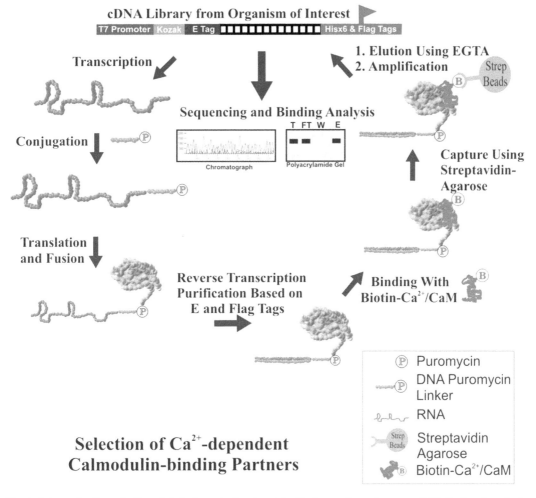

Fig. 1. Scheme for the selection of Ca^{2+}-dependent, calmodulin-binding partners from an mRNA-displayed synthetic peptide or natural proteome library.

2. Materials

2.1. Reagents

1. Expand long template PCR system (Roche).
2. T7 RNA polymerase (NEB).
3. RNase-free DNase (Promega).
4. Retic lysate IVT™ kit (Ambion).
5. [^{35}S]-L-methionine (Perkin Elmer).
6. Oligo(dT) cellulose (Ambion).
7. RNase-free 10 mL poly-prep chromatography column (Biorad).
8. SuperScript II RNase H$^-$ reverse transcriptase (Invitrogen).
9. Anti-FLAG M2 affinity gel (Sigma).
10. FLAG peptide (Sigma).
11. Biotinylated calmodulin (CalBioChem).
12. Binding buffer: 50 mM Tris–HCl, pH 7.5, 150 mM NaCl, 0.05% Tween-20, 1 mg/mL BSA, 5 mM 2-mercaptoethanol, and 0.5 mM $CaCl_2$.
13. Elution buffer: 50 mM Tris–HCl, pH 7.5, 150 mM NaCl, 0.05% Tween-20, 1 mg/mL BSA, 5 mM 2-mercaptoethanol, and 2 mM EGTA.

2.2. Equipment

1. Thermal cycler.
2. Nanodrop spectrophotometer.
3. UV lamp (Black Ray Lamp 365 nm, 0.16 Amps).
4. Barnstead Labquake Shaker/Rotator.
5. Scintillation counter.

3. Methods

3.1. Construction of Synthetic cDNA Library Coding Synthetic Combinatorial Peptide Library

For the synthetic peptide library, a high-quality cDNA library is synthesized and assembled according to the published protocol (5, 10). Each sequence in the cDNA library contains a T7 RNA polymerase promoter, a TMV translation enhancer sequence, an N-terminal FLAG tag coding sequence, a random cassette encoding 20 consecutive random codons, and a His × 6 tag coding sequence. The amino acid compositions in the random region can be designed to be close to the natural proteins or contain any residues of interest at desired positions and levels while the possibility of coding stop codons is minimized (5). The random region can be readily synthesized by oligo synthesizer using three phosphoramidite mixtures with appropriate proportions of each of the four phosphoramidites in the DNA synthesis, corresponding to each of the three positions in the random codons (5). Consensus sequences are flanked on both sides of the random region to facilitate annealing with other oligos and PCR amplification for assembly of the final library.

3.2. Construction of Natural cDNA Library Coding Natural Proteome Library

Natural cDNA libraries suitable for mRNA display-based selection can be generated from any organisms or tissues, where approximately 1.0 μg of high-quality mRNA can be isolated. Specifically, high-quality poly (A⁺) mRNA sequences are prepared by removing genomic DNAs, ribosomal RNAs, and tRNAs through three rounds of stringent oligo(dT) purification. If the mRNAs of interest do not contain poly (A⁺) tail (i.e., bacterial mRNAs), appropriate approaches, such as use of a combination of microExpress bacterial mRNA enrichment kit and RiboPure-Bacteria kit (Ambion), should be employed for at least three cycles to rigorously remove ribosomal RNAs and tRNAs. The resulting mRNAs are reversely transcribed to generate the corresponding cDNAs with a nucleotide mix containing 5-Me dCTP to protect ORF regions from the subsequent restriction digestion of the directional linker. After the generation of blunt ends with T4 DNA polymerase, a directional linker [i.e., *Eco*RI/*Hin*dIII linker (GCTTGAATTCAAGC)] is ligated to both ends that allows for directional ligation of different left and right consensus sequences upstream and downstream of the coding cDNA region. The left consensus sequence contains a T7 promoter and a deletion mutant of TMV 5′-UTR for efficient in vitro transcription and translation, respectively. The right consensus cassette contains a short sequence for hybridizing with a puromycin-containing oligo linker. Sequences that code for various affinity tags can be included on both ends to facilitate the purification of mRNA-displayed peptide or protein library. Depending on the design of the selection scheme, some special sequences can be incorporated. For instance, an N-terminal BirA site allowing for the site-specific introduction of a biotin molecule at each of the mRNA–protein fusion molecules could greatly facilitate the immobilization of mRNA-displayed proteome library onto a solid support matrix. A highly specific protease recognition site (e.g., for TEV protease) could facilitate the gentle release of the fusion molecules at any desired selection step. A 5-amino-acid PKA recognition site (RRASV) should allow for the radiolabeling of each peptide or protein sequence in the library with the same efficiency. The ligated dsDNA is PCR amplified using two primers complementary to the consensus T7 promoter and 3′ linker-hybridization region at the 5′ and 3′ ends, respectively. Finally, the PCR product is carefully fractionated using a spin column or gel electrophoresis to generate a cDNA library with desired length distribution (i.e., 0.5–2 kb). A critical requirement of mRNA display is the removal of constructs that contain stop codons, which prematurely terminate translation and prevent the conjugation between mRNA and its coding protein. This is achieved by generating and purifying the mRNA–protein fusion molecules directly followed by regeneration back of the cDNA. Constructs containing frameshifts or stop codons are not purified during affinity purification, effectively removing them from the library (17, 20).

3.3. Generation of mRNA-Displayed Synthetic Peptide or Natural Proteome Library

After the construction of a cDNA library and prior to the selection, one round of preselection should be performed in order to remove out-of-frame sequences and sequences containing stop codons (5, 17, 20). One round of mRNA display consists of the following steps: in vitro transcription, DNase digestion, conjugation with the puromycin oligo linker, in vitro translation/fusion formation, oligo(dT) mRNA purification, reverse transcription, protein affinity purification, preselection or functional selection, and regeneration of the selected sequences. The details for most of these procedures are described in Chapter 6. Briefly, the cDNA library is in vitro transcribed using T7 RNA polymerase; the mRNA templates with puromycin at the 3′ ends are generated by cross-linking with an oligonucleotide containing a psoralen residue and a puromycin residue at its 5′ and 3′ ends, respectively; and in vitro translation is performed using rabbit reticulocyte lysate and mRNA–protein fusion formation accomplished under optimized conditions (3). After the fusion formation, free mRNA templates and mRNA–protein fusions are first isolated from the translation reaction mixture using an oligo(dT) column, followed by converting into DNA/RNA hybrids through reverse transcription to remove secondary mRNA structures that might interfere with the subsequent selection. The resulting mRNA-displayed synthetic peptide or natural proteome library is purified using an affinity column (i.e., anti-FLAG) and used for selection.

3.4. Functional Selection of Synthetic Peptides or Natural Proteins that Bind to Calmodulin in a Ca^{2+}-Dependent Manner

One critical issue in using mRNA display for functional selection is to design the selection scheme that minimizes the capture of nonspecific sequences while maximizing efficient enrichment of sequences with desired properties. All recombinant protein targets, small molecules, and other components used in the selection should be RNase and DNase free to maintain intact mRNA–protein fusion molecules. Typically, binding buffer should be supplemented with RNase-free tRNA and BSA to reduce nonspecific RNA and protein binding. It is important to optimize the wash steps that are used prior to elution. A stringent wash effectively removes molecules that may bind to the target nonspecifically, but may also result in the loss of desired weakly bound sequences. An appropriate wash volume and stringency should be determined prior to the selection by using both positive- and negative-control sequences.

1. Dilute the purified mRNA-displayed library in 0.5–2 mL of binding buffer.
2. Apply the mRNA-displayed library to a precolumn of 100–300 μl UltraLink Plus Streptavidin Agarose beads to remove sequences that may bind matrix nonspecifically. Collect the flow-through for subsequent selection steps.

3. Incubate the flow-through with 5–25 μg of biotinylated calmodulin for 30–90 min at 4°C.

4. Add 100–300 μl of streptavidin–agarose beads pre-equilibrated with binding buffer to the mixture and incubate for 30 min at room temperature with gentle mixing.

5. Load the slurry mixture to an empty nuclease-free 10 mL poly-prep chromatography column. Retain flow-through for further analysis.

6. Wash unbound fusion molecules from the column with 9–24-column volumes (3×3 to 8×3 CV) of binding buffer. Retain each wash fraction for further analysis.

7. Elute the fusion molecules that bind to the target protein under desired conditions (see Note 2). Elute molecules that bind to calmodulin in a Ca^{2+}-dependent manner with 1-column volume of elution buffer (binding buffer minus $CaCl_2$ plus 2 mM EGTA) four times. Retain each elution fraction for further analysis.

8. Count 1/100 of each fraction, including flow-through, wash, elution, and 1/10 of the beads using liquid scintillation (see Note 3).

9. PCR amplify the eluted mRNA–protein fusion sequences under the conditions that have been previously titrated on a small scale (see Note 4).

10. Perform the next round of selection using the regenerated cDNA library (see Note 5). Enrichment data measuring radioactive counts from the flow-through, wash, and elution steps can be used to monitor the enrichment of the desired sequences as demonstrated in Fig. 2.

3.5. Functional Confirmation

After selection, several hundred clones are typically sequenced to determine the identity of the selected proteins. When the selection is from an mRNA-displayed natural protein library, cDNA microarrays can also be used to reveal the identity of the genes that are present in the pool.

It is of great importance to develop a high-throughput assay that allows for biochemical characterization of the selected clones to confirm which of the sequences indeed possess the desired properties (see Note 6). The nucleic acid portion of the fusion molecules contains all the necessary sequences for efficient in vitro transcription and translation, and therefore can be directly used after PCR amplification as templates to generate radiolabeled proteins. Autoradiography can be used in binding assays against the immobilized target protein, allowing for sensitive analysis of binding between the two molecules. As illustrated in Fig. 3, Ca^{2+}-dependent, calmodulin-binding analysis of individual natural proteins (Fig. 3a) or synthetic peptides (Fig. 3b) isolated from the selection showed that the selected protein or peptide sequences bound to calmodulin in a Ca^{2+}-dependent manner (10, 17).

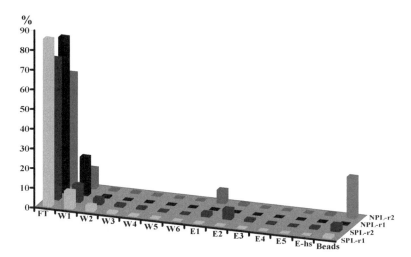

Fig. 2. Elution profile of Ca^{2+}-dependent, calmodulin-binding selection using an mRNA-displayed random peptide library (SPL-r1 and SPL-r2) or human proteome library (NPL-r1 and NPL-r2). The Y axis is the percentage (%) of the radiolabeled fusion molecules present in each fraction. The percentage was calculated based on the input of fusion molecules that were labeled with [^{35}S]-methionine. *FT* flow-through, *W1–W6* washes in the presence of Ca^{2+}, *E1–E5* elution with a buffer containing 2 mM EGTA and 150 mM NaCl, *E-hs* elution with a buffer containing 2 mM EGTA and 1 M NaCl, *Beads* on streptavidin–agarose beads. Round 1 and round 2 are labeled as r1 and r2, respectively.

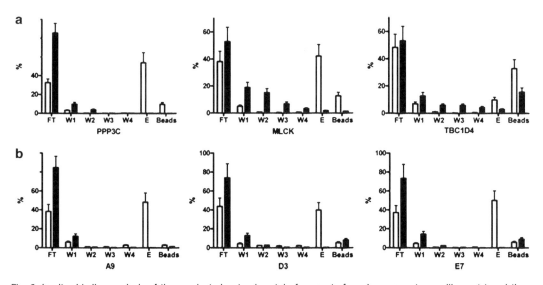

Fig. 3. In vitro binding analysis of three selected natural protein fragments from human proteome library (**a**) and three selected short peptides from combinatorial peptide library (**b**) with biotinylated calmodulin. Radiolabeled proteins or peptides were generated by coupled transcription/translation and purified using Ni-NTA beads. The purified protein product was incubated with biotinylated calmodulin in the presence of 1 mM $CaCl_2$ or 2 mM EGTA. The complex was captured with streptavidin–agarose beads. After extensive wash, the bound proteins or peptides were eluted using a buffer containing 2 mM EGTA. The Y axis is the percentage (%) of a protein or peptide present in each fraction calculated from its radioactive counts. *FT* flow-through, *W1–W4* washes, *E* elution with EGTA. *White bars* binding in the presence of Ca^{2+}, *black bars* binding in the presence of EGTA.

4. Notes

1. Typically, two to five rounds of selection are necessary to isolate peptides or proteins with desired properties for selections using a competitive or conditional elution. If the selection is simply based on nonspecific elution to disrupt the interaction between a target and mRNA-displayed library, the eluted pool should be carefully monitored after each round of selection to prevent enrichment of nonspecific sequences. This can be done through the monitoring of both radioactive counts and sequence analysis of the selected molecules from the eluted pool.

2. For other protein–protein interactions, 10–20-fold excess of unmodified target protein can be used to compete with the immobilized target protein to release the target-binding partners. The milder and more selective the elution conditions, the more efficiently each round of selection enriches for the desired functional sequences.

3. Radioactive counts should return to baseline levels prior to elution of target-binding molecules. The amount of radioactivity present in elution fractions is generally low in the first several rounds of selection. However, after several rounds of enrichment, the percentage of radioactivity present in the elution fractions from the selection dramatically increases.

4. The number of PCR cycles required to effectively regenerate the cDNA library without overamplification is critical for the next round of selection. Too many cycles of PCR have the potential to generate artificial sequences that overwhelm the selected pools. A small-scale diagnostic PCR should be performed at the end of the each selection cycle to determine the appropriate number of cycles necessary to regenerate the library. For a synthetic cDNA library, the length of the PCR products is fixed and therefore the quality of the library is easy to determine. For a natural cDNA library, the distribution of the original library should be used as a benchmark to determine the number of PCR cycles required for regeneration.

5. For the iterative rounds of selection, the general selection procedure remains the same for the generation and purification of the mRNA–protein fusion molecules, but the selection pressure can be gradually increased for subsequent selection cycles. This can be achieved by reducing the amount of target protein used during binding, shortening the incubation time or increasing the incubation temperature, adding binding competitors, and increasing the wash stringency (times and volumes) or elution specificity.

6. One major challenge in the functional selection is how to remove abundant sequences that could dominate the pool as selection goes on. Some peptide or protein sequences might be preferentially enriched, which could interfere with the isolation of other sequences with the same or even better properties. These abundant sequences can be effectively removed at the mRNA level. Specifically, after determining the identity of the abundant molecules by sequencing a couple of hundred clones, biotinylated antisense oligos are designed against the variable region mapped by aligning the abundant sequences (17, 20). Following RNA hydrolysis of mRNA/cDNA hybrids and neutralization, hybridization of the biotinylated oligos with the complementary cDNA, and passage through a streptavidin column, these abundant sequences can be effectively removed. This unique feature significantly increases the chance of discovering nonabundant sequences.

References

1. Roberts, R.W. & Szostak, J.W. (1997) RNA-peptide fusions for the in vitro selection of peptides and proteins. *Proc Natl Acad Sci U S A* **94**, 12297–12302.
2. Nemoto, N., Miyamoto-Sato, E., Husimi, Y. & Yanagawa, H. (1997) In vitro virus: bonding of mRNA bearing puromycin at the 3′-terminal end to the C-terminal end of its encoded protein on the ribosome in vitro. *FEBS Lett.* **414**, 405–408.
3. Liu, R., Barrick, J.E., Szostak, J.W. & Roberts, R.W. (2000) Optimized synthesis of RNA-protein fusions for in vitro protein selection. *Methods Enzymol* **318**, 268–293.
4. Szostak, J., Roberts, R., and Liu, R. in WO/1998/031700(1998);WO/2000/047775 (2000); U.S. Patent 6,207,446 (2001); U.S. Patent 6,214,553 (2001); U.S. Patent 6,258,558 (2001); U.S. Patent 6,261,804 (2001); U.S. Patent 6,281,344 (2001).
5. Cho, G., Keefe, A.D., Liu, R., Wilson, D.S. & Szostak, J.W. (2000) Constructing high complexity synthetic libraries of long ORFs using in vitro selection. *J Mol Biol* **297**, 309–319.
6. Wilson, D.S., Keefe, A.D. & Szostak, J.W. (2001) The use of mRNA display to select high-affinity protein-binding peptides. *Proc Natl Acad Sci U S A* **98**, 3750–3755.
7. Baggio, R. et al. (2002) Identification of epitope-like consensus motifs using mRNA display. *J Mol Recognit* **15**, 126–134.
8. Xu, L. et al. (2002) Directed evolution of high-affinity antibody mimics using mRNA display. *Chem Biol* **9**, 933–942.
9. Getmanova Ev et al. (2006) Antagonists to human and mouse vascular endothelial growth factor receptor 2 generated by directed protein evolution in vitro. *Chem Biol.* **13**, 549–556.
10. Huang, B.C. & Liu, R. (2007) Comparison of mRNA-display-based selections using synthetic peptide and natural protein libraries. *Biochemistry.* **46**, 10102–10112. Epub 12007 Aug 10109.
11. Cho, G.S. & Szostak, J.W. (2006) Directed evolution of ATP binding proteins from a zinc finger domain by using mRNA display. *Chem Biol.* **13**, 139–147.
12. Olson Ca, Liao Hi, Sun R & Roberts RW (2008) mRNA display selection of a high-affinity, modification-specific phospho-IkappaBalpha-binding fibronectin. *ACS Chem Biol.* **3**, 480–485. Epub 2008 Jul 2001.
13. Hammond, P.W., Alpin, J., Rise, C.E., Wright, M. & Kreider, B.L. (2001) In vitro selection and characterization of Bcl-X(L)-binding proteins from a mix of tissue-specific mRNA display libraries. *J Biol Chem* **276**, 20898–20906.
14. Cujec, T.P., Medeiros, P.F., Hammond, P., Rise, C. & Kreider, B.L. (2002) Selection of v-abl tyrosine kinase substrate sequences from randomized peptide and cellular proteomic libraries using mRNA display. *Chem Biol.* **9**, 253–264.

15. McPherson, M., Yang, Y., Hammond, P.W. & Kreider, B.L. (2002) Drug receptor identification from multiple tissues using cellular-derived mRNA display libraries. *Chem Biol* **9**, 691–698.
16. Horisawa, K. et al. (2004) In vitro selection of Jun-associated proteins using mRNA display. *Nucleic Acids Res.* **32**, e169.
17. Shen, X., Valencia, C.A., Szostak, J.W., Dong, B. & Liu, R. (2005) Scanning the human proteome for calmodulin-binding proteins. *Proc Natl Acad Sci U S A.* **102**, 5969–5974. Epub 2005 Apr 5919.
18. Shen X et al. (2008) Ca(2+)/Calmodulin-binding proteins from the C. elegans proteome. *Cell Calcium.* **43**, 444–456. Epub 2007 Sep 2012.
19. Tateyama, S. et al. (2006) Affinity selection of DNA-binding protein complexes using mRNA display. *Nucleic Acids Res.* **34**, e27.
20. Ju, W. et al. (2007) Proteome-wide identification of family member-specific natural substrate repertoire of caspases. *Proc Natl Acad Sci U S A.* **104**, 14294–14299. Epub 12007 Aug 14229.
21. Fukuda, I. et al. (2006) In vitro evolution of single-chain antibodies using mRNA display. *Nucleic Acids Res.* **34**, e127. Epub 2006 Sep 2029.

Chapter 17

Identification of Candidate Vaccine Genes Using Ribosome Display

Liancheng Lei

Abstract

In vitro protein selection methods that are not biased by the context of living organisms allow the screening of genomic expression libraries against a large number of different ligands. As such, ribosome display is a powerful technology for the in vitro selection of proteins or peptides from large PCR-derived libraries. Libraries can be generated from the genomic fragments of pathogens, thus allowing potential vaccine genes and the immunologically relevant proteins of pathogens to be identified and mapped using ribosome display. This chapter describes a methodology for the use of ribosome display for the identification of potential vaccine genes for the bacterial pathogen APP-5.

Key words: Ribosome display, Vaccine genes, Pathogen, In vitro selection, Cell-free translation

1. Introduction

In recent years, several in vitro protein selection technologies have emerged (1–5). One such technology, ribosome display, is based on the generation of ternary complexes consisting of protein–ribosome–mRNA, thereby establishing a linkage between genotype and phenotype that is used for evolutionary experiments. Various in vitro translation systems, including those derived from *Escherichia coli* (1, 2), rabbit reticulocytes (3, 6), and wheat germ (7), have been used for ribosome display. Ribosome stalling, resulting in the production of ternary complexes, can be achieved by the omission of a stop codon from the mRNA (2, 3) or through use of antibiotics (1). After selection, mRNA is recovered and amplified as DNA by reverse transcription polymerase chain reaction (RT-PCR) and is then selected further or subcloned for analysis of selected proteins. In order to identify the genes encoding immunoreactive peptides of a pathogen, for the purpose of vaccine identification, a ribosome display library can be prepared using genomic DNA fragments of the pathogen of interest. This is a powerful approach for fingerprinting the repertoire of immune reactive proteins that could serve as potential candidates for active and passive vaccination against pathogens. Ribosome display has been used to comprehensively identify candidate vaccine genes and map the immunologically relevant proteins of a pathogenic microorganism *Actinobacillus pleuropneumoniae* serotype 5 (APP-5) that causes porcine contagious pleuropneumonia (8). The method for identifying vaccine genes using ribosome display is described here using the example of APP-5.

2. Materials

2.1. Construction of the Pathogen (APP-5) Library for Ribosome Display

2.1.1. Extraction of the Pathogen (APP-5) DNA

1. Serotype 5 reference strain of APP (Control Institute of Veterinary Bioproducts and Pharmaceuticals, China) or alternative pathogen of interest.
2. Brain heart infusion (BHI) broth: 15 mg/mL BHI (Difco Laboratories, Michigan, USA), 10 μg/mL NAD, 1 mg/mL glucose, or alternative growth media suitable for the pathogen of interest. Store at 4°C.
3. TE buffer: 10 mM Tris–HCl, 1 mM EDTA, pH 8.5.
4. 100 mg/mL lysozyme. Store at 4°C.
5. 10 M NaCl.
6. 20% (w/v) SDS.
7. 1 mg/mL protease K. Store at −20°C.
8. Phenol/chloroform/isoamyl alcohol (25:24:1). Store at 4°C.
9. Chloroform/isoamyl alcohol (24:1). Store at 4°C.
10. 3 M sodium acetate, pH 5.2. Store at 4°C.
11. Absolute and 70% (v/v) ethanol. Store at −20°C.
12. Nuclease-free water.

2.1.2. Fragmentation and Cloning of the Pathogen (APP-5) Genomic DNA

1. Shotgun Cleavage Kit.
2. 0.5 mM Ethylenediaminetetraacetic acid (EDTA).
3. Nuclease-free water.
4. 0.5× TBE buffer: 45 mM Tris base, 45 mM boric acid, 1 mM EDTA, pH 8.0.
5. 2% (w/v) Agarose, 0.5× TBE, 0.5 μg/mL ethidium bromide gel.
6. 0.7% (w/v) Agarose, 0.5× TBE, 0.5 μg/mL ethidium bromide gel.

7. 500 mM Na_2EDTA, pH 8.0
8. Phenol/chloroform/isoamyl alcohol (25:24:1). Store at 4°C.
9. Chloroform/isoamyl alcohol (24:1). Store at 4°C.
10. 3 M Sodium acetate. See Subheading 2.1.1, item 10.
11. Absolute and 70% (v/v) ethanol. Store at 4°C.
12. DNA marker in the 100–2,000 bp range. Store at −20°C.
13. 6× DNA gel loading buffer.
14. T4 DNA polymerase (Invitrogen, USA, or alternative supplier).
15. 5× T4 DNA polymerase buffer (Invitrogen, USA, or alternative supplier).
16. 25 mM $MgCl_2$.
17. 0.5 mM dNTP mix: 0.5 mM each of dATP, dGTP, dCTP, and dTTP.
18. 0.5 mM dATP.
19. 10× Taq polymerase buffer: 500 mM NaCl, 100 mM Tris–HCl, pH 7.5, 100 mM $MgCl_2$, 10 mM DTT (provided with Taq DNA polymerase).
20. Taq DNA Polymerase. Store at −20°C.
21. pGEM-T Vector System (Promega, USA). Store at −20°C.
22. *E. coli* JM109 electrocompetent cells (Promega, USA). Store at −80°C.
23. LB medium containing 100 μg/mL ampicillin.
24. LB agar Petri dishes containing 100 μg/mL ampicillin.
25. SOC medium. Store at 4°C.
26. QuickLyse Miniprep Kit (Qiagen, USA) or equivalent.
27. 30% (v/v) glycerol.

2.1.3. Construction of the Pathogen (APP-5) DNA Fragment Library for Ribosome Display

1. 10× Taq DNA polymerase buffer, see Subheading 2.1.2, item 18.
2. Taq DNA polymerase. Store at −20°C.
3. 10 mM dNTP mix: 10 mM each of dATP, dGTP, dCTP, and dTTP.
4. Rid1 primer: 5′-CGAA<u>TAATACGACTCACTATAGGGA</u>**GAC CACAACGGTTTCCC**ACTAGTAATAATTTTGTTT AACTTTAAGAAGGAGATATATCCATG*CCGACGT CGCATG*-3′ (100 bp) (see Notes 1 and 2).
5. Rid2 primer: 5′-CGAA<u>TAATACGACTCACTATAGGGA</u>**GAC CACAACGGTTTCCC**ACTAGTAATAATTTT GTTTAACTTTAAGAAGGAGATATATCCATGC *CGACGTCGCATG*-3′ (101 bp) (see Notes 1 and 2).

6. Rid3 primer: 5'-CGAA<u>TAATACGACTCACTATAGGG</u>A**GAC CACAACGGTTTCCC**ACTAGTAATAATTTT GTTTAACTTTAAGAAGGAGATATATCCATGGC *CCGACGTCGCATG*-3' (102 bp) (see Notes 1 and 2).

7. RDX2 primer: 5'-<u>GGCCCACCCGTGAAGGTGAGCC</u>**CCA CCACCACCAGAACTTC***CAGGCGGCCGCACTA*-3' (54 bp) (see Note 3).

8. RDT primer: 5'-CAACGGTTTCCCACTAGTAATAATTTT GTTTAACTTTAAGAAGGAGATATATCCATG -3' (57 bp) (see Note 2).

9. RDP1 primer: 5'-CGAATAATACGACTCACTATAGGGAGA CCACAACGGTTTCCCAC-3' (42 bp) (see Note 2).

10. LDP21 primer: 5'-CATGCGACGTCGGCATGGATATAT CTCC-3' (28 bp) (see Note 2).

11. LDP22 primer: 5'-CATGCGACGTCGGGCATGGATATAT CTCC-3' (29 bp) (see Note 2).

12. LDP23 primer: 5'-CATGCGACGTCGGGCCATGGATATA TCTCC-3' (30 bp) (see Note 2).

13. 0.5× TBE buffer, see Subheading 2.1.2, item 4.

14. 2% (w/v) Agarose, 0.5× TBE, 0.5 μg/mL ethidium bromide gel.

15. DNA marker, see Subheading 2.1.2, item 12.

16. 6× DNA gel loading buffer.

17. Agarose gel DNA purification kit.

18. Nuclease-free water.

2.2. Ribosome Display Selection of the Pathogen (APP-5) Library

2.2.1. Coupled Transcription and Translation of the Pathogen (APP-5) Fragment Library

1. *E. coli* S30 Extract System for Linear DNA (Promega, USA). Store at −70°C.
2. Transcend Colorimetric Non-Radioactive Translation Detection System (Promega, USA). The kit contains Transcend tRNA, Streptavidin–AP, and Western Blue Stabilized Substrate for Alkaline Phosphatase. Store at −70°C.

2.2.2. Assessment of Luciferase and the Pathogen (APP-5) Control Reactions

1. Luciferase Assay System (Promega, USA).
2. Rnasin Ribonuclease Inhibitor (Promega, USA) (optional).
3. Acetone.
4. 20% (w/v) SDS.
5. 30% acrylamide:bisacrylamide (29:1).
6. 20% (w/v) ammonium persulfate. Make fresh each time.
7. TEMED.
8. 1× TBST: 50 mM Tris–HCl, 150 mM NaCl, 0.1% (v/v) Tween 20, pH 8.0.

9. 4× stacking gel buffer: 0.5 M Tris–HCl, pH 6.8, 0.4% (w/v) SDS.
10. 4× resolving gel buffer: 1.5 M Tris–HCl, pH 8.8, 0.4% (w/v) SDS.
11. 5× SDS-PAGE sample loading buffer: 0.5 M Tris–HCl, pH 6.8, 50% (v/v) glycerol, 0.5% (w/v) bromophenol blue, 10% (w/v) SDS. Prior to use, mix 750 μL of this mix with 250 μL β-mercaptoethanol.
12. PVDF membrane.
13. Transcend colorimetric nonradioactive translation detection system. See Subheading 2.2.1, item 2.

2.2.3. Ribosome Display Selection to Identify Candidate Vaccine Genes

1. DNase I (TaKaRa, China). Store at −20°C.
2. 10× DNase I buffer (TaKaRa, China). Store at −20°C.
3. Antiserum raised against pathogen of interest (see Note 4).
4. ELISA Blocker Blocking Buffer (Thermo Scientific, USA).
5. Quick-Precip Solution (EdgeBio, USA).
6. Dissociation solution: 50 mM Tris–acetic acid, pH 7.5, 150 mM NaCl, 10 mM EDTA in RNase-free water. Store at −20°C.
7. TE buffer, see Subheading 2.1.1, item 3.
8. Access Quick RT-PCR kit (Promega, USA). Store at −20°C.
9. RDS1 primer: 5′-CGAATAATACGACTCAC-3′ (see Note 5).
10. RDS2 primer: 5′-GGCCCACCCGTGAAGGTG-3′ (see Note 5).
11. Nuclease-free water.
12. QIAquick PCR purification kit (Qiagen, USA).
13. 0.5× TBE, see Subheading 2.1.2, item 4.
14. 0.7% (w/v) Agarose, 0.5× TBE, 0.5 μg/mL ethidium bromide gel.
15. DNA marker, see Subheading 2.1.2, item 12.
16. 6× DNA gel loading buffer.
17. Agarose Gel Extraction Kit.
18. 1× PBS: 137 mM NaCl, 2.7 mM KCl, 4.3 mM Na_2HPO_4, 1.4 mM KH_2PO_4, pH 7.4.

2.3. Cloning and DNA Sequence Analysis

1. Agarose Gel DNA Extraction Kit (Qiagen Inc, USA).
2. pGEM-T vector system, see Subheading 2.1.2, item 21.
3. *E. coli* JM109 chemically competent cells or equivalent.
4. SOC medium, see Subheading 2.1.2, item 25.
5. RDS1 primer, see Subheading 2.2.3, item 8.
6. RDS2 primer see Subheading 2.2.3, item 9.
7. 10× Taq DNA polymerase buffer, see Subheading 2.1.2, item 19.

8. Taq DNA polymerase. Store at −20°C.
9. 10 mM dNTP mix, see Subheading 2.1.3, item 3.
10. 0.5× TBE, see Subheading 2.1.2, item 4.
11. 0.7% (w/v) Agarose, 0.5× TBE, 0.5 µg/mL ethidium bromide gel.
12. DNA marker, see Subheading 2.1.2, item 12.
13. 6× DNA gel loading buffer.
14. LB medium containing 100 µg/mL ampicillin.
15. LB agar Petri dishes containing 100 µg/mL ampicillin.
16. Plasmid DNA purification kit.

3. Methods

APP-5 genomic DNA is extracted, digested into fragments, and cloned into a plasmid. The DNA fragment library is converted to the appropriate format for ribosome display by PCR. The APP-5 DNA fragments are then expressed in vitro by ribosome display followed by immunological selection, gene cloning, DNA sequencing, and analysis to identify potential vaccine genes.

3.1. Construction of the Pathogen (APP-5) Library for Ribosome Display

3.1.1. Extraction of the Pathogen (APP-5) DNA

1. Grow a 10-mL culture of APP-5 in BHI broth at 37°C for 6 h with shaking at 150 rpm (or grow the pathogen of interest according to suitable conditions).
2. Centrifuge the culture for 5 min at $8,000 \times g$ and resuspend the bacterial pellet in 9,250 µL of TE buffer. Add 750 µL of 100 mg/mL lysozyme and incubate at 37°C for 2 h.
3. To the 10-mL volume, add 110 µL of 10 M NaCl, 550 µL of 20% (w/v) SDS, 28 µL of 1 mg/mL protease K, and 312 µL of water. Incubate overnight at 50°C.
4. Add 580 µL of 10 M NaCl and then divide the solution (about 6 mL each) between two centrifuge tubes.
5. Add an equal volume of phenol/chloroform/isoamyl alcohol (25:24:1) to each tube. Mix gently for 5 min and centrifuge for 10 min at $13,400 \times g$ at room temperature. Carefully remove the top (aqueous) phase containing the DNA and transfer to a new tube. Repeat this DNA extraction step once. If a white precipitate is still present at the aqueous/organic interface, repeat the extraction once again.
6. Add an equal volume of chloroform:isoamyl alcohol (24:1) and mix gently for 2 min. Centrifuge for 10 min at $9,300 \times g$. Carefully remove the top (aqueous) phase containing the DNA and transfer to a new tube.

7. Add 2.5× the original volume of ice-cold ethanol and 0.1× the original volume of 3 M sodium acetate. Mix gently and incubate overnight at −20°C or for 1 h at −70°C.

8. Pellet the precipitated DNA by centrifuging for 10 min at 4°C in a fixed-angle microcentrifuge at maximum speed. Remove and discard the supernatant. Add 1 ml of room-temperature 70% (v/v) ethanol to wash the pellet. Invert the tube gently to wash and centrifuge for 10 min at maximum speed.

9. Allow the pellet to dry for 5–10 min to remove remaining ethanol and dissolve the DNA in 30–50 µL of nuclease-free water or TE buffer. Measure the DNA concentration by UV spectrophotometry (A260) and store at −20°C.

3.1.2. Fragmentation and Cloning of the Pathogen (APP-5) Genomic DNA

1. Take 10 µg of APP-5 genomic DNA and concentrate by ethanol precipitation as described in Subheading 3.1.1, steps 7–9, to give a concentration of 1.5–5 µg/µL.

2. Digest the APP-5 genomic DNA into fragments of approximately 50–150 bp in size by mild DNase I treatment using a shotgun cleavage kit according to the manufacturer's instructions. Digestion with 1 µL of DNase I diluted 1:200 or 1:300 usually gives the highest quantity of 50–150 bp fragments. Stop the reaction with 2 µL of stop buffer supplied in the kit or with 0.5 M EDTA (see Note 6).

3. Add 0.5 µL of the digestion product to 4.5 µL of nuclease-free water and 1 µL of 6× DNA gel loading buffer and determine the fragmentation efficiency by electrophoresis of this sample on a 2% (w/v) agarose, 0.5× TBE, and 0.5 µg/mL ethidium bromide gel. Run the sample alongside a DNA marker and check that a very thick band of approximately 50–150 bp is present.

4. Purify the DNA fragments from the remainder of the digestion reaction by phenol/chloroform/isoamyl alcohol extraction as described in Subheading 3.1.1, steps 5–9, and dissolve the DNA in nuclease-free water to give a final concentration of approximately 1–2 µg/µL.

5. Blunt end the fragments by mixing 2–4 µg of fragmented DNA, 20 µL of 5× T4 DNA polymerase buffer, 40 µL of 0.5 mM dNTPs, 20 U of T4 DNA polymerase, and nuclease-free water to a final volume of 100 µL. Incubate at 11°C for 15 min and then place on ice.

6. Purify the DNA fragments by phenol/chloroform/isoamyl alcohol extraction as described in Subheading 3.1.1, steps 5–9, and dissolve in 5 µL of nuclease-free water.

7. Convert the blunt ends to 3′ adenine (A) overhangs by mixing 5 µL (2–4 µg) of the DNA fragment, 3 µL of 10× Taq DNA polymerase buffer, 3 µL of 25 mM $MgCl_2$, 12 µL of 0.5 mM

dATP, 15 U of Taq DNA polymerase, and nuclease-free water to a final volume of 30 μL. Incubate at 70°C for 1 h.

8. Purify the DNA fragments by phenol/chloroform/isoamyl alcohol extraction as described in Subheading 3.1.1, steps 5–9, and dissolve the DNA in 10 μL of nuclease-free water. Measure the DNA concentration by UV spectrophotometry (A260).

9. Ligate the DNA fragments into the pGEM-T vector by mixing 1–3 μL (100–200 ng) of APP-5 DNA fragments with 5 μL (250 ng) of pGEM-T Vector. Add 25 μL of 2× rapid ligation buffer, 5 μL of T4 DNA ligase, and nuclease-free water to a final volume of 50 μL. Mix by pipetting and incubate overnight at 4°C.

10. Purify the ligated DNA by phenol/chloroform/isoamyl alcohol extraction as described in Subheading 3.1.1, steps 5–9, and dissolve the DNA in 3 μL of nuclease-free water.

11. Mix 40 μL of *E. coli* JM109 electrocompetent cells (thawed on ice) with the 3 μL of the purified ligated plasmid and incubate on ice for 1 min. Transfer to a pre-chilled 0.2-cm electroporation cuvette and electroporate using a single pulse at 2.5 kV. Immediately add 1 mL of SOC medium to the cuvette and quickly but gently resuspend the cells and transfer to a 17 × 100-mm polypropylene tube (see Note 7). Incubate at 37°C for 1 h with shaking at 225 rpm.

12. Take 0.2 mL of the transformation reaction and plate onto an LB agar Petri dish containing 100 μg/mL ampicillin. Incubate overnight at 37°C to determine the number of transformants.

13. Inoculate the remaining bacteria from step 11 into 10 mL of LB medium containing 100 μg/mL ampicillin and grow at 37°C with shaking at 250 rpm for 12 h to amplify the APP-5 DNA fragment library.

14. Use 3 mL of the culture for plasmid extraction using a QuickLyse Miniprep Kit according to manufacturer's instructions and check the product on a 0.7% (w/v) agarose, 0.5× TBE, and 0.5 μg/mL ethidium bromide gel.

15. Centrifuge the remaining culture at $850 \times g$ for 3 min. Discard half of the supernatant and resuspend the bacterial pellet in the remaining medium. Mix equally with 30% (v/v) glycerol and aliquot into several 100-μL volumes in cryovials and store at −70°C.

3.1.3. Construction of the Pathogen (APP-5) DNA Fragment Library for Ribosome Display

Figure 1 illustrates the process of converting the APP-5 library into the required format for ribosome display by PCR.

1. Set up a PCR containing 10 μL of 10× Taq DNA polymerase buffer, 8 μL of 10 mM dNTP mix, 2 μL of Rids primer pool (10 μM each), 6 μL of primer RDX2 (10 μM), 2 μL of APP-5 genomic fragment plasmids template (50–200 ng), 5 U of Taq

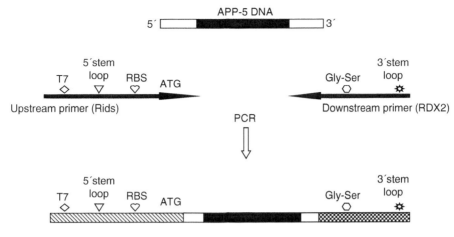

Fig. 1. Preparation of the APP-5 genomic DNA fragment library in ribosome display format by PCR.

DNA polymerase, and nuclease-free water to a final volume of 100 μL. Perform PCR using the following conditions: 95°C for 5 min, 30 cycles of 95°C for 40 s, 60°C for 40 s, 72°C for 1 min, and then 72°C for 10 min.

2. Separate the PCR products on a 2% (w/v) agarose, 0.5× TBE, 0.5 μg/mL ethidium bromide gel. The products should be approximately 230–360 bp. Purify these products using an agarose gel DNA extraction kit according to the manufacturer's instructions (see Notes 8 and 9) and measure the concentration of DNA by UV spectrophotometry (A260). Dilute the DNA concentration 2 μg/μL using RNase-free water and store at −70°C.

3.2. Ribosome Display Selection of the Pathogen (APP-5) Library

3.2.1. Coupled Transcription and Translation of the Pathogen (APP-5) Fragment Library

Several controls are recommended to ensure activity of the reagents and suitable conditions for translation for the pathogen library prior to selection. A positive control is used to confirm activity of the reagents, for example translation of luciferase protein followed by measurement of luciferase activity (see Subheading 3.2.2, step 1). Translation of the pathogen library can be confirmed by inclusion of a biotinylated-lysine loaded tRNA that leads to the production of biotin-containing protein fragments. This can be detected by SDS-PAGE and then western blotting with streptavidin and alkaline phosphatase (see Subheading 3.2.2, steps 2–11). Both of these controls are recommended prior to attempting a ribosome display selection (see Subheading 3.2.3). Other negative controls to check for specificity in parallel are also recommended, for example translation without template DNA.

1. Perform transcription and translation of the APP-5 library and controls using the *E. coli* S30 Extract System for Linear DNA as described in Table 1 (see Notes 10–13).

Table 1
Composition of transcription–translation reactions

Component	APP-5 control	APP-5 selection	Positive control
DNA template	Up to 4 μg of the APP-5 ribosome display library	Up to 4 μg of the APP-5 ribosome display library	8 μL of pBEST*luc* control DNA
Complete amino acid mixture (mix before use)	5 μL	5 μL	5 μL
Transcend tRNA	2 μL	–	–
S30 premix without amino acids (mix gently before use)	20 μL	20 μL	20 μL
S30 extract, linear (mix gently before use)	15 μL	15 μL	15 μL
Nuclease-free water up to	50 μL	50 μL	50 μL

2. Mix gently by pipetting and microcentrifuge for 5 s to collect the reactions at the bottom of the tube.
3. Incubate for 2 h at 37°C (see Note 14).
4. Stop the reactions by placing the tubes in an ice bath for 5 min. Control reactions can be stored at −20°C until use. Proceed with the ribosome display selection using fresh reactions.

3.2.2. Assessment of Luciferase and the Pathogen (APP-5) Control Reactions

1. Determine the successful transcription and translation of the luciferase positive control using the Luciferase Assay System according to manufacturer's instructions. Add 50 μL of one in two serial dilutions of the luciferase standard provided in the kit or of the positive control reaction to wells of a 96-well plate suitable to luminescence detection. Add 50 μL of room-temperature luciferase assay reagent to each well and mix quickly by pipetting. Read the luminescence in a suitable plate reader or imaging system, typically with an exposure of 6–10 min.
2. Prepare a 15% SDS-PAGE gel as described in Table 2 to detect the translation product encoded by the DNA fragment library of the APP-5 positive control reaction tube using western blotting.
3. Precipitate the proteins by adding 10 μL of the translation reaction to 40 μL of acetone in a microcentrifuge tube (see Note 15). Place on ice for 15 min and then centrifuge at $13,400 \times g$ for 5 min. Remove the supernatant and dry the pellet for 15 min under a vacuum. Resuspend the pellet in 5 μL of water.
4. Add 10 μL of SDS-PAGE sample buffer and heat at 100°C for 5 min.

Table 2
Preparation of 15% SDS-PAGE gel for western blotting

Resolving (lower) gel	mL	Stacking (upper) gel	mL
Water	2.4	Water	5.65
30% Acrylamide mix	5	30% Acrylamide mix	1.7
4× Resolving gel buffer	2.5	4× Stacking gel buffer	2.5
20% (w/v) SDS	0.05	20% (w/v) SDS	0.05
20% (w/v) Ammonium persulfate	0.05	20% (w/v) Ammonium persulfate	0.05
TEMED	0.004	TEMED	0.01

5. Separate 5–15 µL of the sample per lane on a 15% SDS-PAGE gel. Typically, electrophoresis is performed at a constant current of 15 mA in the stacking gel and 30 mA in the resolving separating gel until the bromophenol blue dye front has run off the bottom of the gel.

6. After electrophoresis, transfer the proteins to PVDF membrane by electroblotting according to the instrument manufacturer's instructions (see Note 16). For example, electroblotting onto PVDF membranes using a Bio-Rad Trans Blot Plus Cell Tank is performed under wet conditions for approximately 70–80 min at 100 V.

7. Add 15 mL (for a 7×9-cm membrane) of TBST to the membrane and incubate at room temperature for 60 min.

8. Pour off the TBST and add 15 mL of TBST containing 6 µL of streptavidin–AP solution made fresh just before use. Incubate at room temperature for 45–60 min with gentle agitation.

9. Pour off the streptavidin–AP solution and wash two times for 1 min each with 15 mL of TBST, followed by two more 1-min washes with 15 mL of water.

10. Completely cover the membrane with approximately 5 mL of Western Blue Stabilized Substrate for Alkaline Phosphatase. Protect the solution from strong light. Allow the color reaction to develop, typically for 1–15 min, until the bands of interest have reached the desired intensity.

11. Stop the reaction by washing the membrane in water for several minutes, changing the water at least once. Air dry the membrane and capture a digital image of the blot if desired. Purple bands in the sample lanes indicate that the APP-5 fragments were translated and expressed efficiently (see Note 17).

3.2.3. Ribosome Display Selection to Identify Candidate Vaccine Genes

1. Add 2 μL of RNase-free DNase and 6 μL of 10× DNase I buffer to the transcription–translation reaction and incubate for 30 min at 37°C. This step is included to digest the DNA template.

2. Incubate at 65°C for 10 min to activate the DNase.

3. Dilute the antiserum as appropriate (for example, 4,000-fold) with PBS. Add 100 μL per well into wells of a 96-well plate and coated for 2.5 h at 37°C or overnight at 4°C. Prepare two wells per selection and leave one well uncoated for a negative control selection.

4. Wash the wells three times with PBS.

5. Add 100 μL of blocking buffer to each well and incubate for 1 h at 37°C.

6. Wash the wells three times with PBS.

7. Add 20 μL of the translation reaction to each of the antiserum-coated wells and to the negative-control well. Seal the plate and incubate for 45–60 min at 37°C.

8. Wash the wells 15–20 times with PBS, including a 2-min incubation for each wash.

9. Add 100 μL of dissociation solution to each well and incubate on ice with shaking for 5–10 min.

10. Recover the dissociation solution that now contains the eluted mRNA. Replicate selections can be pooled if desired.

11. Purify the mRNA by phenol/chloroform/isoamyl extraction as described in Subheading 3.1.1, steps 5–6, and then precipitate the mRNA using Quick-Precip™ Solution according to manufacturer's instructions. Dissolve the recovered mRNA in 5 μL of TE buffer.

12. Amplify the candidate vaccine genes using the AccessQuick RT-PCR kit by mixing 25 μL of AccessQuick master mix, 0.3 μL of primer RDS1 (10 μM), 0.3 μL of primer RDS2 (10 μM), 5 μL of RNA, nuclease-free water to a final volume of 50, and 1 μL of AMV reverse transcriptase. Amplify as follows: 45–60 min at 45°C for reverse transcription, and then for PCR; 95°C for 2 min; 30 cycles of 95°C for 1 min and 70°C for 3 min, and then 72°C for 10 min.

13. Purify the amplified DNA using a QIAquick PCR purification kit according to the manufacturer's instructions. Elute the purified product in 10 μL nuclease-free water and use for an additional round of selection beginning at Subheading 3.2.1, step 1, performing three rounds of selection in total.

14. After three rounds of ribosome display selection, run the final RT-PCR product on a 0.7% (w/v) agarose, 0.5× TBE, and 0.5 μg/mL ethidium bromide gel and purify using a DNA gel

extraction kit. Use this RT-PCR product for cloning and DNA sequence analysis to identify potential genes encoding for APP-5 antigens.

3.3. Cloning and DNA Sequence Analysis

1. Ligate 25 ng of the purified RT-PCR product from the third round of selection with 50 ng of pGEM-T vector in a total volume of 10 µL according to the manufacturer's instruction.

2. Thaw *E. coli* JM109 chemically competent cells on ice and distribute 100-µL aliquots into chilled, 1.5-mL microcentrifuge tubes on ice. Add 2 µL of ligation product and incubate on ice for 30 min. Heat shock for 45 s at 42°C and replace on ice immediately. Add 900 µL of SOC medium and culture at 37°C for 1.5 h with shaking at 150 rpm.

3. Plate 100 µL of each transformation reaction onto duplicate LB agar plates containing 100 µg/mL ampicillin and incubate the plates overnight (16–24 h) at 37°C.

4. Pick individual colonies and identify whether they contain an insert by PCR with RDS1 and RDS2 as primers. Perform colony PCR in a 25-µL reaction volumes by mixing 2.5 µL of 10× Taq polymerase buffer, 2 µL of 0.2 mM dNTP mix, 0.2 µL of each primer (10 µM), and 2.5 U of Taq polymerase. Mark the Petri dish to allow identification of the clones later, and then pick a small part from each clone directly into the PCR tube as template. Amplify as follows: 95°C for 5 min; 30 cycles of 95°C for 1 min, 54°C for 45 s, and 72°C for 1 min, and then 72°C for 10 min.

5. Check a 5-µL sample of each PCR product on a 0.7% (w/v) agarose, 0.5× TBE gel containing 0.5% ethidium bromide.

6. Identify the clones that contain insert and then pick from the appropriate colony into 5 mL of LB medium supplemented with 100 µg/mL ampicillin. Grow overnight at 37°C with shaking at 250 rpm.

7. Prepare plasmid DNA using a mini-prep kit according to manufacturer's instructions and obtain the DNA sequence using T7 and/or M13 primers.

8. Compare the sequences obtained from the selection to those in the GenBank database by using the BLAST program to identify the genes that are potential vaccines for APP-5.

4. Notes

1. The primers Rid1, Rid2, and Rid3 contain the required upstream sequences for ribosome display; a T7 RNA polymerase promoter (underlined), a 5′ stem loop (bold), and a ribosome-binding site (shaded) and a start codon (ATG).

At the 3′-end, the primers contain a region complementary to the pGEM-T vector (italic) for amplification of the library from this vector. Each of Rid1, Rid2, and Rid3 primers has one additional base after the start codon, and the combination of these three primers together ensures that the APP-5 fragment library is displayed in the correct frame.

2. These Rid primers can be synthesized directly (recommended) or alternatively they can be generated by PCR using shorter oligonucleotides LDP21 (for Rid1), LDP22 (for Rid2), LDP23 (for Rid3), RDT, and RDP1. Prepare three reaction mixes as follows: 5 μL of 10× Taq polymerase buffer, 4 μL of dNTP mix (0.2 mM each), 1 μL of LDP21, LDP22, or LDP 23 (10 μM), 1 μL of RDP1(10 μM), 2 μL of RDT (25 μM), 2.5 U of Taq DNA polymerase, and nuclease-free water to a final volume of 50 μL. Perform PCR using the following conditions: 95°C for 3 min, 30 cycles of 94°C for 40 s, 56°C for 30 s, 72°C for 1 min, and then 72°C for 10 min. Purify each product using a PCR product purification kit and mix equally to a final concentration of 10 μM each Rid. The Rid1, Rid2, and Rid3 pool is termed Rids.

3. The downstream primer RDX2 includes a 3′ stem loop (underlined), a spacer sequence (bold), and a region complementary to the pGEM-T vector (italics).

4. Prepare antiserum for use as the target in selections using appropriate standard methods for immunization of rabbits. Alternatively, polyclonal antibodies prepared from the antiserum can be used, for example coated onto microbeads. The latter method may result in a more efficient selection.

5. RDS1 and RDS2 primers are the same as the 5′-end of the Rids and 5′-end of RDX2, respectively, and are used for RT-PCR recovery of the ribosome display construct after selection.

6. The target DNA must be fairly concentrated (approximately 1.5–5 μg/μL) and free of all traces of Mg^{2+} as this would cause single-strand cleavage by the enzyme. Under these conditions, 1 μL of DNase I diluted 1:200 or 1:300 usually gives the highest quantity of 50–150 bp fragments in preparation for cloning into the T-Vector.

7. The period between applying the pulse and transferring the cells to outgrowth medium is crucial for recovering the maximum number of *E. coli* transformants. Delaying this transfer by even 1 min causes a threefold drop in transformation efficiency and this decline continues to a 20-fold drop by 10-min delay.

8. Template DNA purity is extremely important. If efficiencies are low, examine the quality of the template DNA used. Avoid

adding excessive salts or glycerol with the DNA template. Precipitate the DNA template with sodium acetate rather than ammonium acetate.

9. It is important to maintain purity of this sample to avoid contaminating the S30 reaction with the other PCR products or primer-dimers. Primer-dimers can be removed by ethanol precipitation with sodium acetate.

10. Water purity is extremely important and nuclease-free water must be used in all procedures involving mRNA (and ideally DNA). If efficiencies are low, consider whether the quality of the water used is sufficient.

11. To reduce the chance of RNase contamination, wear gloves throughout the experiment and use microcentrifuge tubes and pipette tips that have been autoclaved and handled only with gloves. Addition of a ribonuclease inhibitor (such as RNAsin) to the translation reaction may be useful but is not required for preventing degradation of mRNA.

12. In general, reactions should not contain more than 4 µg of sample DNA. Higher amounts of linear DNA template can result in higher incorporation of label, but can also increase the number of internal translational starts or prematurely arrested translation products.

13. S30 extract and premix activity are easily decreased at room temperature. Keep on ice, do not warm significantly to thaw, and work as quickly as possible.

14. There may be enhanced expression at lower temperatures (for example, 30°C) for longer times. This appears to be gene specific and may be examined if the standard reaction at 37°C for 1 h does not produce the desired results. Control reactions using Transcend tRNA are useful for establishing the most efficient conditions for in vitro expression of the library.

15. Up to 1 µL of the translation reaction can be run on the gel without acetone precipitation; however, larger volumes can result in a high background on the blot, in which case acetone precipitation is recommended.

16. Coomassie staining is typically not sensitive enough to detect the cell-free translation products, so this is not performed prior to blotting.

17. *E. coli* S30 extract contains one endogenous biotinylated protein, migrating at 22.5 kDa. This may, therefore, be detected when translation products are analyzed by streptavidin detection. Comparison to a no-template control is recommended to distinguish between the endogenous biotinylated protein(s) and the newly synthesized biotinylated translation product.

References

1. Mattheakis LC, Bhatt RR, Dower WJ (1994) An in vitro polysome display system for identifying ligands from very large peptide libraries. Proc Natl Acad Sci USA 91:9022–9026.
2. Hanes J, Plückthun A (1997) In vitro selection and evolution of functional proteins by using ribosome display. Proc Natl Acad Sci USA 94:4937–4942.
3. He M, Taussig MJ (1997) Antibody-ribosome-mRNA (ARM) complexes as efficient selection particles for in vitro display and evolution of antibody combining sites. Nucleic Acids Res 25:5132–5134.
4. Nemoto N, Miyamoto-Sato E, Husimi Y et al (1997) In vitro virus: bonding of mRNA bearing puromycin at the 3-terminal end to the C-terminal end of its encoded protein on the ribosome in vitro. FEBS Lett 414:405–408.
5. Roberts RW, Szostak JW (1997) RNA-peptide fusions for the in vitro selection of peptides and proteins. Proc Natl Acad Sci USA 94:12297–12302.
6. Douthwaite JA, Groves MA, Dufner P et al (2006) An improved method for an efficient and easily accessible eukaryotic ribosome display technology. Protein Eng Des Sel 19:85–90.
7. Takahashi F, Ebihara T, Mie M, et al (2002) Ribosome display for selection of active dihydrofolate reductase mutants using immobilized methotrexate on agarose beads. FEBS Lett 514:106–110.
8. Lei L, Du C, Yang P (2009) Screening of strain-specific *Actinobacillus pleuropneumoniae* genes using a combination method. J Microbiol Methods 77:145–151.

Chapter 18

Ribosome Display for the Selection of Sac7d Scaffolds

Barbara Mouratou, Ghislaine Béhar, Lauranne Paillard-Laurance, Stéphane Colinet, and Frédéric Pecorari

Abstract

Combinatorial libraries of Sac7d have proved to be a valuable source of proteins with favorable biophysical properties and novel ligand specificities, so-called Nanofitins. Thus, Sac7d represents a promising scaffold alternative to antibodies for biotechnological and potentially clinical applications. We describe here the methodology for the construction of a library of Sac7d and its use for selection by ribosome display.

Key words: Ribosome display, In vitro selection, Sac7d, Nanofitin

1. Introduction

Historically, antibodies were the molecules of choice when affinity and specificity were required in a reagent as they could be obtained quite easily via animal immunization. However, their complex molecular organization with several polypeptide chains and disulfide bridges requires the use of eukaryotic hosts for their expression, which leads to high production costs. Antibody fragments were designed to overcome these problems while keeping the binding function. The combination of the creation of synthetic libraries and in vitro selections has enabled antibody fragments to be generated with an affinity for a given target while bypassing the animal immune system. However, these fragments often have weak stability and a tendency to aggregate.

Thus, other proteins called scaffolds, which combine all the advantages of antibodies without their drawbacks in one molecule, have recently been developed. We have previously described the use of the Sac7d protein (1), a dsDNA-binding protein from the hyperthermophilic archaeon *Sulfolobus acidocaldarius*, as such a scaffold (see Fig. 1). It has a simple molecular organization

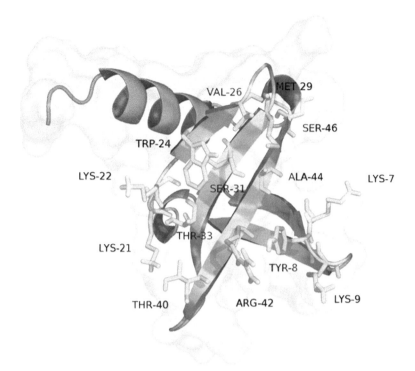

Fig. 1. Schematic representation of wild-type Sac7d (pdb code 1AZP). Side chains of residues involved in DNA binding, and which were substituted, are represented by *sticks*.

with only one polypeptide chain of 66 amino acids and lacks disulfide bridges. Sac7d shows high expression yields in the cytoplasm of *E. coli*. In addition to its high solubility, it is chemically and thermally stable and resistant to extreme pH.

By coupling variations of the Sac7d sequence and ribosome display selections, we have generated binders, called Nanofitins, with sub-nanomolar affinities and high specificities for the bacterial protein PulD (1–3). Anti-PulD binders have been shown to keep the favorable biophysical properties of Sac7d. These results demonstrate that Sac7d can tolerate significant modifications of its sequence via random mutagenesis schemes without altering its fold. Binders remained functional when fused to a reporter such as alkaline phosphatase and green fluorescent protein, the latter being used as a reagent for immunolocalization. In addition, when exported to the *E. coli* periplasm, binders fused to alkaline phosphatase inhibited PulD oligomerization, thereby blocking the type II secretion pathway of which PulD is part. We also demonstrated that a Nanofitin with a low nanomolar dissociation constant for lysozyme could be used to detect labeled lysozyme in a protein chip format (4), as well as their use as a detection or capture reagent of human IgG (unpublished data).

The following sections describe the generation of the Sac7d library in a format compatible with ribosome display and its use for selections against a given target.

2. Materials

2.1. Generation of the Library in Ribosome Display Format

1. SClib1: GGAGATATATCCATGAGAGGATCGCATCACCA TCACCATCACGGATCCGTCAAGGTGAAATTC.
2. SClib2: GGATCCGTCAAGGTGAAATTCNNSNNSNNSGG CGAAGAAAAAGAAGTGGACACTAGTAAGATC.
3. SClib3: CTTGCCGTTGTCGTCGTASNNAAASNNCACS NNTTTGCCSNNACGSNNAACSNNSNNGATCTT ACTAGTGTCCACTTC.
4. SClib4: TAATAACTCTTTCGGGGCATCTTTCTCSNNCA CSNNGCCSNNGCCSNNCTTGCCGTTGTCGTCGTA.
5. SClib5: CCATATAAAGCTTTTTCTCGCGTTCCGCACGC GCTAACATATCTAATAACTCTTTCGGGGCATC.
6. T7B: ATACGAAATTAATACGACTCACTATAGGGAGACC ACAACGG.
7. T7C: ATACGAAATTAATACGACTCACTATAGGGAGACC ACAACGGTTTCCCTC.
8. tolAk: CCGCACACCAGTAAGGTGTGCGGTTTCAGTTG CCGCTTTCTTTCT.
9. SClink: GCGGAACGCGAGAAAAAGCTTTATATGGCCTC GGGGGCC.
10. SDA_MRGS: AGACCACAACGGTTTCCCTCTAGAAATAA TTTTGTTTAACTTTAAGAAGGAGATATATCCA TGAGAGGATCG.
11. tolA_coli_F: GAGAAAGGATCCCTTTATATGGCCTCGGG GGCCGAGTTCGAATCTGGTGGCCAGAAGCAAGCT GAAGAGGCGGCAGCG.
12. tolA_coli_R: TGCATTAAGCTTTTTTTCAGCAGCTTCAG TTGCCGCTTTCTTTC.
13. Qe30for: CTTTCGTCTTCACCTCGAG.
14. Qe30rev: GTTCTGAGGTCATTACTGG.
15. Nuclease-free water (see Note 1).
16. Vent DNA polymerase with reaction buffer and 100 mM $MgSO_4$ (New England Biolabs). Store at –20°C.
17. Taq DNA polymerase with reaction buffer (various suppliers). Store at –20°C.
18. dNTP solution: 10 mM of each dNTP. Aliquot and store at –20°C.
19. Dimethyl sulfoxide (DMSO, molecular biology grade).
20. 1.5% (w/v) ultrapure agarose, 0.5× TBE gel containing 1× SYBR Safe DNA gel stain (Invitrogen) or equivalent.
21. 0.5×TBE: 44.5 mM Tris base, 44.5 mM Boric acid, 1 mM EDTA (pH 8.0).

22. DNA ladder in the range 100–1,000 bp (various suppliers).
23. 6× DNA Gel Loading Dye (various suppliers).
24. DNA Gel Extraction and PCR Cleanup kit (various suppliers).
25. T4 DNA ligase with reaction buffer (various suppliers). Store at −20°C.
26. *Bam*HI and *Hin*dIII restriction enzymes with reaction buffers (various suppliers). Store at −20°C.
27. DNA Plasmid Miniprep kit (various suppliers).
28. pQE30 expression vector (Qiagen) or equivalent (see Note 11).
29. Competent DH5α cells (Invitrogen or other supplier).
30. Luria–Bertani (LB) agar petri dishes without antibiotic.

2.2. In Vitro Transcription

1. TranscriptAid T7 High Yield Transcription kit (Fermentas) or equivalent.
2. RNA purification kit (various suppliers).

2.3. Ribosome Display Selection

1. SCepRev: TCGGCCCCCGAGGCCATATAAAGCTTTTTCTC.
2. αssrA: TTAAGCTGCTAAAGCGTAGTTTTCGTCGTTT GCGACTA.
3. *E. coli* in vitro translation system (homemade S30 extract or a commercially available reagent) (see Note 2).
4. 200 mM L-methionine in nuclease-free water. Aliquot and store at −20°C.
5. Premix for in vitro translation system: 5.3 mM adenosine triphosphate (ATP), 13.4 mM guanosine triphosphate (GTP), 2.7 mM cyclic adenosine monophosphate (cAMP), 80 mM acetyl phosphate, 1.33 μg/μL *E. coli* tRNA, 133 mM Tris–acetate pH 7.4 at 4°C, 6.7% (w/v) PEG 8000, 0.93 mM of all natural amino acids except methionine, 53 μg/mL folinic acid, 534 mM L-glutamic acid monopotassium, 15 mM magnesium acetate, 9.7 μM αssrA oligonucleotide in nuclease-free water. Aliquot and store at −80°C.

Prepare all the solutions listed below using nuclease-free water.

6. 1× TBS: 20 mM Tris–HCl pH 7.4 at 4°C, 150 mM NaCl. Sterile (0.22 μm) filter.
7. 66 nM streptavidin (Sigma) in 1× TBS. Prepare from a stock solution of 20 μM streptavidin in 1× TBS, aliquoted and stored at −20°C.
8. WBT: 50 mM Tris–acetate pH 7.4 at 4°C, 150 mM NaCl, 50 mM Magnesium-acetate, 0.1% (v/v) Tween-20. Sterile (0.22 μm) filter before adding Tween-20.
9. TBS-BSA: 1× TBS containing 0.5% (w/v) BSA.

10. WBT-BSA: WBT containing 0.5% (w/v) BSA.
11. Elution Buffer: 50 mM Tris–acetate pH 7.4 at 4°C, 150 mM NaCl, 20 mM ethylenediaminetetraacetic acid (EDTA). Sterile (0.22 μm) filter.
12. 25 μg/μL solution of *S. cerevisiae* RNA (various suppliers). Aliquot and store at −20°C.
13. H minus reverse transcriptase with reaction buffer (various suppliers).
14. RiboLock RNase Inhibitor (Fermentas) or equivalent. Store at −20°C.
15. MaxiSorp plates (Nunc).
16. Biotinylated target of interest (see Note 3).
17. 66 nM NeutrAvidin (Thermo Scientific) in 1× TBS. Prepare from a stock solution of 20 μM NeutrAvidin 1× TBS, aliquoted and stored at −20°C.

2.4. Analysis of Selection Outputs (ELISA and DNA Sequencing)

1. Biotinylated target of interest.
2. 66 nM streptavidin in 1× TBS (see Subheading 2.3 item 7).
3. 66 nM NeutrAvidin in 1× TBS (see Subheading 2.3 item 17).
4. *E. coli* in vitro translation system (see Subheading 2.3 item 3).
5. 200 mM L-methionine in nuclease-free water (see Subheading 2.3 item 4).
6. Premix for in vitro translation system (see Subheading 2.3 item 5).
7. TBST: 1× TBS containing 0.1% (v/v) Tween-20.
8. Anti-RGS-His6-HRP conjugate (Qiagen).
9. 1 mg/mL OPD (*o*-Phenylenediamine dihydrochloride) substrate (various suppliers).
10. DH5α F′IQ cells (Invitrogen or other supplier).
11. LB agar petri dishes containing ampicillin (100 μg/mL) and kanamycin (25 μg/mL).
12. LB liquid media containing 100 μg/mL ampicillin, 25 μg/mL kanamycin, and 1% (w/v) glucose.
13. 2YT liquid media containing 100 μg/mL ampicillin, 25 μg/mL kanamycin, and 0.1% (w/v) glucose.
14. Isopropyl β-D-1-thiogalactopyranoside (IPTG).
15. Lysis solution: 1× TBS containing 1× BugBuster (Merck) and 5 μg/mL DNAse.
16. Qe30for: CTTTCGTCTTCACCTCGAG.
17. Qe30rev: GTTCTGAGGTCATTACTGG.

3. Methods

The Sac7d library is created by randomization of 14 amino acid residues (K7, Y8, K9, K21, K22, W24, V26, M29, S31, T33, T40, R42, A44, and S46) known to interact with double-stranded DNA (see Fig. 1 and Note 4). As the gene encoding Sac7d is quite short (~200 bp), the library can be synthesized with standard oligonucleotides and degenerated nucleotides in a single-step PCR. The PCR product obtained is then assembled with the 5′ and 3′ parts necessary for selection by ribosome display, including a T7 promoter, a ribosome binding site, a tolA linker, and stem loops at the ends to stabilize the construct (see Fig. 2 and Notes 5 and 6). To generate a reliable source of tolA spacer, we designed a plasmid derived from pFP1001 which encodes the part of the tolA gene that is needed for the ribosome display construct. The tolA gene is directly amplified from *E. coli* DH5α cells. The final PCR product is then transcribed in vitro, and after translation of RNAs, the ternary complexes are used for selections. Depending on the desired characteristics for binders, four to seven rounds of selection are usually necessary.

3.1. Generation of the Library

3.1.1. Generation of the Sac7d Library Fragment

1. In the first step, prepare a DNA product by assembly PCR using a combination of four standard (T7C, SDA_MRGS, SClib1, SClib5) and three degenerate oligonucleotides encoding NNS triplets (SClib2, SClib3, SClib4) (see Notes 7 and 8). Prepare three PCR reaction mixes in tubes containing 2 pmol

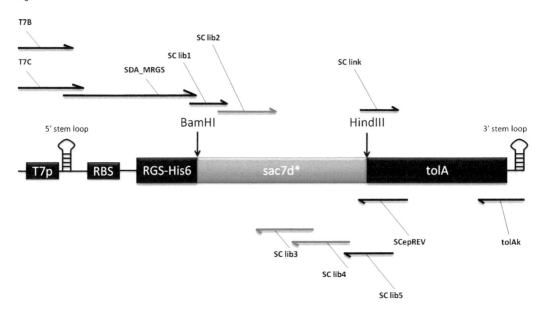

Fig. 2. Scheme of the ribosome display construct used for selection. *T7p* T7 promoter, *RBS* ribosome binding site, *RGS-His6* Arg-Gly-Ser-His6 are the starting amino acid residues of the coding sequence and constitute a tag for the detection of binders with anti-RGS-His6-HRP conjugate antibody, *sac7d** randomized gene of Sac7d, *tolA*: TolA spacer. Standard and degenerated oligonucleotides used to generate the library in this format are indicated by *black* and *gray horizontal arrows*, respectively. According to Fig. 1, the following amino acid residues were randomized: 7, 8, 9, 21, 22, 24, 26, 29, 31, 33, 40, 42, 44, 46.

of each internal primer (0.2 μL of 10 μM primer SDA_MRGS, SClib1–4), 10 pmol of each external primer (1 μL of 10 μM primer T7C and SClib5), 3 mM $MgSO_4$ (1.5 μL of 100 mM), 2 μL of dNTP mix (containing 10 mM of each dNTP), 5 μL of 10× Vent buffer, and 1 U of Vent polymerase then add nuclease-free water to 50 μL.

2. Place the tubes in a thermocycler and preheat the reaction to 95°C for 5 min. Perform 5 cycles of PCR of 95°C for 30 s, 50°C for 30 s, 72°C for 30 s, followed by 30 cycles of PCR of 95°C for 30 s, 64°C for 30 s, 72°C for 30 s, and then 1 cycle of PCR of 72°C for 5 min.

3. Mix 10 μL of 6× loading buffer with each 50 μL PCR product and load into wells of a 1.5% (w/v) agarose, 0.5× TBE, 1× SYBR Safe gel. Also load 5 μL of 1,000 bp DNA ladder in an adjacent lane and run the gel at 150 V for approximately 25 min.

4. Excise the bands corresponding to the expected size of 330 bp (see Note 9) and purify the DNA product using a DNA gel extraction kit (elute DNA with 30 μl of Elution Buffer) according to the manufacturer's specifications.

5. Load 5 μl of the purified library sample on a 1.5% (w/v) agarose, 0.5× TBE, 1× SYBR Safe gel and quantify by comparison to a DNA ladder run alongside. A good yield is approximately 15 ng/μL for each PCR, thus about 1.3 μg of purified DNA should be obtained if three reactions are performed (see Note 10).

6. In order to enable the quality control of the library, clone the purified DNA into the pFP1001 vector or equivalent (see Note 11) using *Bam*HI and *Hin*dIII restriction enzymes. Digest 100 ng of purified library fragment with *Bam*HI and *Hin*dIII enzymes for 2 h at 37°C.

7. Mix the digested product with the appropriate volume of 6× loading buffer and gel purify the DNA (200 bp) as described in steps 3 and 4.

8. Ligate the purified DNA product into *Bam*HI and *Hin*dIII digested vector. Mix 20 ng of the fragment library, 100 ng of the digested vector, 1 μL of 10× T4 DNA ligase buffer and nuclease-free water to 10 μL. Add 2.5 U of T4 DNA ligase and incubate at room temperature for 60 min.

9. Transform *E. coli* DH5α cells to obtain individual clones.

10. DNA sequence of approximately 48–96 colonies (a convenient number in a 96-well plate) using the Qe30for and Qe30rev primers to ensure that the library was synthesized as designed and there is no strong bias in the nucleotide composition (see Note 12).

3.1.2. Generation of the pFP1001_tolA Vector

1. Spread a dilution of an overnight DH5α culture on an LB agar petri dish and incubate overnight at 37°C to obtain isolated colonies.

2. Using a sterile pipette tip, pick a colony and inoculate a PCR mix containing 10 pmol of each primer (1 μL of 10 μM primer tolA_coli_F and tolA_coli_R), 4 mM $MgCl_2$ (4 μL of 50 mM), 1 μL of dNTP mix (containing 10 mM of each dNTP), 5 μL of DMSO, 5 μL of 10× Taq buffer, and 2 U of Taq polymerase, then add nuclease-free water to 50 μL.

3. Place the tubes in a thermocycler and preheat the reaction to 95°C for 5 min. Perform 30 cycles of PCR of 95°C for 30 s, 50°C for 30 s, 72°C for 30 s, followed by 1 cycle of 72°C for 5 min.

4. Check that the PCR reaction gives a band corresponding to the expected size of 327 bp by agarose gel electrophoresis.

5. Purify the remaining DNA product using a PCR purification kit according to the manufacturer's specifications.

6. Subclone the purified DNA into the FP1001 vector (or equivalent) as described in Subheading 3.1.1 steps 6–9.

7. Obtain DNA sequence using the Qe30for and Qe30rev primers for several colonies and from one clone with the correct sequence, prepare a stock of purified pFP1001_tolA vector by carrying out a plasmid miniprep according to the manufacturer's specifications.

3.1.3. Generation of the tolA Fragment

1. From the pFP1001_tolA vector produced in Subheading 3.1.2, large quantities of tolA spacer can be obtained quickly by PCR amplification. For this purpose, prepare a series of at least ten PCR reaction mixes containing 25 pmol of each primer (0.25 μL of 100 μM primer SClink and tolAk), 20 ng of pFP1001_tolA vector (0.2 μL of 100 ng/μL), 1 μL of dNTP mix (containing 10 mM of each dNTP), 5 μL of 10× Vent buffer, and 1 U of Vent polymerase, then add nuclease-free water to 50 μL.

2. Place the tubes in a thermocycler and preheat the reaction to 95°C for 5 min. Perform 30 cycles of PCR of 95°C for 30 s, 55°C for 30 s, 72°C for 30 s, followed by 1 cycle of 72°C for 5 min.

3. Pool the reactions and gel purify as described in Subheading 3.1.1 steps 3 and 4. The PCR reaction should give a band of 333 bp.

4. Determine the concentration of the tolA linker by measuring the optical density at 260 nm and considering that a 50 μg/ml double strand DNA solution has an optical density of 1 (for 1 cm light path). Adjust the concentration to 80 ng/μl. This is the stock that will be used for the construction of the library and during selection.

3.1.4. Generation of the Ribosome Display Construct

1. Prepare a series of 20 PCR reaction mixes in tubes containing 25 pmol of each primer (0.25 µL of 100 µM primer T7B and tolAk), 40 ng of Sac7d library (2.7 µL of 15 ng/µL), 53 ng of tolA linker (0.66 µL of 80 ng/µL), 1 µL of dNTP mix (containing 10 mM of each dNTP), 5 µL of 10× Vent buffer, and 1 U of Vent polymerase, then add nuclease-free water to 50 µL.

2. Place the tubes in a thermocycler and preheat the reaction to 95°C for 5 min. Perform 8 cycles of PCR of 95°C for 30 s, 45°C for 30 s, 72°C for 50 s, followed by 30 cycles of PCR of 95°C for 30 s, 55°C for 30 s, 72°C for 50 s and by 1 cycle of PCR of 72°C for 5 min.

3. Pool the tubes and check that the PCR reaction gives a band corresponding to the expected size of 635 bp by agarose gel electrophoresis. Purify the PCR product using a PCR purification kit according to the manufacturer's specifications.

4. Determine the concentration of the PCR product by measuring the optical density at 260 nm. The final concentration should be around 100 ng/µL yielding about 100 µg DNA library construct (see Note 13). The library can be stored at −80°C for several months or years.

3.2. In Vitro Transcription

1. Prepare the transcription reaction as follows. To 9 µg of the PCR product from Subheading 3.1.4 step 4, add 36 µL of 5× TranscripAid reaction buffer, 72 µL of NTP mix, 18 µL of TranscriptAid enzyme mix, and nuclease-free water to 180 µL (see Notes 14 and 15).

2. Incubate at 37°C for 2 h (see Note 16). The transcription mix should be turbid.

3. Purify the mRNA using a RNA purification kit according to the manufacturer's specifications.

4. Determine the concentration of mRNA by measuring the optical density at 260 nm and considering that a 40 µg/ml RNA solution has an optical density of 1 (for 1 cm light path). Dilute the mRNA with nuclease-free water to a final concentration of 2.5 µg/µL. Aliquot and store at −80°C.

3.3. Ribosome Display Selection

1. Add 100 µL of a 66 nM solution of streptavidin in 1× TBS into the wells of a MaxiSorp microtiter plate (see Notes 17 and 18), including wells for prepanning to avoid selection of streptavidin binders (see Note 19). Incubate overnight at 4°C or for 1 h at room temperature.

2. Wash with 3 × 300 µL of 1× TBS.

3. Add 300 µL of TBS-BSA to block any remaining surfaces of the plate, and incubate at room temperature for 1 h.

4. Discard the remaining buffer from the plate and tap the plate upside down onto adsorbent paper to remove any residual

buffer. Add 100 μL of a 150 nM solution of biotinylated target to the wells to be used for target selection and add TBS-BSA to the streptavidin only wells. Incubate the plate at room temperature for 1 h or overnight at 4°C.

5. Wash with 4 × 300 μL of 1× TBS and once with 300 μL WBT. Fill the wells with WBT and keep the plate at 4°C until use.

6. During or after the plate immobilization of the target, prepare the required number of translation reactions (see Note 17) where each 50 μL reaction contains 5.9 μL of nuclease-free water, 0.9 μL of 200 mM methionine, 18.6 μL of Premix, 22.7 μL of S30 extract, and 1.9 μL of mRNA from Subheading 3.2 step 4. Translate for 7 min at 37°C (see Note 20). Add 200 μL of ice-cold WBT-BSA to stop the translation reaction. Centrifuge at 20,000 × g for 5 min at 4°C and place on ice until use.

7. For the depletion of streptavidin binders, remove the WBT only from the wells lacking the target protein and add 240 μL of the stopped translation reaction in its place to perform the preplanning step. Incubate for 60 min with gentle shaking at 4°C.

8. Remove WBT from the target-coated wells and transfer the 240 μL of the stopped translation reaction to the target-coated wells to perform the selection step. Incubate for 30 min with gentle shaking at 4°C.

9. Tap the plate upside down onto adsorbent paper to remove any residual buffer and wash eight times for a few seconds each with 300 μL of ice-cold WBT-BSA to remove unbound complexes.

10. To elute the bound mRNA, add 200 μL of ice-cold Elution Buffer containing 50 μg/mL S. *cerevisiae* RNA. Incubate for 10 min with gentle shaking at 4°C.

11. Purify the mRNA using a RNA purification kit (elute mRNA with 25 μl Elution Buffer) according to the manufacturer's specifications.

12. Prepare a PCR tube per eluted well containing 24 μL of eluted purified mRNA and primer SCepRev (0.44 μL of 100 μM). Incubate at 70°C for 5 min and immediately chill on ice.

13. Prepare the reverse transcription mix; add 8.7 μL of 5× Reverse Transcription buffer, 4.4 μL of dNTP mix (containing 10 mM of each dNTP), 2.7 μL of Ribolock (40 U/μL), 2.2 μL of Minus H reverse transcriptase, and nuclease-free water to 20.70 μL. Add the denatured mRNA (from step 10) to the reverse transcription mix and incubate at 42°C for 1 h in a PCR machine. Stop the reaction by heating at 70°C for 10 min. Chill on ice.

14. To amplify the reverse transcription products, prepare a series of four PCR reaction mixes in tubes containing 25 pmol of each primer (0.25 μL of 100 μM primer SCepRev and SDA_MRGS), 5 μL of the reverse transcription template (from step 11), 1 μL of dNTP mix (containing 10 mM of each dNTP), 1.5 μL of $MgSO_4$, 2.5 μL of DMSO, 5 μL of 10× Vent buffer, and 1 U of Vent polymerase then add nuclease-free water to 50 μL (see Notes 21 and 22).

15. Place the tubes on a thermocycler and preheat the reaction to 95 °C for 5 min. Perform 5 cycles of PCR of 95 °C for 30 s, 55 °C for 30 s, 72 °C for 30 s, followed by 40 cycles of PCR of 95 °C for 30 s, 64 °C for 30 s, 72 °C for 30 s and by 1 cycle of 72 °C for 5 min.

16. Pool the replicate reactions and check that the PCR reaction gives a band corresponding to the expected size of 314 bp by agarose gel electrophoresis (see Note 23). Purify the PCR product using a PCR purification kit according to the manufacturer's specifications and determine its concentration by measuring the optical density at 260 nm. The DNA can be stored at −80 °C for several months.

17. To proceed with an additional round of selection, replace the promoter and the spacer by PCR. Prepare a PCR reaction mix in a tube containing 25 pmol of each primer (0.25 μL of 100 μM primer T7B and tolAk), 300 ng of PCR template (from step 16), 320 ng of tolA linker (4 μL of 80 ng/μL from Subheading 3.1.3 step 4), 2 μL of dNTP mix (containing 10 mM of each dNTP), 5 μL of DMSO, 10 μL of 10× Vent buffer, and 2 U of Vent polymerase then add nuclease-free water to 100 μL.

18. Place the tubes in a thermocycler and preheat the reaction to 95 °C for 5 min. Perform 8 cycles of PCR of 95 °C for 30 s, 45 °C for 30 s, 72 °C for 50 s, followed by 32 cycles of PCR of 95 °C for 30 s, 55 °C for 30 s, 72 °C for 50 s and by 1 cycle of 72 °C for 5 min.

19. Check that the PCR reaction gives a band corresponding to the expected size of 635 bp by agarose gel electrophoresis. Purify the PCR product using a PCR purification kit according to the manufacturer's specifications and determine its concentration by measuring the optical density at 260 nm. The final concentration should be around 150–200 ng/μL. The PCR product is now ready for in vitro transcription as described in Subheading 3.2 (see Note 24) and a further round of selection as described in Subheading 3.4 (see Notes 25–27). The PCR product can be stored it at −80 °C for several months or years.

3.4. Analysis of Selection Outputs by ELISA

Enrichment of a selection can be observed by the decreasing number of PCR cycles needed to obtain the same quantity of PCR product from round to round. Usually a decrease of five cycles for every round of selection is required (see Note 27). To test the specificity of the material selected, the last rounds of selection can be done in parallel with the presence and absence of the target in the wells of selection. No PCR product after the RT step should be obtained in the absence of the target, while a PCR product should be obtained in its presence. The most convincing way to check that the selection is successful is to perform an ELISA to detect the binding of Nanofitins on the target via their RGS-His6 tag.

1. Coat wells of a MaxiSorp plate with biotinylated target or streptavidin/NeutrAvidin as described in Subheading 3.3 step 1.
2. Perform in vitro translation of the mRNA from the last round of selection as described in Subheading 3.3 step 6, except incubate the translation of 1 h at 37°C rather than for 7 min.
3. Dilute the translation reaction eight times with TBST and add 100 μl into each well to be tested. Incubate for 1 h at RT with gentle shaking.
4. Wash the wells six times with 300 μL of TBST.
5. Add 100 μl of anti-RGS-His6-HRP conjugate (1/4,000) into each well. Incubate for 1 h at room temperature with gentle shaking.
6. Wash the wells six times with 300 μL of TBST.
7. Distribute 100 μl of a 1 mg/ml OPD substrate and read the absorbance at 450 nm. The ratio of OD450 for the target to OD450 for the negative control should be at least ten for a successful selection.

3.5. Analysis of Individual Clones

1. To investigate individual clones that have been selected, subclone the gel purified PCR product from a positive selection into pFP1001 using *Bam*H1 and *Hin*dIII as in described Subheading 3.1.1 steps 6–8.
2. Transform ligation reactions into *E. coli* DH5αF'IQ cells and plate onto an LB agar petri dish containing ampicillin (100 μg/ml) and kanamycin (25 μg/ml). Incubate overnight at 37°C to obtain individual colonies.
3. Add 1.8 ml of LB media containing ampicillin (100 μg/ml), kanamycin (25 μg/ml) and 1% (w/v) glucose per well of a deep 96-well plate.
4. Inoculate each well by picking an isolated colony from the petri dish using a sterile pipette tip and incubate overnight at 37°C with shaking. This is the master plate.
5. Add 1.2 ml of 2YT/ampicillin (100 μg/ml)/kanamycin (25 μg/ml)/0.1% glucose in each well of a new deep 96-well plate.

6. Transfer 200 μl of culture from the master plate to the fresh plate using a multi-channel pipette.

7. Incubate for 2–3 h at 37°C with shaking at 450 rpm until the medium is turbid.

8. Induce Nanofitin expression by addition of IPTG to each well to a final concentration of 0.5 mM.

9. Incubate overnight at 30°C with shaking at 450 rpm.

10. Centrifuge the deep-well plate at $2,500 \times g$ for 10 min and discard the supernatant in each well, for example using a vacuum system connected by a flexible tube to a pipette tip.

11. Add 50 μl of Lysis Solution per well and shake the plate vigorously at 1,000 rpm for 30 min.

12. Add 250 μl of 1× TBS per well and shake the plate vigorously at 1,000 rpm for 5 min.

13. Centrifuge the deep-well plate at $2,500 \times g$ for 10 min and determine the binding of individual clones by ELISA as described in Subheading 3.5, using 100 μl of supernatant instead of the in vitro translated pool.

14. Determine the DNA sequence of positive clones using the primers Qe30for and Qe30rev.

4. Notes

1. We recommend using nuclease-free water for reactions and preparations of solutions to avoid degradation of nucleic acids in general and of mRNAs that are particularly labile. We use a water purification unit (Elga or other suppliers) that delivers ultra filtered water that has a resistivity of 18.2 MΩ-cm, and that is free from RNase and DNase contamination. We usually autoclave it before use and store it at 4°C.

2. The S30-Extract is prepared according to Amstutz et al. (5). The extract is a cytoplasmic fraction (from MRE600, an *E. coli* strain that is deficient for RNase I) that has been depleted of endogenous DNA and mRNA. As an alternative to using a homemade S30 extract and premix, we also have used the commercial PUREsystem Classic II with success.

3. The target protein can be chemically biotinylated with the lysine reactive EZ-Link Sulfo-NHS-LC-LC-Biotin (Thermo Scientific) reagent following the manufacturer's instructions. This reagent has a spacer arm of about 30 Å that ensures a sufficient degree of liberty for the target to interact with binders once immobilized. We use find conditions that lead to an average of 3 biotins per protein. This ratio can be quickly determined

using the HABA/Avidin reagent (Sigma). A disadvantage of this approach is that the labeling occurs mostly randomly on the surface of the protein and thus could affect interesting epitopes. Another consequence is that target molecules are not identically presented to binders. Alternatively, one can fuse the target of interest to an AviTag and biotinylate in vivo or in vitro using the biotinylation enzyme BirA (Avidity). This approach leads to an oriented and homogeneous presentation of the target of interest without interfering with potential epitopes to bind, but can be only used for recombinant proteins.

4. The binding area of about 1,000 A^2 is quite large and slightly concave to match the spherical shape of globular proteins.

5. The TolA spacer ensures that the displayed proteins have a sufficient degree of liberty to interact with the target.

6. The ribosome display construct used here is based on that described by Amstutz et al. (5) but with slight modifications to fit our needs, such as different detection tags and restriction sites.

7. We recommend using highly purified (HPLC or gel purification) oligonucleotides for the construction of the library to avoid as far as possible unwanted sequences due to $n-1$ products.

8. NNS codons (N = A, C, T or G and S = C or G) are one of the mutagenic schemes that encode all 20 amino acids while minimizing the number of stop codons.

9. This DNA product corresponds to the gene of Sac7d, flanked by the 5′ sequence necessary for ribosome display and an additional 3′ sequence necessary for subsequent step PCR assembly with the tolA spacer.

10. A quantity of 100 ng is equivalent to about 2.8×10^{11} molecules, thus the upper limit for the size of the library at this step is roughly 10^{11} to 10^{12} variants.

11. pFP1001 is an expression vector derived from the pQE30 vector Qiagen. We replaced the suppressible stop from the original vector by two nonsuppressible stops to allow expression in the strain DH5αF′IQ. However, pQE30, or any other vector that has *Bam*HI and *Hin*dIII unique restriction sites, can be used in place of pFP1001.

12. In our hands, DNA sequencing confirms that the percentage of correct clones without any frame shifts or deletions was about 65%. The quality of the library is important since this will determine the "useful diversity" of the library and thus the chance of successfully selecting for binders.

13. For this large PCR, we use 2.2×10^{12} molecules from the Sac7d library (800 ng) as DNA input. The aim is to retain as much as possible the large diversity of the library. Each microgram of DNA

product obtained is equivalent to about 1.5×10^{12} molecules. The quality of the PCR product is crucial for the efficiency of the subsequent in vitro transcription. A sharp band without any smear by agarose gel electrophoresis can be used without further gel purification as the template for in vitro transcription. If the band is smeary or if there are other major bands present, try to improve the PCR, for example by changing magnesium concentration, annealing temperature, primers, etc.

14. The volume of the transcription reaction is calculated to retain a high library complexity. 9 μg of DNA is equivalent to about 1.4×10^{13} molecules and corresponds to a 10× oversampling of a library of 1.4×10^{12}. Thus the probability that each sequence of the initial sample is represented at least one time is about 100%.

15. Working with RNA requires several basic precautions including wearing gloves, using disposable materials and tips and tubes purchased RNase and DNase-free. The water quality is also important (see Note 1).

16. Incubations can be performed from 2 h to overnight. Generally, most of the product is generated within 1–2 h; however, longer incubations will yield more product if necessary.

17. We usually perform the first round multiple times (at least four wells per target) and combine the resulting pools for the second round of selection.

18. If the target is not biotinylated, it can be directly immobilized as described for streptavidin. Note, however, that this can lead to partial denaturation of the target molecule and affect the selection against specific epitopes that are not accessible in the native form of the target.

19. Prepare wells that are identically coated but lack the target to perform a prior negative selection step (prepanning) before the positive selection, in an attempt to reduce the isolation of matrix-binding sequences. This is done for all rounds of selection.

20. The duration of the translation should be optimized for every new batch of *E. coli* S30 extract and premix. Usually translation times between 5 and 10 min give optimal yields. To determine this optimal time, perform one round of selection with a known binder with translation times of 5, 6, 7, 8, 9 and 10 min and comparison of the PCR yields (with 12 PCR cycles) by agarose gel electrophoresis to indicate the optimal time of translation.

21. This reaction should be performed at least in quadruplicate to obtain the variability contained in half the amount of RT. It is important to prepare a negative control reaction by replacing the mRNA template with water to detect any contamination of reagents.

22. It is important to prepare a negative control reaction by replacing the RT template with water to detect any contamination of PCR solutions. Don't forget to also test the RT negative control prepared in the previous step (see Note 21). If a band is observed in the negative reaction, discard all PCR solutions and repeat the selection.

23. When a diffuse band or other side-products appear on the agarose gel, gel purify the band of interest and use this product as template for a second PCR. This will normally yield a high quality DNA.

24. As the diversity drops after the first round of selection, it is not necessary to carry out large transcriptions of selection for later rounds. We perform the following transcription reaction after the first round: to 1 μg of DNA template, add 4 μL of 5× TranscriptAid reaction buffer, 8 μL of NTP mix, 2 μL of TranscriptAid enzyme mix, and nuclease-free water to 20 μL.

25. We recommend alternating the use of streptavidin and NeutrAvidin to avoid the generation of binders against both molecules. Sometimes, it can be advantageous to perform selections in solution by using magnetic beads (coated with avidin or streptavidin).

26. The stringency of the selection can be increased by (a) decreasing the concentration of the target protein to 10 nM and/or (b) increasing the number of washing steps and the duration of the washing steps to 1 h.

27. Adjust the number of PCR cycles after the RT step according to the selection round. Performing too many cycles can normalize the relative proportions of different pool members, reducing the selective enrichment due to binding. We recommend reducing by about 5 cycles per round.

Acknowledgment

The authors thank previous members of the laboratory who helped to develop this protocol. This work was supported by "La Région des Pays de la Loire".

References

1. Mouratou B, Schaeffer F, Guilvout I et al (2007) Remodeling a DNA-binding protein as a specific in vivo inhibitor of bacterial secretin PulD. Proc Natl Acad Sci USA 104, 17983–17988
2. Krehenbrink M, Chami M, Guilvout I et al (2008) Artificial binding proteins (Affitins) as probes for conformational changes in secretin PulD. J Mol Biol 383, 1058–1068
3. Buddelmeijer N, Krehenbrink M, Pecorari F et al (2009) Type II secretion system secretin PulD localizes in clusters in the *Escherichia coli* outer membrane. J Bacteriol 191, 161–168

4. Cinier M, Petit M, Williams MN et al (2009) Bisphosphonate adaptors for specific protein binding on zirconium phosphonate-based microarrays. Bioconjugate Chemistry 20, 2270–2277

5. Amstutz P, Binz HK, Zahnd C et al (2006) Ribosome Display: In Vitro Selection of Protein-Protein Interactions. In Celis J (ed) Cell Biology – A Laboratory Handbook, Elsevier Academic Press

Part V

Incorporation of Non-natural Amino Acids for Selection by Ribosome Display and Related Methods

Chapter 19

Charging of tRNAs Using Ribozymes and Selection of Cyclic Peptides Containing Thioethers

Patrick C. Reid, Yuki Goto, Takayuki Katoh, and Hiroaki Suga

Abstract

In vitro selection methods represent a powerful approach toward identifying high-affinity peptide ligands from highly diverse peptide libraries against a desired target. We herein describe a method for the display and selection of cyclic thioether peptide libraries. Reprogramming the initiation event from fMet to an N-chloroacetyl-amino acid by utilizing flexizyme to rapidly and efficiently prepare the aa-tRNA can be effectively used to initiate translation, upon which the thiol group of an inserted cysteine at the C terminus of the designed library spontaneously reacts to yield a nonreducible cyclic thioether peptide readily compatible with any in vitro display methods. Thus, cyclic peptides already in a nonreducible stable form can be selected directly against the target of interest.

Key words: Flexizyme, Cyclic thioether peptides, In vitro selection, Ribosome display, Genetic code reprogramming

1. Introduction

Cyclic peptides have long represented an intriguing therapeutic class of molecules, with some well-known natural product peptides adopting a macrocyclic structure, such as cyclosporin A, ADEP1, and chlorofusin (1–3). Cyclic peptides exhibit a rigid structure, high affinity for their targets, and often increased resistance to proteases/degradation compared to their linear counterparts (4–6). Numerous reports of the selection of cyclic peptides by in vitro display methods can be found in the literature (7–10). Unfortunately, the vast majority of these cyclic peptides are linked via a disulfide linkage, as alternative chemical methods to catalyze peptide cyclization after translation are predominantly incompatible with current in vitro display methods. These disulfide-linked cyclic peptides are easily reduced in vivo rendering these peptides poor therapeutics. In order to develop such peptides toward the clinic, the disulfide linkage must be replaced with a nonreducible linkage, often a laborious process itself, as modifications of the linkage often alter the peptide structure resulting in significant loss in activity and/or target affinity (11). Therefore, the ability to translate and display peptides cyclized via a nonreducible linkage (e.g., in a therapeutically usable form) in a manner compatible with in vitro display methods represents a potentially significant advancement in the discovery of cyclic peptides for therapeutic use.

The following sections describe a protocol for the expression and selection (in this protocol, Ribosome Display, although any in vitro display method can be used) of cyclic peptides linked via a nonreducible thioether linkage that forms spontaneously upon translation, without the need of any exogenous reagents (12).

2. Materials

1. *RNase-free water*. Water generated by an ultrapure water system.

2. *10× PCR buffer*. 500 mM KCl, 100 mM Tris–HCl (pH 9.0), and 1% (v/v) Triton X-100. For preparation of 50 ml solution, mix 12.5 ml of 2 M KCl, 5 ml of 1 M Tris–HCl (pH 9.0), and 0.5 ml of Triton X-100, and then add water up to 50 ml. Store it in aliquots at −20°C (stable for years).

3. *5 mM dNTPs*. Mix 1 ml of 100 mM dATP, 1 ml of 100 mM dCTP, 1 ml of 100 mM dGTP, and 1 ml of 100 mM dTTP with 16 ml of water. Store it in aliquots at −20°C (stable for at least a year).

4. *Taq DNA polymerase*. Heterogeneously express *Taq* DNA polymerase in *Escherichia coli* and purify as previously reported (13).

5. *Phenol/chloroform/isoamyl alcohol solution (25:24:1)*. Mix 50 ml of TE-saturated phenol, 48 ml of chloroform, and 2 ml of isoamyl alcohol. Can be stored at 4°C for at least a year.

6. *Chloroform/isoamyl alcohol solution (24:1)*. Mix 96 ml of chloroform and 4 ml of isoamyl alcohol. Can be stored at room temperature for at least a year.

7. *10× T7 buffer*. 400 mM Tris–HCl (pH 8.0), 10 mM spermidine, and 0.1% (v/v) Triton-X. For preparation of 10 ml solution, add 0.485 g of Tris–HCl, 50 μl of 2 M spermidine, and 10 μl of Triton X-100 to 6 ml of RNase-free water. Adjust the pH of solution to 8.0 by adding 1 M HCl, and then add water up to 10 ml. Store it in aliquots at −20°C (stable for years).

8. *25 mM NTPs.* First, make 100 mM stock solution of each NTP (ATP, CTP, GTP, and UTP). Dissolve 600 mg of each NTP in 10 ml of RNase-free water. Dilute a small portion of the solution with 100 mM MOPS–KOH buffer (pH 7.0), and measure the absorbance to determine the concentration (the molar absorption coefficients at pH 7.0 for ATP, CTP, GTP, and UTP are 15,400 at 259 nm, 9,000 at 271 nm, 13,700 at 253 nm, and 10,000 at 262 nm, respectively). Adjust the concentration to 100 mM by adding RNase-free water. Then, mix equal volume of 100 mM ATP, CTP, GTP, and UTP to make 25 mM NTPs solution. Store it in aliquots at −20°C (stable for years).

9. *T7 RNA polymerase.* Express His_6-tagged T7 RNA polymerase in *E. coli* and purify it according to standard methods.

10. *Acrylamide gel solution (8, 12, or 20%).* Mix 14.4 g of urea, 4 ml of 5× TBE, and proper amount of 40% acrylamide/bisacrylamide (19/1) solution (8, 12, and 20 ml to make 8, 12, and 20% gel, respectively), and add RNase-free water up to 40 ml. Mix gently until urea dissolves completely. Add 400 μl of 10% APS and 30 μl of TEMED to the solution right before pouring to a slab-gel equipment.

11. *2× RNA loading buffer.* 8 M urea, 2 mM EDTA, 2 mM Tris, and 0.004% BPB. For preparation of 50 ml solution, mix 24 g of urea, 372 mg of EDTA·2Na·$2H_2O$, 12 mg of Tris, and 250 μl of 2% bromophenol blue (BPB), and then add water up to 50 ml. pH of the resulting solution should be around 7.5. Can be stored at room temperature for at least a year.

12. *Acid PAGE loading buffer.* Mix 50 μl of 3 M NaOAc (pH 5.2), 20 μl of 0.5 M EDTA (pH 8.0), 930 μl of formamide, and 8 μl of 2% BPB. Can be stored at room temperature for at least a year.

13. *Acid–acrylamide gel solution (20%).* Mix 1.8 g of urea, 83 μl of 3 M NaOAc (pH 5.2), and 2.5 ml of 40% acrylamide/bisacrylamide (19/1) solution, and add RNase-free water up to 5 ml. Mix gently until urea dissolves completely. Add 50 μl of 10% APS and 4 μl of TEMED to the solution right before pouring to a mini-gel equipment.

14. *E. coli ribosome.* Purify ribosome from *E. coli* A19 strain by following a method reported previously (14).

15. *5× E. coli wash buffer.* Prepare 50 ml by combining 10× TBS (25 ml), 3 M $MgCl_2$ (4.15 ml), 10% Tween20 (2.5 ml), q.s. to 50 ml in MilliQ.

16. *1× HB buffer.* Can be prepared from 2× HB buffer below.

17. *2× HB buffer.* Prepare in water, per 100 μl – 40 μl 5× *E. coli* wash buffer, 57.5 μl water, and 2.5 μl 20% heparin.

18. *Elution buffer*. Prepare in water, 25 mM EDTA (pH 8.0), 100 mM KCl, 2.5 µg/µl RNA.

19. *Reverse transcription mixture*. Water, 0.875 µl; 0.3 M NaCl, 2.5 µl; 5 mM dNTPs, 2.5 µl (final 0.5 mM); 100 µM reverse transcription primer, 0.625 µl (final 5 µM); 5× RT buffer (supplied with RT enzyme), 5 µl; and 1 µl of RT enzyme (H−).

20. *RT enzyme (H−)*. Promega M-MLV Reverse Transcriptase (RNase H−).

21. *1× PCR buffer*. 10× PCR buffer, $MgCl_2$ 2.5 mM final, dNTPs 0.25 mM final, with the desired forward and reverse primers at final concentrations of 0.25 µM each.

22. *10× TS solution*. 500 mM HEPES–KOH (pH 7.6), 1 M potassium acetate, 120 mM magnesium acetate, 20 mM ATP, 20 mM GTP, 10 mM CTP, 10 mM UTP, 200 mM creatine phosphate, 1 mM 10-formyl-5,6,7,8-tetrahydrofolic acid, 20 mM spermidine, 10 mM DTT, 15 mg/ml *E. coli* total tRNA. Use and store it in low-binding tubes. Store it in aliquots at −20°C.

23. *10× RP solution*. 3 mM magnesium acetate, 12 µM *E. coli* ribosome, 6 µM MTF, 27 µM IF1, 4 µM IF2, 15 µM IF3, 2.6 µM EF-G, 100 µM EF-Tu, 6.6 µM EF-Ts, 2.5 µM RF2, 1.7 µM RF3, 5 µM RRF, 40 µg/ml creatine kinase, 30 µg/ml myokinase, 1 µM inorganic pyrophosphatase, 1 µM nucleotide diphosphate kinase, and 1 µM T7 RNA polymerase. Use and store it in low-binding tubes. Store it in aliquots at −80°C.

24. *10× ARS solution*. Mix the ARSs; 7.3 µM AlaRS, 0.3 µM ArgRS, 3.8 µM AsnRS, 1.3 µM AspRS, 0.2 µM CysRS, 0.6 µM GlnRS, 2.3 µM GluRS, 0.9 µM GlyRS, 0.2 µM HisRS, 4.0 µM IleRS, 0.4 µM LeuRS, 1.1 µM LysRS, 0.3 µM MetRS, 6.8 µM PheRS, 1.6 µM ProRS, 0.4 µM SerRS, 0.9 µM ThrRS, 0.3 µM TrpRS, 0.2 µM TyrRS, and 0.2 µM ValRS.

25. *5 mM each amino acid solution*. Mix the 19 proteinogenic amino acids, Met is excluded, to make a 5 mM stock solution.

26. *Glycogen*. Prepared at 20 mg/ml in RNase-free water.

27. *Required primers*. All primers are 5′–3′.

    Fx-F: GTAATACGACTCACTATAGGATCGAAAGATTT CCGC.

    Fx-R1: ACCTAACGCTAATCCCCTTTCGGGGCCGCGG AAATCTTTCGATCC.

    T7-F: GGCGTAATACGACTCACTATAG.

    T7-F2: GTAATACGACTCACTATAGGATCGAAAGATTT CCGC.

    Fx-R2: ACCTAACGCTAATCCCCT.

    I-F1:   GTAATACGACTCACTATAGGCGGGGTGGAGCA GCCTGGTAGCTCGTCGG.

I-R1: GAACCGACGATCTTCGGGTTATGAGCCCGACGAGCTACCAGGCT.

I-R2: TGGTTGCGGGGGCCGGATTTGAACCGACGATCTTCGGG.

I-R3: TGGTTGCGGGGGCCGGATTT.

28. *Streptavidin magnetic beads.* Invitrogen Dynal Streptavidin M280 Magnetic Beads commercially available from Invitrogen.
29. *TA cloning kit.* Promega pGEM-T cloning kit.
30. *Sequencing kit.* Applied Biosystem BigDye Terminator Sequencing Kit.
31. *Transcription Kit.* HiScribe T7 In Vitro Transcription Kit (E2030; New England Biolabs).

3. Methods

3.1. Preparation of DNA Template for Flexizyme (eFx) (15)

1. Prepare a master mix solution for the following extension and PCR reactions on ice by mixing 120 µl of 10× PCR buffer, 12 µl of 250 mM $MgCl_2$, 60 µl of 5 mM dNTPs, and 9 µl of *Taq* DNA polymerase with 1,000 µl of RNase-free water. Keep the master mix solution on ice.

2. Mix 0.5 µl of 200 µM Fx-F primer and 0.5 µl of 200 µM Fx-R1a reverse primer in a 200-µl PCR tube with 100 µl of the master mix solution.

3. Set the sample in a PCR thermal cycler and carry out the extension reaction with 95°C for 1 min; five cycles of 50°C for 1 min and 72°C for 1 min. The extension product is stable at −20°C for years.

4. Mix 5 µl of the extension product (without any purification) with 2.5 µl of 200 µM T7-F primer, 2.5 µl of 200 µM Fx-R2, and 1,000 µl of the master mix solution. Aliquot the resulting solution into five 200-µl PCR tubes. Divide the resulting solution to five aliquots in 200-µl PCR tubes.

5. Set the tubes in a PCR thermal cycler and run it with 95°C for 1 min; 12 cycles of 95°C for 40 s, 50°C for 40 s, and 72°C for 40 s.

6. Check the amplified DNA by agarose gel electrophoresis. If the band corresponding to the objective band is faint, run 2–3 additional PCR cycles.

7. Combine the DNA solutions and add 1 ml of phenol/chloroform/isoamyl alcohol solution to the resulting sample. Shake the tube intensely.

8. Centrifuge the sample at $15,000 \times g$ for 5 min at 25°C.

9. Recover the water layer (upper phase) and mix it with 1 ml of chloroform/isoamyl alcohol solution. Shake the tube intensely.
10. Centrifuge the sample at 15,000×g for 5 min at 25°C.
11. Recover the water layer and add 100 μl of 3 M NaCl and 2.2 ml of ethanol. Mix the sample well.
12. Centrifuge the sample at 15,000×g for 15 min at 25°C.
13. Remove the supernatant and add 500 μl of 70% ethanol to the tube.
14. Centrifuge the sample at 15,000×g for 15 min at 25°C.
15. Remove the supernatant completely. Open the tube lid and cover it with tissues, and then dry the DNA at room temperature for 10 min.
16. Add 100 μl of RNase-free water and resuspend the DNA pellet (see Note 1).

3.2. Preparation of DNA Template for Initator tRNA$_{CAU}^{fMetE}$ (12)

1. Prepare a master mix solution for the following extension and PCR reactions on ice by mixing 132 μl of 10× PCR buffer, 13.2 μl of 250 mM MgCl$_2$, 66 μl of 5 mM dNTPs, and 9.9 μl of *Taq* DNA polymerase with 1,100 μl of RNase-free water. Keep the master mix solution on ice (see Note 2).
2. Mix 0.5 μl of 20 μM I-F1 and 0.5 μl of 20 μM I-R1 in a 200-μl PCR tube with 10 μl of the master mix solution.
3. Set the sample in a PCR thermal cycler and carry out the extension reaction with 95°C for 1 min; five cycles of 50°C for 1 min and 72°C for 1 min.
4. *First PCR*: Mix 10 μl of the extension product (without any purification) with 0.5 μl of 200 μM T7-F2 primer, 0.5 μl of 200 μM I-R2, and 190 μl of the master mix solution in a 200-μl PCR tube.
5. Set the tubes in a PCR thermal cycler and run it with 95°C for 1 min; five cycles of 95°C for 40 s, 50°C for 40 s, and 72°C for 40 s.
6. Second PCR: Mix 5 μl of the PCR product (without any purification) with 2.5 μl of 200 μM T7-F2 primer, 2.5 μl of 200 μM I-R3, and 1,000 μl of the master mix solution. Aliquot the resulting solution into five 200-μl PCR tubes. Divide the resulting solution to five aliquots in 200-μl PCR tubes.
7. Set the tubes in a PCR thermal cycler and run it with 95°C for 1 min; 12 cycles of 95°C for 40 s, 50°C for 40 s, and 72°C for 40 s.
8. Check the amplified DNA by agarose gel electrophoresis. If the band corresponding to the objective band is faint, run 2–3 additional PCR cycles.

9. *Purification of the PCR product.* Combine the DNA solutions and add 1 ml of phenol/chloroform/isoamyl alcohol solution to the resulting sample. Shake the tube intensely.

10. Centrifuge the sample at $15{,}000 \times g$ for 5 min at 25°C.

11. Recover the water layer and mix it with 1 ml of chloroform/isoamyl alcohol solution. Shake the tube intensely.

12. Centrifuge the sample at $15{,}000 \times g$ for 5 min at 25°C.

13. Recover the water layer and add 100 μl of 3 M NaCl and 2.2 ml of 100% ethanol. Mix the sample well.

14. Centrifuge the sample at $15{,}000 \times g$ for 15 min at 25°C.

15. Remove the supernatant and add 500 μl of 70% ethanol to the tube.

16. Centrifuge the sample at $15{,}000 \times g$ for 15 min at 25°C.

17. Remove the supernatant completely. Open the tube lid and cover it with tissues, and then dry the DNA at room temperature for 10 min.

18. Add 100 μl of RNase-free water and resuspend the DNA pellet (see Note 3).

3.3. Transcription of eFx and tRNA$_{CAU}^{fMetE}$

1. Prepare in vitro transcription reaction mixture. This step can be performed using option A or option B depending on RNA molecules to be made (see Note 4).

    (a) Preparation of eFx.
    Mix 100 μl of 10× T7 buffer, 100 μl of 100 mM DTT, 120 μl of 250 mM MgCl$_2$, 200 μl of 25 mM NTPs, 15 μl of 2 M KOH, 100 μl of DNA template, and 20 μl of T7 RNA polymerase with 345 μl of RNase-free water.

    (b) Preparation of tRNA$_{CAU}^{fMetE}$
    Mix 100 μl of 10× T7 buffer, 100 μl of 100 mM DTT, 90 μl of 250 mM MgCl$_2$, 150 μl of 25 mM NTPs, 11.25 μl of 2 M KOH, 50 μl of 100 mM GMP, 100 μl of DNA template, and 20 μl of T7 RNA polymerase with 303 μl of RNase-free water.

2. Incubate the transcription reaction mixture in an air incubator at 37°C for 5 h. White precipitations of inorganic pyrophosphate can be observed if the transcription reaction proceeds accordingly.

3. Add 20 μl of 100 mM MnCl$_2$ and 4 μl of DNase I to the reaction mixture. Incubate the solution at 37°C for an additional 30 min.

4. Add 75 μl of 500 mM EDTA (pH 8.0), 100 μl of 3 M NaCl, and 1 ml of isopropanol. Mix the sample well and stand it at room temperature for 5 min.

5. Centrifuge the sample at 15,000×*g* for 5 min at 25°C.

6. Remove the supernatant completely. Open the tube lid and cover it with saran wrap, and then dry the RNA at room temperature for 10 min. The RNA pellet can be stored at −20°C for at least a week.

3.4. Purification of Flexizyme and Initiator tRNA by PAGE

1. Add 100 μl of RNase-free water and resuspend the RNA pellet. Mix the RNA solution with 100 μl of 2× RNA loading buffer.

2. Incubate the sample on a heat block at 95°C for 1 min.

3. Apply the resulting sample onto a denaturing polyacrylamide gel and run. (Use 12% polyacrylamide gel with 250 V for 1 h to purify eFx. Use 8% polyacrylamide gel with 250 V for 1 h to purify tRNA$_{CAU}^{fMetE}$.)

4. Remove the gel from the gel plates and put it on a TLC plate containing a fluorescent indicator covered with a plastic wrap. Cover the gel with another plastic wrap. Visualize RNA band by irradiating with 260-nm UV lamp in a dark room. Trace the pattern of RNA band with a marker. Mark the band promptly (see Note 5).

5. Cut the gel by a razor along the mark. Crush the gel pieces containing RNA finely in a 50-ml tube. Recovery yield of RNA can be improved by crushing the gel into a paste.

6. Add 3 ml of 0.3 M NaCl to the resulting gel paste, and then shake the tube at room temperature for 1 h.

7. Centrifuge the sample at 15,000×*g* for 5 min at 25°C. Recover the supernatant carefully.

8. Repeat steps 6 and 7 one more time.

9. Combine all supernatants, and filter through a 0.45-μm syringe filter.

10. Add twofold volume of ethanol to the resulting RNA solution. Mix the sample well.

11. Centrifuge the sample at 15,000×*g* for 15 min at 25°C.

12. Remove the supernatant, add 1,000 μl of 70% ethanol to the tube, and wash the pellet.

13. Centrifuge the sample at 15,000×*g* for 3 min at 25°C.

14. Remove the supernatant completely. Open the tube lid and cover it with tissues, and then dry the RNA at room temperature for 10 min.

15. Add 50 μl of RNase-free water and resuspend the RNA pellet.

16. Determine the concentration of RNA by a UV spectrometer. (The length of tRNA$_{CAU}^{fMetE}$ and eFx are 76 and 45, respectively.)

17. Dilute the RNA solution with RNase-free water to make a 250 μM stock solution of flexizyme. The RNA solution can be stored at −20°C for at least 2 years. Anticipated results of 15,000–45,000 pmol of RNA can be obtained from a 1-ml-scale transcription reaction.

3.5. Preparation of N-Chloroacetyl-Trp-CME

25 mM N-chloroacetyl-Trp-CME in DMSO-d_6. Synthesize the acid substrate for flexizyme reaction in two steps from *tryptophan* according to the previously reported method (12). To a solid of N-chloroacetyl-Trp-CME, add DMSO-d_6 to make a 200 mM stock solution. Keep it at −20°C (generally stable for a year). To make 25 mM working solution, mix 10 μl of 200 mM stock solution with 70 μl of DMSO-d_6. 25 mM working solution can be stored in aliquots at −20°C (generally stable for a year) (see Note 6).

3.6. Acylation of Initiator tRNA with Flexizyme

1. Mix 2 μl of 500 mM HEPES–KOH buffer (pH 7.5), 2 μl of 250 μM eFx, and 2 μl of 250 μM tRNA$_{CAU}^{fMetE}$ with 6 μl of RNase-free water.

2. Heat the sample at 95°C for 2 min, and then slowly cool it at room temperature over 5 min.

3. Add 4 μl of 3 M MgCl$_2$ into the sample, and then incubate it at room temperature for 5 min followed by incubation on ice for 3 min.

4. Add 4 μl of 25 mM acid substrate in DMSO-d_6 into the sample and mix well.

5. Incubate the acylation reaction mixture on ice for 2 h.

6. Add 80 μl of 0.3 M NaOAc (pH 5.2) and 200 μl of ethanol into the reaction mixture to quench the acylation reaction.

7. Centrifuge the sample at 15,000×g for 15 min at 25°C. Then, remove the supernatant completely (see Note 7).

8. Add 50 μl of 70% ethanol containing 0.1 M NaOAc (pH 5.2) to the tube and vortex the tube well to break the RNA pellet into pieces (see Note 8).

9. Centrifuge the sample at 15,000×g for 5 min at 25°C. Then, remove the supernatant completely.

10. Repeat steps 8–9 one more time.

11. Add 50 μl of 70% ethanol to the tube.

12. Centrifuge the sample at 15,000×g for 3 min at 25°C.

13. Open the tube lid and cover it with saran wrap, and then dry the RNA at room temperature for 5 min (see Note 9).

3.7. Preparation of Desired Ribosome Display Library and Transcription to mRNA

1. Prior to implementation of this protocol for the selection of cyclic thioether peptide libraries using ribosome display, significant consideration should be given to library design. The library should contain the necessary upstream 5′ ribosome-binding

elements and the necessary 3′ tether sequence required for ribosome display. Expression of the thioether-closed cyclic peptide library only requires an initiation ATG site in which the chloroacetyl-Trp initiates the sequence translation, and a C-terminal cysteine codon (either TGT or TGC may be used) to which the thioether linkage forms between the chloroacetyl-Trp and the cysteine spontaneously after translation. The codons between the N-terminal chloroacetyl-Trp and C-terminal cysteine can be randomized through the appropriate use of NNN codons (NNK, NNW, NNS, or NNT, NNA can be used depending on the desire for library diversity, to which consideration should be given). DNA libraries of the format ATG-NNN_{4-15}-TGT/TGC have been successfully explored, with all yielding the spontaneous cyclic thioether peptides upon translation.

2. Transcription of the dsDNA library is performed prior to translation expression according to the manufacturer's instructions using the HiScribe T7 In Vitro Transcription Kit (E2030; New England Biolabs). The mRNA can be aliquoted and stored at −80 C.

3.8. Selection of Target

1. Significant consideration should be given to target selection. The better the quality of target protein (e.g., purity and proper folding/native/active structure), the greater the likelihood of selected peptides to exhibit the desired affinity/activity.

3.9. In Vitro Translation

1. To facilitate the reconstitution of translation systems (16, 17), three solutions are prepared separately; one is composed of total tRNAs and small molecules, including buffer, nucleotide triphosphates, and amino acids, referred to as the TS solution; the second contains ribosome and essential protein factors, referred to as the RP solution; the other consists of ARSs, referred to as the ARS solution (see Note 10).

2. Set up the translation reaction as follows (see Note 11):

10× TS solution	0.5 μl
10× RP solution	0.5 μl
10× ARS solution	0.5 μl
5 mM each amino acid solution	0.5 μl
Library mRNA	1.6 μM
acyl–tRNAs synthesized by flexizyme	500 pmol each (see Note 12)
RNase-free water	To 5 μl

3. Incubate the reaction mixture at 37°C in a water bath for 30 min.

4. Add an equal volume of 2× HB buffer to the translation mixture and place on ice (see Note 13).

3.10. Selection

1. All selection steps should be performed in low-binding Eppendorf tubes and at 4°C to ensure stability of the ribosome–peptide–mRNA complex.

2. The following selection steps assume the availability of a biotin target and the use of streptavidin magnetic beads (SA beads; Invitrogen DYNAL beads are recommended) although affinity tag combinations can also be used.

3. Perform negative selection (Rd1: 1 time, Rd2+: 3–6 times) (see Note 14) by adding the same volume of beads (prewashed in TBS-T) as used in the positive selection step to the translation mixture and incubating on a rotator at 4°C for 20 min.

4. Perform positive selection by adding the desired target protein to a final concentration of 250 nM (see Notes 15 and 16) by first determining the binding capacity of the biotin target protein to the SA beads, and then calculating the amount of beads required to reach 250 nM in the translation volume.

5. Incubate on rotator at 4°C for 1 h.

6. Wash beads four times in 100 μl of 1× HB buffer.

7. Add elution buffer to the beads and incubate at 4°C for 30 min (see Note 17).

8. Add (in Rd1 only) $MgCl_2$ (final 30 mM) and NaCl (final 0.3 M) and tap to mix (see Note 18).

9. Add glycogen (1 μl of 20 mg/ml).

10. Add 2.5 volumes of 100% EtOH.

11. Centrifuge at $15,000 \times g$ for 15 min at room temperature.

12. Wash pellet twice with 70% EtOH and centrifuge again.

13. Dry for 5 min at room temperature.

14. Resuspend pellet in 12.5 μl of water.

3.10.1. Reverse Transcription, Real-Time PCR, and PCR

1. Add an equal volume of the reverse transcription mixture and incubate for 30 min in a water bath at 42°C.

2. The resulting reverse transcription mixture is diluted 20-fold in 1× PCR buffer. Real-time PCR is then performed on a sample of this mixture to determine the amount of recovered DNA (see Note 19).

3. Perform amplification of selected DNA by PCR by simply adding Taq DNA polymerase to the mixture and aliquoting to the appropriate PCR tubes (see Note 20). Example conditions: 95°C for 40 s; 58°C for 40 s; 72°C for 40 s.

3.10.2. Additional Rounds of Selection	The number of rounds required to observe positive binding varies depending on the target protein, library diversity, and selection conditions employed. Normally, greater than a 1% recovery of DNA should be observed by the fifth to sixth rounds.
3.10.3. Cloning and Sequencing	PCR products are cloned by way of TA cloning using Promega's pGEM-T cloning kit, according to the manufacturer's protocol. Properly cloned in sequences can then be prepared for sequence analysis using Applied Biosystem's BigDye Terminator Sequencing Kits according to the included manufacturer's protocol, and sent for sequencing to any number of vendors/core facilities.

4. Notes

1. The DNA template for flexizyme can be aliquoted and stored at −20°C for years.
2. The initiator tRNA used for reprogramming is $tRNA_{CAU}^{fMetE}$ derived from *E. coli* initiator tRNA with a single mutation of C1G and no base modifications. However, omitting Met is sufficient enough to deplete the background initiation by fMet (12, 18, 19).
3. The DNA template for $tRNA_{CAU}^{fMetE}$ can be aliquoted and stored at −20°C for years.
4. All the following steps should be performed in an RNase-free manner. Use RNase-free tubes, pipettes, pipette tips, and water. Wear gloves at all times.
5. Irradiation of UV at a short range for a long time may cause RNA damage.
6. In the case where the quality of the acid substrate stock solution needs to be checked, measure its NMR spectrum.
7. Ethanol precipitation and the following washing steps should be carried out at around room temperature, not at lower temperature, to avoid undesirable precipitation of salts.
8. The washing steps (steps 8–9) are critical for the following translation reaction. Carryover of magnesium and sodium ions would decrease the efficiency of translation reaction.
9. Dried acyl–tRNA pellets can be stored at −80°C for up to 2 weeks with no significant loss in initiation efficiency.
10. The in vitro translation system has been customized for optimal efficiency of initiation suppression and library expression. The use of other translation systems may result in a significant loss in efficiency.
11. The volume of translation depends on the round of the selection. We commonly use the following: Rd1 (100 μl), Rd2 (20 μl),

Rd3+ (5 μl). The translation mixture setup is described at the 5-μl scale, and can be multiplied by 20 to get the 100-μl-scale equivalent.

12. Resuspend the acyl–tRNAs in proper amount of 1 mM NaOAc (pH 5.2) just before mixing them with the translation reaction mixture because they are unstable and easily hydrolyzed in solution (especially under alkaline conditions).

13. The HB buffer serves to stabilize the ribosome–peptide–mRNA complex prior to selection.

14. Negative selection is a critical step in the selection process. Careful consideration should be given to match the number of negative selection steps with the desired positive selection target. Magnetic beads are highly recommended. The same beads used in the positive selection step should be also used in the negative selection steps.

15. Target protein must have a tag. Biotinylated target protein is the preferred tag of choice where possible. Target protein can be first captured to a streptavidin magnetic bead (after determining the binding capacity) and added directly to the display mixture in a one-step process, or tagged protein can be first added to the display mixture, incubated for 30 min, and then the streptavidin bead added subsequently in a two-step process. Other tag and magnetic bead combinations are also possible.

16. The final concentration of target protein can be varied as desired. We normally use 250 nM in Rd1 and Rd2, and then decrease the target protein concentration in subsequent rounds.

17. Elution buffer serves to break the complexes to release the template mRNA. The amount of elution buffer to use varies, but we commonly use as follows: Rd1 (25 μl), Rd2 (10 μl), Rd3+ (5 μl).

18. The described precipitation step is only performed in round one. Subsequent rounds can proceed directly to reverse transcription by adding the reverse transcription mixture (which is at 2×) directly to the bead–elution buffer mixture.

19. Real-time PCR should be used to not only measure/monitor the enrichment of the selected DNA, but also to monitor the potential enrichment of nonspecific sequences by measuring the amount of sequences that bind in the negative selection steps. This information should then be used to increase the number of negative selections if needed. Real-time PCR results should be used to calculate the number of PCR cycles required to amplify the DNA templates. We normally use CP + 5 cycles.

20. Exact PCR conditions vary depending on primers and PCR machine. Please consult previous protocols of ribosome display for more information (20).

Acknowledgments

We would like to thank Dr. Hiroshi Murakami and all of the members of the Suga lab and PeptiDream for their helpful suggestions.

References

1. Starzl, T.E. et al. Liver transplantation with use of cyclosporin A and prednisone. *N. Engl. J. Med.* **305** (5): 266–9 (1981).
2. Brötz-Oesterhelt, H. et al. Dysregulation of bacterial proteolytic machinery by a new class of antibiotics. *Nature Med.* **11**, 1082–1087 (2005).
3. Duncan, S.J. et al. Isolation and structure elucidation of Chlorofusin, a novel p53-MDM2 antagonist from a Fusarium sp. *J. Am. Chem. Soc.* **123**(4), 554–560 (2001).
4. Bogdanowich-Knipp, S. et al. Solution stability of linear vs. cyclic RGD peptides. *J. Pept. Res.* **53**, 530–541(1999).
5. Gudmundsson, O. S. et al. Phenylpropionic acid-based cyclic prodrugs of opioid peptides that exhibit metabolic stability to peptidases and excellent cellular permeation. *Pharm. Res.* **16**, 16–22 (1999).
6. Borchardt, R. T. Optimizing oral absorption of peptides using prodrug strategies. *J. Control Rel.* **62**, 231–238 (1999).
7. Hall, P.R. et al. Phage display selection of cyclic peptides that inhibit Andes virus infection. *J. Virol.* **17**, 8965–8969 (2009).
8. Litovchick, A. et al. Selection of cyclic peptide aptamers to HCV IRES RNA using mRNA display. *Proc. Natl. Acad. Sci. USA* **105**(40), 15293–15298 (2008).
9. Torregrossa, P. et al. Selection of poly-alpha 2,8-sialic acid mimotopes from a random phage peptide library and analysis of their bioactivity. *J. Biol. Chem.* **279**(29), 30707–30714 (2004).
10. Meiring, M.S. et al. In vitro effect of a thrombin inhibition peptide selected by phage display technology. *Thromb. Res.* **107**(6), 365–371 (2001).
11. Pero, S.C. et al. Identification of novel non-phosphorylated ligands, which bind selectively to the SH2 domain of Grb7. *J. Biol. Chem.* **277**, 11918–11926 (2002).
12. Goto, Y. et al. Reprogramming the translation initiation for the synthesis of physiologically stable cyclic peptides. *ACS Chem. Biol.* **3**, 120–129 (2008).
13. Pluthero, F.G. Rapid purification of high-activity Taq DNA polymerase. *Nucleic Acids Res.* **21**, 4850–4851 (1993).
14. Clemons, W.M., Jr. et al. Crystal structure of the 30 S ribosomal subunit from Thermus thermophilus: purification, crystallization and structure determination. *J. Mol. Biol.* **310**, 827–843 (2001).
15. Murakami, H., Ohta, A., Ashigai, H. & Suga, H. A highly flexible tRNA acylation method for non-natural polypeptide synthesis. *Nat. Methods* **3**, 357–359 (2006).
16. Kung, H.F. et al. DNA-directed in vitro synthesis of beta-galactosidase. Studies with purified factors. *J. Biol. Chem.* **252**, 6889–6894 (1977).
17. Shimizu, Y. et al. Cell-free translation reconstituted with purified components. *Nat. Biotechnol.* **19**, 751–755 (2001).
18. Goto, Y., Murakami, H. & Suga, H. Initiating translation with D-amino acids. *RNA* **14**, 1390–1398 (2008).
19. Goto, Y. & Suga, H. Translation initiation with initiator tRNA charged with exotic peptides. *J. Am. Chem. Soc.* **131**, 5040–5041 (2009).
20. Zahnd, C. et al. Ribosome Display: selecting and evolving proteins in vitro that specifically bind to a target. *Nature Methods* **4**, 269–279 (2007).

Chapter 20

Update on Pure Translation Display with Unnatural Amino Acid Incorporation

R. Edward Watts and Anthony C. Forster

Abstract

The identification of peptide and protein ligands by directed evolution in vitro has been of enormous utility in molecular biology and biotechnology. However, the translation step in almost all polypeptide selection methods is performed in vivo or in crude extracts, restricting applications. These restrictions include a limited library size due to transformation efficiency, unwanted competing reactions in translation, and an inability to incorporate multiple unnatural amino acids (AAs) with high fidelity and efficiency. These restrictions can be addressed by "pure translation display" where the translation step is performed in a purified system. To date, all pure translation display selections have coupled genotype to phenotype in a ribosome display format, though other formats also should be practical. Here, we detail the original, proof-of-principle, pure-translation-display method because this version should be the most suitable for encoding multiple unnatural AAs per peptide product toward the goal of "peptidomimetic evolution." Challenges and progress toward this ultimate goal are discussed and are mainly associated with improving the efficiency of ribosomal polymerization of multiple unnatural AAs.

Key words: Pure translation, Ribosome display, mRNA display, In vitro directed evolution, In vitro selection, Ligand and drug discovery, Unnatural amino acid, tRNA, Peptidomimetic, Peptide, Protein synthesis

1. Introduction

1.1. Successes and Limitations of the Progenitors: Displays Using In Vivo and Crude Translation Systems

The encoding, synthesis, selection and directed evolution of vast peptide or protein libraries has been enabled by monoclonal antibody technology, phage display, ribosome display (Fig. 1), mRNA display and emulsions (e.g. 1–8). Library members are usually selected or evolved to identify ligands for an immobilized target, but selections for more complicated functions such as catalysis are also feasible.

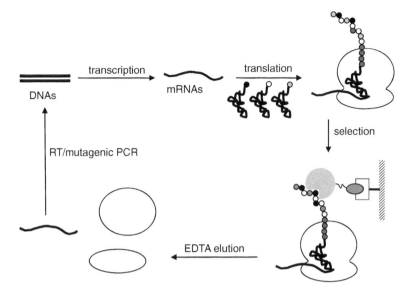

Fig. 1. Evolution of a ligand for a target by ribosome display. A library of DNAs encoding random sequences of AAs is transcribed into mRNAs. The mRNAs are translated into polypeptides and the ribosome stalls prior to release of the peptide and mRNA due to a deficiency in stop codons. The complexes are stabilized and then screened for their ability to bind an immobilized target. Nonbinders are washed away and the mRNAs encoding the binders are released and amplified by reverse transcription-mutagenic PCR. The cycle is repeated until the library converges on a consensus sequence(s) for the highest affinity ligands.

All of these methods were developed using either in vivo or crude translation systems. Despite the enormous utility of these approaches, there are some limitations. Monoclonal antibody technology is obviously limited to antibodies, while phage display is limited in diversity by cellular transformation efficiency. The in vitro translation systems of ribosome display and mRNA display are adversely affected by unwanted competing reactions. These include degradation of DNA templates and mRNAs by various nucleases, translation of endogenous mRNAs, competition from natural aminoacyl-tRNAs or release factors when trying to incorporate unnatural AAs, dissociation of the ternary complex that links genotype to phenotype by transfer-messenger RNA (tmRNA) and release factors, degradation of the displayed peptide or protein products by proteases, and interference of cellular components with the binding and washing steps of the selection.

1.2. Pure Translation Displays

The primary motivation for our method was the proposal that directed evolution could be engineered for identification of peptide ligands containing unnatural AAs for drug discovery (9). Unnatural AAs are known to impart upon peptides drug-like properties such as protease resistance and membrane permeability. As a first step, we reconstituted a purified translation system (9) which

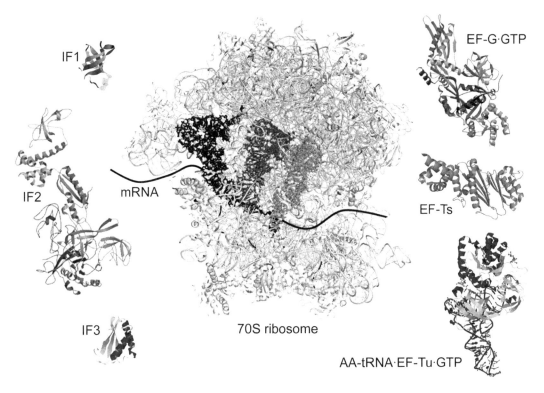

Fig. 2. Components of the purified translation system for our pure translation display. The system is based upon that of *E. coli* except that release factors and RSs are omitted. Images were created from coordinates deposited in the Protein Data Bank (PDB entries 3I4O, 3CW2, 2IFE, 2XEX, 1XB2, 1OB2, 2WDG and 2WDI).

enabled the second step: programming the ribosome to incorporate multiple unnatural AAs per product (10) (discussed below; Fig. 2). Accomplishing the second step in cells or impure cell extracts with high fidelity was impractical due to competition from natural components (11, 12). Our third step, based upon ribosome display (2), was development of the first translation display in a purified, reconstituted, translation system (13).

Though our "pure translation display" system was rudimentary, it nevertheless demonstrated the capability of preventing all of the unwanted competing reactions described in Subheading 1.1, suggesting utility for selections even with natural AAs alone. Indeed, more complex versions of pure translation display have been developed that contain all 20 aminoacyl-tRNA synthetases (RSs), in contrast to our RS-free system. All of these systems use versions of the "PURE" protein synthesis system (14), commercially available in recent years, in a format like ribosome display. The PURE translation system is currently less efficient than crude systems in terms of protein yield (15) but has found utility in several labs. The first selection of protein ligands in a purified translation system used both IL-13 and human growth hormone receptor as targets (16). Ligands have also been selected against hen egg white lysozyme (17), and epitopes of anti-FLAG M2 antibody and anti-β-Catenin antibodies have been mapped in purified systems (18). In addition, compensatory, second-site mutations have been selected that restore binding to proline-rich sequences by a mutated WW domain (19).

1.3. Practical Considerations for Performing Pure Translation Display with Unnatural AAs

Our method described below was developed toward the goal of peptidomimetic ligand evolution. Only three precharged aminoacyl-tRNAs and no RSs or release factors were used in the translation, thus leaving 58 of the 64 codons open for future reassignment to unnatural AAs (13). Genotype was noncovalently coupled to phenotype like in ribosome display with two main differences: the spacer sequence traversing the ribosome exit tunnel was a polymer of only two different AAs encoded by a highly repetitive mRNA sequence, and ribosomal pausing was induced near the ends of the mRNAs rather than at their 3′ ends by omission of cognate aminoacyl-tRNAs (Fig. 3). In the proof-of-principle selection for optimal spacer length for binding to avidin, biotinyl-methionine prepared by chemical modification of Met-tRNA$_i^{fMet}$ (see Subheading 3.3) was displayed at the amino terminus of the polypeptide. As expected, mRNAs that encoded polypeptides long enough to traverse the ribosome tunnel and hence be accessible to avidin were selected from shorter mRNAs (Fig. 3).

It should be mentioned that this 2004 method is still the only published selection from a library of mRNAs by an unnatural AA version of pure translation display and that the method was reviewed

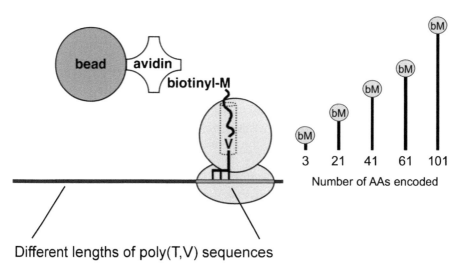

Fig. 3. Schematic of selection of mRNAs from our model mRNA library by pure translation display. The mRNA encoding 101 AAs was selected most efficiently, the mRNAs encoding 61 and 41 AAs were moderately selected, and the mRNAs encoding 21 and 3 AAs were not recovered from the selection.

in detail in the following year (20). Thus, some discussion of impediments to major progress since then and efforts to address these impediments is warranted here. The main difficulty is low product yields when incorporating multiple unnatural AAs (see Note 3 below). Nevertheless, purified systems give higher yields of unnatural AA incorporation than crude systems and our method is compatible with all types of unnatural AA technology. So this method is currently being pursued by several laboratories including ours with the goal of peptidomimetic evolution. With this goal in mind, there are three main practical decisions to be made:

1. Choice of tRNA acylation method for completely reassigning multiple codons to unnatural AAs (so that the sequence of each selected mRNA specifies a unique full-length peptidomimetic).
2. Whether or not to include some RSs.
3. Choice of method for coupling genotype to phenotype.

For decision (1), there are several means of charging tRNAs with unnatural AAs. The four most promising charging methods have all now been shown to incorporate multiple unnatural AAs per ribosomal product with high fidelity (Table 1, left column). All of these methods were optimized initially for a single unnatural AA incorporation per product, which is more straightforward and efficient than multiple unnatural AA incorporations. One is a chemoenzymatic method that chemically couples an amino-protected AA to the dinucleotide pdCpA, which is then ligated to a tRNA^{minusCA} in vitro transcript using T4 RNA ligase (21). A second chemoenzymatic method chemically activates an AA, which is then charged onto a full-length tRNA by a specially evolved ribozyme (22). A third chemoenzymatic method charges total tRNA with RSs and supplied natural AAs, then chemically monomethylates the aminoacyl-tRNAs (23). The fourth method mischarges a full-length tRNA using its cognate RS and an unnatural AA that had been determined empirically to be an efficient substrate for the RS (24, 25).

For decision (2), inclusion of RSs in the translation can be counter-productive because at least seven of the 20 RSs can hydrolyze certain mischarged AAs off their cognate tRNAs in a proofreading reaction known as posttransfer editing (26), and RSs charge tRNAs with trace contaminating cognate natural AAs in marked preference to unnatural AAs (25). Only the fourth mischarging method described above is performed in situ by including certain RSs in the translation mixture (25, 27), though it is theoretically possible to leave out the RSs by precharging the tRNAs. In this method, contaminating AAs do get incorporated, though this problem has been eliminated in many cases (25, 27). Translations performed with unnatural AAs precharged by ribozymes also generally include certain RSs, presumably to decrease the concentration of total tRNA necessary to incorporate

Table 1
Multiple incorporations of elongator unnatural AAs per product in purified translation systems

Method of tRNA acylation	Unnatural AA type					No. of unnatural AAs incorporated	Notes	References
	Side chain analog	N-methyl AA	N-alkyl glycine (peptoid)	α-Hydroxy acid	Cyclization (nondisulfide)			
Ligation of pdCpA-protected AA to tRNA^minusCA transcript	X					3–5	Five consecutive allylglycines incorporated at 30% yield relative to wild type; three consecutive different unnatural AAs incorporated at 55% yield	(10)
	X					3–5	Similar to lower yields; slow incorporation rate	(28)
	X					2–5	Effects of tRNA modifications, anticodon mutations, dCA and unnatural AA defined	(36)
Mischarging of native tRNA with unnatural AA using RS	X					10	Ten different unnatural AAs incorporated into a single peptidomimetic at 37% yield	(27)
	X					13	Thirteen unnatural AAs incorporated into a single peptidomimetic	(25)

Method				Description	Ref
Chemical mono-methylation of native AA-tRNA	X			2, 3	Two consecutive N-Me-AAs incorporated at 23–68% yield; three consecutive gave lower yields	(44)
Mischarging of tRNA transcript using activated AA and ribozyme	X			6	Three different unnatural AAs incorporated in six consecutive positions	(22)
		X		2–12	Four consecutive different α-hydroxy acids incorporated into polyesters	(31)
	X		X	5–10	Three different N-methyl AAs incorporated into 5–10 consecutive positions at ~10% yield	(30)
		X	X	2–6	Three different N-alkyl glycines incorporated in 2–6 consecutive positions at 50–1% yields	(29)

the natural AAs into the peptidomimetics. (The concentration of total tRNA is important because very high tRNA concentrations can inhibit translation (28).) Though the extent of proofreading of unnatural AA-tRNAs by RSs is uncertain, proofreading and recharging in situ was apparently not problematic in translations performed with aminoacyl-tRNAs precharged by ribozymes based on identification of many correct products by mass spectrometry (29–31).

For decision (3), the theoretical use of pure translation display in an mRNA display format has conceivable advantages and disadvantages over the ribosome display format. mRNA Display has the disadvantage of three extra steps: coupling of the mRNA library members to puromycin, coupling these mRNA–puromycin conjugates to their polypeptide products, and purification prior to selection (8). Furthermore, the coupling of mRNA–puromycin to polypeptide is slow and provides additional time for the competing reaction of peptidyl-tRNA drop off to occur; drop off may be exacerbated in peptidomimetic synthesis because it is predicted to be faster for short unnatural versus natural peptidyl-tRNAs (20). Nevertheless, a peptide containing multiple unnatural AAs has been successfully fused to an mRNA–puromycin conjugate in a proof-of-principle experiment (27). The advantages of the mRNA display format would be a more stable (covalent) coupling of mRNA to polypeptide, obviating the presence of ribosomes in the selection step (perhaps reducing nonspecific binding) and obviating the need to synthesize a spacer peptide to span the ribosome tunnel. Coupling of genotype to phenotype for pure translation display could also be done in theory by tRNA display (32) or emulsions (33), but we will focus on the ribosome display format here because pure translation display selections have only been done in the ribosome display format to date. In the ribosome display format, the spacer codon sequence should ideally be translated by natural aminoacyl-tRNAs because multiple such incorporations are generally more efficient than with unnatural aminoacyl-tRNAs (28). Minimizing the number of different spacer codons is desirable to maximize the number of codons left available for reassignment to unnatural AAs. However, cloning, translation and display of highly repetitive sequences has proved challenging (see Note 7 below). Pure translation display is illustrated in the protocol below for incorporation of biotinylated methionine and selection on avidin beads.

2. Materials

2.1. In Vitro Transcription for mRNA Synthesis

1. 5× Milligan transcription buffer: 200 mM Tris–acetate, pH 8.0, 5 mM spermidine, 0.25 µg/µl BSA, 100 mM $MgCl_2$.
2. 0.5 M DTT.

3. 0.2% Triton X-100.
4. 30% PEG 8000.
5. Mix of ATP/CTP/GTP at 16.7 mM each, pH 8.0.
6. 2.5 mM UTP, pH 8.0.
7. α-^{32}P-labeled UTP (10 µCi/µl; 3,000 Ci/mmol).
8. 0.25 µg/µl Plasmid DNA template, linearized by a restriction enzyme.
9. T7 RNA polymerase (USB, 80 U/µl, 250 U).

2.2. Charging of tRNA with a Natural AA

1. 5× Charging buffer. 500 mM HEPES–KOH, pH 7, 75 mM MgCl$_2$, 75 mM DTT, 25 mM neutralized ATP.
2. Purified RSs (see Subheading 3.4).
3. AAs (Sigma).
4. Purified tRNA isoacceptors (Sigma).

2.3. Charging of tRNA with Unnatural AA

1. tRNA$_i^{fMet}$ (Sigma).
2. Methionine (Sigma).
3. Purified MetRS (see Subheading 3.4).
4. Sulfo-NHS-LC-biotin (Pierce).
5. Sodium acetate buffer, pH 5.
6. Microcon 10 ultrafiltration device (Amicon).

2.4. Preparation of Escherichia coli Translation Factor Proteins

1. Lysis buffer: 50 mM Tris–HCl, pH 8.0, 300 mM NaCl, 1 mM MgCl$_2$, 10 mM imidazole, 10 mM β-mercaptoethanol.
2. Wash buffer: 50 mM Tris–HCl, pH 8.0, 300 mM NaCl, 1 mM MgCl$_2$, 20 mM imidazole, 10 mM β-mercaptoethanol.
3. Elution buffer: 50 mM Tris–HCl, pH 8.0, 300 mM NaCl, 1 mM MgCl$_2$, 250 mM imidazole, 10 mM β-mercaptoethanol.
4. Dialysis buffer: 10 mM Tris–HCl, pH 7.4, 1 mM MgCl$_2$, 1 mM DTT.
5. *E. coli BL21(DE3)* (Novagen).
6. *E. coli BL21(DE3)*pLysS (Novagen).
7. Ni–NTA His-binding resin (Novagen).
8. Slide-A-Lyzer dialysis cassette (Pierce).
9. pET24a vector (Novagen).
10. IPTG (Sigma).

2.5. Purification of Ribosomes

1. Buffer BIII: 10 mM Tris–acetate, pH 8.2, 14 mM magnesium acetate, 60 mM potassium acetate, 1 mM DTT.
2. Wash buffer: 10 mM Tris–acetate, pH 8.2, 14 mM magnesium acetate, 60 mM potassium acetate, 1 mM DTT, 1 M NH$_4$Cl.

3. Storage buffer: 1 mM Tris–HCl, pH 7.4, 10 mM magnesium acetate, 1 mM DTT.
4. SOLR *E. coli* host strain (Stratagene).
5. Thick-wall polycarbonate centrifuge tubes (Beckman Coulter, 38 ml).

2.6. In Vitro Translation

1. 5× Translation Buffer, pH 6.73: 180 mM Tris–acetate, pH 7.5, 50 mM sodium 3,3-dimethyl-glutarate, pH 5.69, 180 mM ammonium acetate, 10 mM DTT, 140 mM potassium phosphoenolpyruvate, pH 6.55, 200 mM potassium acetate, 4 mM spermidine HCl.
2. 0.5 M magnesium acetate.
3. 0.1 M GTP, pH 7.1.
4. Pyruvate kinase (Sigma, 2.6 U/μl, prepared from rabbit muscle).
5. Purified recombinant IF1, IF2, IF3, EF-Tu, EF-Ts, EF-G, washed ribosomes.
6. Aminoacyl-tRNAs charged with natural and unnatural AAs.
7. Wash Buffer (WB). 50 mM magnesium acetate, 150 mM NaCl, 50 mM HEPES–KOH, pH 7.0, 10 mM DTT, 0.1% Tween 20.

2.7. Preparation of Target

1. Avidin-coated beads (Invitrogen).
2. Biotin (Pierce).

2.8. In Vitro Selection

1. Elution buffer (EB). 100 mM EDTA, 150 mM NaCl, 50 mM HEPES–KOH, pH 7.0, 10 mM DTT, 0.1% Tween 20.

3. Methods

3.1. In Vitro Transcription for mRNA Synthesis

1. The transcription reaction mixture of 25 μl contains:
   5.0 μl of 5× Milligan transcription buffer
   0.25 μl of 0.5 M DTT
   1.25 μl of 0.2% Triton X-100
   6.7 μl of 30% PEG 8000
   0.75 μl of ATP/CTP/GTP mix at 16.7 mM each, pH 8.0
   0.25 μl of 2.5 mM UTP, pH 8.0
   2.5 μl of α-^{32}P-labeled UTP
   4.0 μl of cut plasmid DNA (0.25 μg/μl)

1.17 μl of water

3.13 μl of T7 RNA polymerase

The mixture is incubated at 37°C for 3 h.

2. Transcripts are purified by 7 M urea/8% PAGE and detected by autoradiography. The bands are excised and mRNA eluted overnight with elution buffer. Transcripts are then precipitated with ethanol (see Note 1).

3.2. Charging of tRNA with a Natural AA

1. Plasmids encoding His-tagged RSs for methionine (MetRS), valine (ValRS), and threonine (ThrRS) have been published and are available from ref. (14).

2. His-tagged RSs are overexpressed in *E. coli*, the cells lysed by sonication and RSs purified from Ni–NTA columns according to standard manufacturer's protocols (Novagen).

3. Purified RSs are dialyzed with a Slide-A-Lyzer dialysis cassette.

4. A typical reaction to charge a tRNA isoacceptor with an AA contains:

   16 μM of tRNA isoacceptor

   60 μM of AA (if nonradioactive; 16 μM if radioactive)

   3 μM purified RS

   1× charging buffer

   The charging reaction is incubated at 37°C for 30 min.

5. Upon completion, the aminoacyl-tRNA is extracted with phenol saturated with sodium acetate, pH 5, precipitated with ethanol at −80°C, and the supernatant is removed.

6. The aminoacyl-tRNA is washed by resuspending in potassium acetate, pH 5.1, followed by centrifugal concentration in a tube with a 10,000 MW cut-off membrane (Vivaspin).

7. The aminoacyl-tRNA is stored at −80°C (see Note 2).

3.3. Charging of tRNA with an Unnatural AA

1. Met-tRNA$_i^{fMet}$ is biotinylated in a reaction whose final concentrations are: 19 μM Met-tRNA$_i^{fMet}$, 1.1 mg/ml sulfo-NHS-LC-biotin, 50 mM NaHCO$_3$.

   The reaction is incubated at 0°C for 40 min or less (to limit hydrolysis of the aminoacyl bond).

2. The pH is adjusted by addition of excess sodium acetate, pH 5, and the biotinylated aminoacyl-tRNA product separated by centrifugation/buffer exchange in a Microcon 10 ultrafiltration device.

   See Note 3.

3.4. Preparation of E. coli Translation Factor Proteins

1. Plasmids for expression of His-tagged translation factor proteins are prepared by PCR from published plasmids and subcloned into a pET24a vector (Novagen) or obtained from published sources: IF1, IF2, and IF3 in ref. (9); EF-Tu, EF-Ts in ref. (34); EF-G in ref. (35).

2. Factors were overexpressed, cells lysed by sonication, and factors purified from Ni-NTA columns according to standard manufacturer's protocols (Novagen) except that EF-Tu required addition of 10 μM GDP in all but the last dialysis step.

3. Purified factors are dialyzed with a Slide-A-Lyzer dialysis cassette. Precipitated IF3 was recovered by redissolving in 5 M urea, diluted and then dialyzed against dialysis buffer containing 100 mM NH_4Cl.

4. All factors are stored at −80°C and none but EF-Tu loses activity with numerous freeze-thaw cycles. EF-Tu is stored at 4°C after thawing and is active for several weeks (see Note 4).

3.5. Purification of Ribosomes

1. SOLR cells are grown to mid-log phase in 50 μg/ml kanamycin and are resuspended in buffer BIII (1 g/ml).

2. Cells are sonicated and spun at $10,000 \times g$ at 4°C for 15 min to remove cell debris.

3. The supernatant is removed and spun at $30,000 \times g$ at 4°C for 30 min.

4. 5× Buffer BIII is added to the resulting supernatant (S30) to fill the polycarbonate centrifuge tube, and the solution centrifuged at 23,600 rpm ($100,000 \times g$) for 15 h at 4°C (Beckman L-70 ultracentrifuge, rotor Sw28).

5. The supernatant (S100) is stored at −80°C.

6. The ribosome pellet is washed 3 times by stirring in wash buffer (0.1 ml × weight of pellet) at 4°C overnight and followed by repelleting at 23,600 rpm ($100,000 \times g$).

7. The 3× washed ribosome pellet is stirred in wash buffer overnight a fourth time and then centrifuged at 23,600 rpm ($100,000 \times g$) for 1 min to remove the particulate contaminants.

8. The supernatant is removed and then spun for 15 h at 23,600 rpm ($100,000 \times g$). The pellet is resuspended in a volume of storage buffer giving 1 OD_{260} U/μl ribosome solution. The ribosome solution is stored at −80°C (see Note 5).

3.6. In Vitro Translation

1. The translation reaction mixture equivalent to ten translation reaction volumes is made by combining the following in order on ice:

   10 μl of 5× Translation buffer

   0.96 μl of 0.5 M magnesium acetate

0.72 µl of pyruvate kinase (0.48 U/µl, diluted with 25 mM HEPES–KOH, pH 7.0)

0.34 µl 0.7 µg/µl IF1

0.43 µl 5.6 µg/µl IF2

0.34 µl 1.6 µg/µl IF3

0.5 µl 0.1 M GTP, pH 7.1

0.98 µl washed ribosomes (1.0 A_{260} U/µl)

4.80 µl 3.2 µg/µl EF-Tu

0.30 µl 2.6 µg/µl EF-Ts

1.12 µl 1.7 µg/µl EF-G

1.32 µl 19 µM biotinyl - Met - $tRNA_i^{fMet}$

2. Translation reactions (5 µl) are performed by adding in order:

    2.18 µl translation mix

    Aminoacyl-tRNAs to a final concentration of ~500 nM in each tRNA

    mRNA to a final concentration of 500 nM

    Water to a final volume of 5 µl

3. Translations are run for 30 min at 37°C without preincubation.

4. mRNA–ribosome–peptidyl-tRNA ternary complexes are stabilized by placing on ice for 2 min followed by addition of 95 µl cold WB to adjust the Mg^{2+} concentration to 48 mM (see Note 6).

3.7. Preparation of Target

1. Avidin beads are soaked in WB.

2. The beads are blocked by a negative-control translation mix deficient in biotinyl - Met - $tRNA_i^{fMet}$.

3. The blocked beads are divided into aliquots. Aliquots for biotin-blocked controls were also blocked with a 25% volume of 5 mM biotin for 5 min at 37°C.

3.8. In Vitro Selection

1. In a cold room, the chilled, stabilized, ternary complexes from the translation reaction are incubated for 5 min and then added to the tubes containing the avidin beads.

2. The mixture is incubated for 40 min at 4°C with intermittent mixing by tapping.

3. The selection mixture is spun gently and the supernatants removed.

4. The beads are washed 3× with 200 µl WB.

5. To elute mRNAs, 40 µl EB is added, the mixture is incubated for 5 min, and then the supernatant is recovered (see Notes 1 and 7).

4. Notes

1. Our proof-of-principle selection used radioactive mRNAs to measure washing and binding efficiencies essentially in real time with a Geiger counter and to analyze library composition directly by PAGE and autoradiography. However, practical applications of the method should use mRNAs prepared without radiolabel using higher NTP concentrations.

2. Aminoacyl-tRNAs are notoriously unstable to hydrolysis, especially under physiological to alkaline pH values. However, we find aminoacyl-tRNA stock solutions can be stored for years with many freeze-thaw cycles without substantial deacylation. $tRNA_i^{fMet}$ and $tRNA^{Val}$ are commercially available from Sigma, along with a few other tRNA isoacceptors. The $tRNA^{Thr}$ isoacceptor that we used in the past is no longer commercially available, but it should be possible to substitute it with total tRNA.

3. Our system incorporates virtually any single elongator unnatural L-AA almost quantitatively, but it incorporates multiple unnatural L-AAs with much lower yields (10, 28, 36). These inefficiencies currently limit the ability to synthesize and evolve a peptidomimetic library because of the likely production of too few, different, full-length, library members containing several unnatural AAs. We are investigating the mechanisms responsible for these inefficiencies. Single N-alkyl AAs are incorporated very slowly (37) due to decreased chemical reactivity engendered by increased steric bulk on the amine nucleophile (38, 39). Incorporation of a single hydroxy acid derivative of Phe is also very slow (40). Consecutive unnatural L-AAs are incorporated very slowly too (28), and it has been hypothesized that nascent peptidyl-tRNA drop off rates may be competitive (20). While a lack of posttranscriptional modifications in the engineered in vitro tRNA transcripts is tolerated well (36, 37), the penultimate dC and anticodon mutations can be very inhibitory when templating consecutive unnatural AA incorporations (36). Interestingly, very high concentrations of unnatural aminoacyl-tRNAs show promising improvements in incorporation efficiencies (30). The state of the art for multiple, specific, unnatural AA incorporations per ribosomal product using various charging methods is summarized in Table 1.

4. Factors were assayed by factor-specific assays and by dependencies in translations (9, 20).

5. A new method that purifies His-tagged ribosomes on Ni has been reported to be faster, cheaper and less detrimental to ribosome activity (41).

6. Problems due to loss of activity in the translation assay were generally solved faster by going back to stocks of most of the components rather than by trouble shooting the numerous components individually.

7. Only one round of selection was performed, but multiple rounds of selection and amplification (Fig. 1) should be carried out when using diverse libraries. The efficiency of recovery of the main selected mRNA from a single round of selection was estimated by PAGE to be only ~0.2% (13), although this is similar to the efficiency of crude translation ribosome displays (2–5). In contrast to crude systems, optimization to virtually 100% efficiency should be possible in a purified system (42). One possible explanation for our low yield may be aggregation due to the use of a highly hydrophobic poly(T,V) spacer sequence. This presumably explains why our analysis of the translation products by various gel systems proved impractical. Another reason to avoid hydrophobic spacers is that they can bind nonspecifically to targets, as observed by the polyPhe spacer (43). We are, therefore, testing various lengths of two other highly repetitive spacer sequences designed to improve hydrophilicity (while minimizing the number of different AAs to maximize reassignable codons): polyHis and poly(Glu$_3$Ala$_4$) (Gao and Forster, unpublished data). However, these engineered spacers have been translated with only modest efficiencies so far, requiring further optimization of translation conditions or spacer sequences.

Acknowledgments

We thank Tarjani Thaker for help with Fig. 2, and Craig Goodwin, Seth Villarreal, and the editors for comments on the manuscript. This work was supported by the National Institutes of Health and the American Cancer Society.

References

1. Smith GP, Petrenko VA. (1997) Phage Display. *Chem. Rev.* **97**, 391–410.
2. Mattheakis LC, Bhatt RR, Dower WJ. (1994) An in vitro polysome display system for identifying ligands from very large peptide libraries. *PNAS* **91**, 9022–9026.
3. Hanes J, Pluckthun A. (1997) In vitro selection and evolution of functional proteins by using ribosome display. *PNAS* **94**, 4937–4942.
4. Schaffitzel C, Zahnd C, Amstutz P, Luginbühl B, Plückthun A. (2005) In vitro selection and evolution of protein-ligand interactions by ribosome display. In Golemis E, Adams P, eds : *Protein-protein interaction: A molecular cloning manual* 2nd edn. Cold Spring Harbor Laboratory Press, Cold Spring Harbor, NY 517–548.
5. Lipovsek D, Pluckthun A. (2004) In-vitro protein evolution by ribosome display and mRNA display. *Journal of Immunological Methods* **290**, 51–67.
6. Nemoto N, Miyamoto-Sato E, Husimi Y, Yanagawa H. (1997) In vitro virus: Bonding of

mRNA bearing puromycin at the 3′-terminal end to the C-terminal end of its encoded protein on the ribosome in vitro. *FEBS Lett* **414**, 405–408.

7. Roberts RW, Szostak JW. (1997) RNA-peptide fusions for the in vitro selection of peptides and proteins. *PNAS* **94**, 12297–12302.

8. Takahashi TT, Roberts RW. (2009) In Vitro Selection of Protein and Peptide Libraries Using mRNA Display. In Mayer, G., ed.: *Nucleic Acid and peptide Aptamers: Methods and Protocols* 1st edn. **535**. pp 293–314.

9. Forster AC, Weissbach H, Blacklow SC. (2001) A Simplified Reconstitution of mRNA-Directed Peptide Synthesis: Activity of the Epsilon Enhancer and an Unnatural Amino Acid. *Analytical Biochemistry* **297**, 60–70.

10. Forster AC, Tan Z, Nalam MNL, Lin H, Qu H, Cornish VW, Blacklow SC. (2003) Programming peptidomimetic syntheses by translating genetic codes designed de novo. *PNAS* **100**, 6353–6357.

11. Li S, Millward S, Roberts R. (2002) In Vitro Selection of mRNA Display Libraries Containing an Unnatural Amino Acid. *JACS* **124**, 9972–9973.

12. Frankel A, Millward SW, Roberts RW. (2003) Encodamers: Unnatural Peptide Oligomers Encoded in RNA. *Chemistry and Biology* **10**, 1043–1050.

13. Forster AC, Cornish VW, Blacklow SC. (2004) Pure Translation Display. *Analytical Biochemistry* **333**, 358–364.

14. Shimizu Y, Inoue A, Tomari Y, Suzuki T, Yokogawa T, Nishikawa K, Ueda T. (2001) Cell-free translation reconstituted with purified components. *Nature Biotechnology* **19**, 751–755.

15. Jewett MC, Forster AC. (2010) Update on designing and building minimal cells. *Curr. Op. Biotech.* **21**, 697–703.

16. Villemagne D, Jackson R, Douthwaite JA. (2006) Highly efficient ribosome display selection by use of purified components for in vitro translation. *Journal of Immunological Methods* **313**, 140–148.

17. Ohashi H, Shimizu Y, Ying BW, Ueda T. (2007) Efficient protein selection based on ribosome display system with purified components. *Biochemical and Biophysical Research Communications* **352**, 270–276.

18. Osada E, Shimizu Y, Akbar BK, Kanamori T, Ueda T. (2009) Epitope Mapping Using Ribosome Display in a Reconstituted Cell-Free Protein Synthesis System. *Journal of Biochemistry* **145**, 693–700.

19. Yanagida H, Matsuura T, Yomo T. (2007) Compensatory Evolution of a WW Domain Variant Lacking the Strictly Conserved Trp Residue. *Journal of Molecular Evolution* **66**, 61–71.

20. Tan Z, Blacklow SC, Cornish VW, Forster AC. (2005) De novo genetic codes and pure translation display. *Methods* **36**, 279–290.

21. Robertson SA, Ellman JA, Schultz PG. (1991) A General and Efficient Route for Chemical Aminoacylation of Transfer RNAs. *JACS* **113**, 2722–2729.

22. Murakami H, Ohta A, Ashigai H, Suga H. (2006) A highly flexible tRNA acylation method for non-natural polypeptide synthesis. *Nature Methods* **3**, 357–359.

23. Merryman C, Green R. (2004) Transformation of Aminoacyl tRNAs for the In Vitro Selection of "Drug-like" Molecules. *Chemistry and Biology* **11**, 575–582.

24. Hartman MCT, Josephson K, Szostak JW. (2006) Enzymatic aminoacylation of tRNA with unnatural amino acids. *PNAS* **103**, 4356–4361.

25. Hartman MCT, Josephson K, Lin CW, Szostak JW. (2007) An Expanded Set of Amino Acid Analogs for the Ribosomal Translation of Unnatural Peptides. *PLoS ONE* **2**, e972.

26. Minajigi A, Francklyn CS. (2010) Aminoacyl Transfer Rate Dictates Choice of Editing Pathway in Threonyl-tRNA Synthetase. *Journal of Biological Chemistry* **285**, 23810–23817.

27. Josephson K, Hartman MCT, Szostak JW. (2005) Ribosomal Synthesis of Unnatural Peptides. *JACS* **127**, 11727–11735.

28. Forster AC. (2009) Low modularity of aminoacyl-tRNA substrates in polymerization by the ribosome. *Nucleic Acids Research* **37**, 3747–3755.

29. Kawakami T, Murakami H, Suga H. (2008) Ribosomal Synthesis of Polypeptoids and Peptoid-Peptide Hybrids. *JACS* **130**, 16861–16863.

30. Kawakami T, Murakami H, Suga H. (2008) Messenger RNA-Programmed Incorporation of Multiple N-Methyl-Amino Acids into Linear and Cyclic Peptides. *Chemistry and Biology* **15**, 32–42.

31. Ohta A, Murakami H, Higashimura E, Suga H. (2007) Synthesis of Polyester by Means of Genetic Code Reprogramming. *Chemistry and Biology* **14**, 1315–1322.

32. Merryman C, Weinstein E, Wnuk SF, Bartel DP. (2002) A Bifunctional tRNA for In Vitro Selection. *Chemistry and Biology* **9**, 741–746.

33. Griffiths AD, Tawfik DS. (2006) Miniaturising the laboratory in emulsion droplets. *Trends in Biotechnology* **24**, 395–402.

34. Hwang YW, Sanchez A, Hwang MCC, Miller DL. (1997) The Role of Cysteinyl Residues in the Activity of Bacterial Elongation Factor Ts, a Guanosine Nucleotide Dissociation Protein. *Arch. Biochem. Biophys.* **348**, 157–162.

35. Semenkov YP, Rodnina MV, Wintermeyer W. (1996) The "allosteric three-site model" of elongation cannot be confirmed in a well-defined ribosome system from Escherichia coli. *PNAS* **93**, 12183–12188.

36. Gao R, Forster AC. (2010) Changeability of individual domains of an aminoacyl-tRNA in polymerization by the ribosome. *FEBS Letters* **584**, 99–105.

37. Pavlov MY, Watts RE, Tan Z, Cornish VW, Ehrenberg M, Forster AC. (2008) Slow peptide bond formation by proline and other N-alkylamino acids in translation. *PNAS* **106**, 50–54.

38. Zhang B, Tan Z, Dickson LG, Nalam MNL, Cornish VW, Forster AC. (2007) Specificity of Translation for N-Alkyl Amino Acids. *JACS* **129**, 11316–11317.

39. Watts RE, Forster AC. (2010) Chemical Models of Peptide Formation in Translation. *Biochemistry* **49**, 2177–2185.

40. Bieling P, Beringer M, Adio S, Rodnina MV. (2006) Peptide bond formation does not involve acid-base catalysis by ribosomal residues. *Nature Structural and Molecular Biology* **13**, 423–428.

41. Ederth J, Mandava CS, Dasgupta S, Sanyal S. (2009) A single-step method for purification of active His-tagged ribosomes from a genetically engineered *Escherichia coli*. *Nucleic Acids Research* **37**, e15.

42. Zavialov AV, Buckingham RH, Ehrenberg M. (2001) A Posttermination Ribosomal Complex Is the Guanine Nucleotide Exchange Factor for Peptide Release Factor RF3. *Cell* **107**, 115–124.

43. Cochella L, Green R. (2004) Isolation of antibiotic resistance mutations in the rRNA by using an in vitro selection system. *PNAS* **101**, 3786–3791.

44. Subtelny AO, Hartman MCT, Szostak JW. (2008) Ribosomal Synthesis of N-Methyl Peptides. *JACS* **130**, 6131–6136.

Chapter 21

In Vitro Selection of Unnatural Cyclic Peptide Libraries via mRNA Display

Zhong Ma and Matthew C.T. Hartman

Abstract

The ribosomal synthesis of drug-like peptides containing unnatural amino acids is possible due to the broad substrate specificity of the ribosome. In this protocol, a reconstituted *Escherichia coli* ribosomal translation system (PURE) is adapted to incorporate unnatural amino acids into mRNA-displayed peptide libraries, which are used in in vitro selection.

Key words: Unnatural amino acids, Peptide library, mRNA display, In vitro selection, mRNA-peptide fusion, Ligand discovery, PURE system, Ribosomal translation

1. Introduction

Vast peptide libraries can now be reliably created and subjected to in vitro selection in order to find novel ligands for a wide variety of biomolecules and materials. The techniques for the creation of the most diverse peptide libraries involve the biochemical machinery of translation. As a result, until recently, the peptide libraries created using these methods contained linear peptides composed of natural amino acids. These peptides often suffer from poor biostability. The recent development of translation systems reconstituted entirely from purified components (1, 2) as well as an ever-expanding list of unnatural aminoacyl-tRNAs (3–5) have enabled the ribosomal synthesis of cyclic peptides (6–10) composed primarily of unnatural amino acids (11, 12). Cyclization and incorporation of unnatural building blocks are two common methods to make peptides more drug-like (13).

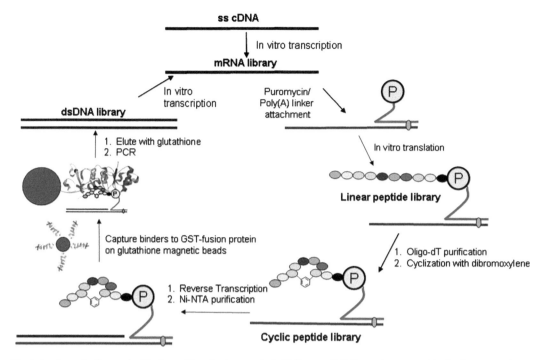

Fig. 1. In vitro selection cycle. Starting with a single-stranded cDNA, an mRNA library is prepared by in vitro transcription (Subheading 3.6). This mRNA is then photo-crosslinked to a puromycin-containing linker (Subheading 3.7). In vitro translation leads to a linear peptide (Subheading 3.8) which is purified on Oligo(dT)-Cellulose (Subheading 3.9). Cyclization is performed on the oligo(dT) column (Subheading 3.9). After reverse transcription and Ni-NTA purification (Subheading 3.10) the library of mRNA peptide fusions is incubated with the desired immobilized target (Subheading 3.10). Binders are eluted and the resulting mRNA-peptide fusions are amplified by PCR (Subheading 3.11).

The following section describes a protocol for the creation of cyclic, unnatural peptide libraries using mRNA display (14, 15). The protocol is summarized in Fig. 1. Although more sophisticated methods for attachment of unnatural amino acids onto tRNA have been developed, this protocol focuses on unnatural amino acids that have already been described as efficient substrates for in vitro translation (12). In a typical mRNA display selection, the initial troubleshooting may take 1–2 months. Once established, each round of selection will take ~3 days. In vitro selections typically converge on a small number of sequences within ten rounds. Thus the entire selection process may take 3–5 months of focused work. This may seem daunting at first, but when one considers that during this process over 10^{13} different unnatural, cyclic peptides have been created and sorted, the overall efficiency is quite remarkable.

Compared to a standard mRNA display experiment which utilizes cell extracts (16), this protocol requires the use of the PURE translation system. This system is commercially available from a number of sources and is suitable for mRNA display for peptides containing all natural amino acids. It is recommended that the user who is unfamiliar with mRNA display begins with this system before investing the time in developing the customizable versions of the system which are required for unnatural amino acid mRNA display.

2. Materials

2.1. Urea PAGE Purification of Library DNA

1. Owl P10DS Dual Gel System (Thermo Scientific).
2. Elutrap Electroelution System (includes Elutrap chamber and BT1 and BT2 membranes) (Whatman). Store the membranes at 4°C (see Note 1).
3. 20 × 20 cm glass-backed TLC plate with fluorescent dye excitable at 254 nm.
4. UVP Compact Lamp, 254 nm, 4 W (UVP).
5. EC Apparatus electrophoresis power supply EC 600 or other power supply with constant watt capabilities.
6. Urea PAGE loading buffer: 8 M Urea, 2 mM Tris–HCl, pH 7.5, 20 mM EDTA, 0.25% bromophenol blue (w/v), 0.25% xylene cyanol.
7. SequaGel sequencing system (National Diagnostics).
8. Ammonium persulfate (APS): prepare a 10% solution (w/v) in ddH$_2$O freshly (see Note 2).
9. $N'N'N'N'$-tetramethylethlyenediamine (TEMED). Store at 4°C.
10. 5× TBE buffer: 445 mM Tris base, 445 mM boric acid, 10 mM EDTA (see Note 3).
11. 3 M NaOAc, pH 5.2. Store at −20°C.
12. 100% Ethanol and 70% ethanol. Store at −20°C.

2.2. PCR and TOPO Cloning

1. Taq DNA polymerase (various suppliers). Store at −20°C.
2. 10× ThermoPol PCR buffer (New England Biolabs) (see Note 4). Store at −20°C.
3. dNTP solution (various suppliers). Store at −20°C.
4. Agarose (various suppliers). Choose molecular biology grade.
5. 5× Agarose gel loading buffer with Orange G as the dye (0.25% Orange G (w/v), 30% glycerol, 6 mM EDTA) (see Note 5). Store at room temperature.
6. 100 bp DNA ladder (various supplies). Store at −20°C.
7. Ethidium bromide solution (10 mg/mL). Store at room temperature. Ethidium bromide is a mutagen. Wear gloves and avoid exposure. Dispose of ethidium bromide accordingly.

8. QIAquick Gel Extraction Kit (Qiagen). Store at room temperature.
9. TOPO TA Cloning Kit (Invitrogen). Store the components according to manufacturer's instructions.
10. LB-Agar plates with 50 μg/mL ampicillin. Store at 4°C.

2.3. Transcription and Purification of mRNA

1. The equipment necessary for Urea PAGE purification of nucleic acids (see Subheading 2.1).
2. T7 RNA polymerase (various suppliers). Store at −20°C.
3. 10× Transcription buffer (400 mM Tris–HCl, pH 7.8, 0.1% Triton X-100). Store at 4°C.
4. NTP solutions: 100 mM each NTP (Sigma-Aldrich), pH adjusted to 7–8 with NaOH (check pH with a pH paper). Sterile filter through 0.22 μm syringe filter and aliquot. Store at −80°C.
5. 300 mM spermidine. Store at −20°C.
6. 1 M $MgCl_2$. Store at room temperature.
7. 1 M DTT, freshly made (see Note 6).
8. 0.1 mg/mL Inorganic pyrophosphatase from *Escherichia coli* (Sigma-Aldrich) (see Note 7). Store at −20°C.
9. RNase inhibitor (various suppliers). Usually supplied as 40 U/μL. store at −20°C.
10. Solid urea.
11. 0.5 M EDTA, pH 8.0. Store at room temperature.
12. Turbo DNase (Applied Biosystems/Ambion). Store at −20°C.

2.4. Psoralen Photo-Crosslinking

1. 365 nm handheld UV lamp (UVP).
2. 96-Well round bottom crosslinking plate (Costar).
3. XL-PSO oligonucleotide: 5′-PsoC6-(uagccggug)$_2$′$_{-OMe}$-15xA-2xSpacer 9-ACC-Puro-3′ (see Note 8). Dissolve in ddH_2O to a final concentration of 125 μM. Aliquot and store at −20°C.
4. 1 M HEPES-KOH, pH 7.6. Store at room temperature.
5. 1 M KCl. Store at room temperature.
6. 25 mM spermidine. Store at −20°C.
7. 3 M KOAc, pH 5.5. Store at −4°C.

2.5. Translation

1. Econo-Pac column (Bio-Rad).
2. 50 mL Nalgene Oakridge centrifuge tubes (Thermo Fisher Scientific).
3. B-PER reagent (in phosphate buffer) (Thermo Scientific/Pierce). Store at room temperature.
4. Ni-NTA Agarose (Qiagen). Store at 4°C.

5. 100 mg/mL ampicillin and 25 mg/mL kanamycin stocks. Store at −20°C.

6. 1 M IPTG. Store at −20°C.

7. Slide-A-Lyzer dialysis cassettes (Thermo Scientific/Pierce) (see Note 9).

8. His compatible protease inhibitor set VII (Merck/EMD) (see Note 10). Store at −20°C.

9. Ni-NTA wash buffer (I): 50 mM NaH_2PO_4, 300 mM NaCl, 20 mM imidazole. Add 5 mM 2-mercaptoethanol before use. Store at 4°C.

10. Ni-NTA elution buffer (I): 50 mM NaH_2PO_4, 300 mM NaCl, 250 mM imidazole. Add 5 mM fresh 2-mercaptoethanol before use. Store at 4°C.

11. Enzyme dialysis buffer with glycerol: 50 mM HEPES-KOH, pH 7.6, 100 mM KCl, 10 mM $MgCl_2$, 30% glycerol. Add 2-mercaptoethanol to 7 mM before use. Store at 4°C.

12. BeadBeater with a medium blender chamber and glass beads (BioSpec Products).

13. Ribosome buffer A: 10 mM Tris–HCl pH 7.5, 10 mM $Mg(OAc)_2$, 100 mM NH_4Cl, 0.25 mM EDTA. Add 2-mercaptoethanol to 7 mM before use. Store at 4°C.

14. Ribosome buffer B: 10 mM Tris–HCl pH 7.5, 10 mM $Mg(OAc)_2$, 500 mM NH_4Cl. Add 2-mercaptoethanol to 7 mM before use. Store at 4°C.

15. Ribosome buffer B with 30% sucrose: 10 mM Tris–HCl pH 7.5, 10 mM $Mg(OAc)_2$, 500 mM NH_4Cl, 30% sucrose (w/v). Add 2-mercaptoethanol to 7 mM before use. Store at 4°C.

16. Ribosome buffer C: 10 mM Tris–HCl pH 7.5, 10 mM $Mg(OAc)_2$, 60 mM NH_4Cl, 0.5 mM EDTA. Add 2-mercaptoethanol to 3 mM before use. Store at 4°C.

17. Ultracentrifuge with Type Ti70.1 rotor (Beckman Coulter).

18. Quick-Seal ultracentrifuge tubes and tube rack (Beckman Coulter).

19. Standard 3× polymix buffer: in a 15-mL Falcon tube, add 4 mL ddH_2O and add final concentrations of 8 mM putrescine, 1 mM spermidine, 5 mM K_2HPO_4, 95 mM KCl, 5 mM NH_4Cl. Adjust pH to 7.70–7.80 with 600 mM HCl. Add $Mg(OAc)_2$ to a final concentration of 5 mM. In a second 15-mL Falcon tube, prepare 3 mL solution of 0.5 mM $CaCl_2$. Mix the two solutions well by pouring one tube of solution into another. Repeat this at least ten times. Add ddH_2O to a total volume of 9.75 mL. Filter the solution through 0.22 μm syringe filter and place at room temperature until use.

20. Creatine kinase: dissolve lyophilized creatine kinase (Roche Applied Science) in 10 mM Tris–HCl, pH 7.5, 10 mM Mg(OAc)$_2$, 100 mM NH$_4$Cl and adjust OD$_{280}$ to 10. This solution should be stored at 4°C for not more than 1 week.

21. Creatine phosphate: dissolve creatine phosphate potassium salt (Merck/EMD) in ddH$_2$O to make 0.5 M solution. Store at −80°C.

22. Nucleoside 5′-diphosphate kinase from bovine liver (Sigma-Aldrich). Store at 4°C.

23. Myokinase from rabbit muscle (Sigma-Aldrich). Store at 4°C.

24. Putrescine dihydrochloride (Sigma-Aldrich) 2.4 M. Store at −20°C.

25. (6R,S)-5,10-formyl-5,6,7,8-tetrahydrofolic acid (Schircks Laboratory): dissolve at 2 mg/mL in 20 mM DTT and adjust pH to 7 with KOH. Store at −80°C.

26. 25 mM ATP and 25 mM GTP potassium salt solution: because sodium inhibits translation, the sodium salt form of ATP and GTP should be exchanged to potassium. Mix thoroughly 200 μL of 100 mM ATP or GTP from Subheading 2.3 with 20 μL 3.81 M KCl and 600 μL ethanol from −20°C. Centrifuge immediately at 16,000×g for 5 min at 4°C. Wash the pellets with 70% ethanol twice and once with 100% ethanol. Aspirate all the remaining liquid and air-dry the pellets. Resuspend the pellets in 200 μL ddH$_2$O. Dilute the solutions to 1:5,000 and measure the OD$_{260}$. Calculate the concentration of ATP or GTP by using 13,700 OD$_{260}$ of GTP = 1 M and 15,400 OD$_{260}$ of ATP = 1 M. Make up a solution containing both 25 mM ATP and 25 mM GTP based on the calculations. Store the 100 μL aliquots at −80°C.

27. Deacylated *E. coli* total tRNA (Roche Applied Science): dissolve tRNA at 100 mg/mL in 1 M Tris–HCl (pH 9). Incubate at 37°C for 2 h. Dialyze overnight against 50 mM Tris–HCl (pH 9). Precipitate with 0.1 volume of KOAc, pH 5.5 and 3 volumes of ethanol. Wash the pellet with 70% ethanol. Air-dry at room temperature. Resuspend the tRNA in ddH$_2$O and adjust the concentration to 100 mg/mL. Store at −80°C in small aliquots (see Note 11).

28. 2.85 M KCl/1 M MgCl$_2$. Store at room temperature.

29. All amino acids and analogs are dissolved in ddH$_2$O at 1, 10, or 50 mM, and the pH was adjusted to 7.0–7.5 with 1 M KOH. Filter through 0.22 μm syringe filter and aliquot. Store at −20°C.

30. Isotopically labeled ^{35}S-Met [specific activity: >1,000 Ci (37.0 TBq)/mmol] (Perkin-Elmer). Upon receiving, thaw and store in 10 μL aliquots at −80°C.

2.6. Oligo(dT) Purification and Cyclization

1. Millipore UltraFree-MC centrifugal filter devices with Durapore membrane (Millipore).
2. Beckman Coulter scintillation counter.
3. Oligo(dT)-Cellulose Type 7 (GE Healthcare). Store at −20°C.
4. 5 mg/mL glycogen (Applied Biosystems/Ambion). Store at −20°C.
5. Oligo(dT) binding buffer: 20 mM Tris–HCl, pH 7.8, 10 mM EDTA, 1 M NaCl, 0.2% Triton X-100. Add 0.5 mM fresh TCEP before use (see Note 12). Store at 4°C.
6. Oligo(dT) wash buffer: 20 mM Tris–HCl, pH 7.8, 0.3 M NaCl, 0.1% Triton X-100. Add fresh TCEP to 0.5 mM before use. Store at 4°C.
7. Cyclization buffer: 20 mM Tris–HCl, pH 7.8, 0.66 M NaCl, 3 mM α,α′-dibromo-m-xylene (Sigma-Aldrich/Fluka), 33% acetonitrile (v/v), 0.5 mM TCEP (fresh). To make the cyclization buffer: prepare a stock solution of 30 mM Tris–HCl, pH 7.8 and 1 M NaCl. Store at room temperature. Before use, prepare a 10 mM solution of dibromo-m-xylene in acetonitrile. Mix 2.64 mL of 30 mM Tris–HCl, pH 7.8 and 1 M NaCl with 1.32 mL of 10 mM dibromo-m-xylene in acetonitrile. Discard any unused dibromo-m-xylene.

2.7. Reverse Transcription and Ni-NTA Purification

1. Superscript III First Strand Synthesis Kit (Invitrogen). Store at −20°C.
2. Denaturing Ni-NTA binding buffer: 100 mM NaH_2PO_4, 10 mM Tris–HCl, 6 M guanidinium hydrochloride, 0.2% Triton X-100. Adjust pH to 8.0. Add 2-mercaptoethanol to 5 mM before use. Store at 4°C.
3. Ni-NTA wash buffer (II): 100 mM NaH_2PO_4, 300 mM NaCl, 0.2% Triton X-100. Adjust pH to 8.0. Add 2-mercaptoethanol to 5 mM right before use. Store at 4°C.
4. Ni-NTA elution buffer (II): 50 mM NaH_2PO_4, 300 mM NaCl, 250 mM imidazole, 0.2% Triton X-100. Adjust pH to 8.0. Add 2-mercaptoethanol to 5 mM before use. Store at 4°C.

2.8. Selection

1. Pierce magnetic glutathione beads (Thermo Fisher Scientific).
2. Magnetic separation stand (Invitrogen).
3. Fixed speed Labquake tube rotators (Thermo Fisher Scientific).
4. GST beads wash buffer: 125 mM Tris–HCl, pH 8.0, 150 mM NaCl. Store at 4°C.
5. Selection buffer: 50 mM Tris–HCl, pH 8.0, 150 mM NaCl, 4 mM $MgCl_2$, 0.25% Triton X-100. Store at 4°C.

6. GSH elution buffer: 250 mM Tris–HCl, pH 9.0, 500 mM NaCl, 100 mM reduced glutathione, 1% Triton X-100. Prepare fresh and filter through 0.22 μm syringe filter.
7. Highly purified protein selection target (see Note 13).
8. 10 mg/mL BSA (New England Biolabs). Store at −20°C.
9. Dialysis buffer: 0.1% Triton X-100 in ddH$_2$O. Pre-cool at 4°C.

2.9. PCR Amplification of Selected Fusions

1. PCR primers for mRNA-peptide fusions. The 5′ primer sequence is: TAATACGACTCACTATAGGGTTAACTTT AGTAAGGAGG. The 3′ primer sequence is: CTAGCTACC TATAGCCGGTGGTGATG.
2. Phenol:chloroform:isoamyl alcohol (25:24:1) (Sigma-Aldrich). Store at 4°C.

3. Methods

Prior to beginning an in vitro selection there are several factors that need to be considered.

3.1. Target

For the entire selection ~1 mg of a particular protein target is required. We typically overexpress proteins in bacteria as GST fusions.

3.2. Unnatural Building Blocks

A random peptide library composed of all natural amino acids has been successfully used in mRNA display and has been described in a previous protocol (16). However, there are a wide variety of unnatural amino acid analogs that are efficient substrates for translation (12). Many of these can simply be substituted for their natural amino acid counterparts. For this example, we have replaced six of the natural amino acids with unnatural counterparts (Fig. 2), but we have in other cases replaced as many as 12 natural amino

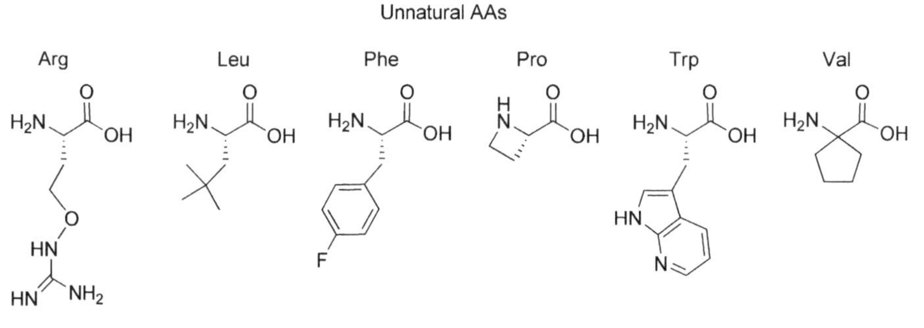

Fig. 2. Unnatural amino acids used in this peptide library. Written above each unnatural amino acid is the natural amino acid it is replacing in the library.

acids. One consideration is that once the in vitro selection has been completed, it is typically necessary to synthesize the unnatural peptides on the solid phase, so it is wise to consider the availability of the corresponding Fmoc unnatural amino acids.

3.3. Library Creation

The flexibility of mRNA display allows the synthesis of linear or cyclized peptides. This library includes fixed cysteines at both termini for cyclization. However, it is possible to encode a mixture of linear and cyclic peptides by omitting the second cysteine. In this case, the cyclization size is not fixed and the peptides can be linear (see Note 14) when no cysteine is present in the library region. One can encode for an enhanced percentage of cysteine codons in the random region by using the codon NNB (B=T, G, or C) (see Note 15).

The DNA sequence of the random library is: TAATACGACTCACTATAGGGTTAACTTTAGTAAGGAGG ACAGCTAAATGTGCNNSNNSNNSNNSNNSNNSNN SNNSNNSNNSNNSNNSTGCGGCTCCGGTAGCTTAGG CCACCATCACCATCACCACCGGCTATAGGTAGCTAG.

The detailed sequence composition of the library is as follows:

1. TAATACGACTCACTATA: T7 promoter, followed by GGG with the first "G" as the transcription start.
2. TTAACTTTAG: Epsilon enhancer (1).
3. TAAGGAGG: Shine Dalgarno sequence, also known as the ribosome binding site.
4. ACAGCTAA: the Spacer between ribosome binding site and the start codon, with "AA" at end (17).
5. ATGTGC(NNS)$_{12}$TGC: translation start codon for methionine (ATG) and the codon for the fixed cysteine (TGC), followed by 12 random NNS codons. N denotes A, T, C, or G, and S denotes G, or C. The translated peptides have the sequence of MCX$_{12}$CGSGSLGHis$_6$, where X can be one of the natural or unnatural amino acids.
6. GGCTCCGGTAGCTTAGGC: codons for GlySerGlySer-LeuGly, the flexible linker with two out-of-frame stop codons.
7. CACCATCACCATCAC: codons for His5 tag. The sixth His of His6 tag is in the following sequence.
8. CACCGGCTAT: hybridization region for the Psoralen cross-linker (XL-PSO oligonucleotide, see below). The CAC encodes the sixth His of the intact His6 tag.
9. AGGTAGCTAG: 3′-UTR to allow those non-crosslinked peptides to release at the in-frame TAG stop codons.

To construct the mRNA library to be used in the PURE system, DNA oligonucleotides with the antisense sequence of the above library DNA are synthesized and purified by Urea PAGE. Two hundred-nanomole scale synthesis with good quality is usually enough, although we prefer 1 μmol scale.

The antisense sequence of the library DNA is named as CX12REV: CTAGCTACCTATAGCCGGTGGTGATGGTGAT GGTGGCCTAAGCTACCGGAGCCGCASNNSNNSNNS NNSNNSNNSNNSNNSNNSNNSNNSNNSNNSNN SNNGCACATTTAGCTGTCCTCCTTACTAAAGTTAACCC TATAGTGAGTCGTATTA.

A DNA oligonucleotide with the T7 promoter (TAATAC GACTCACTATA) is annealed to the 3′ region of the antisense library DNA. The formation of the DNA duplex in the T7 promoter is necessary to initiate the transcription reaction and library mRNA is then transcribed.

The sequence of the XL-PSO oligonucleotide (18) is:
5′-PsoC6-(uagccggug)$_{2'\text{-OMe}}$-15xA-2xSpacer9-ACC-Puro-3′.

The 5′ end of the XL-PSO oligonucleotide is a psoralen moiety which is attached through a C6 alkyl chain to the 5′-phosphate of the linker. The linker hybridization sequence is prepared from 2′-O-methyl-RNA phosphoramidites to enhance the pairing stability between mRNA and the XL-PSO oligonucleotide. Flexible spacers are used to tether 5′-dAdCdC-puromycin to the 3′-end of a stretch of Poly(A). The Poly(A) is designed for purification of the cross-linked mRNA by oligo(dT)-cellulose.

3.4. Urea PAGE Purification of Library DNA

3.4.1. Urea PAGE

1. Set up a large 8% SequaGel (20 cm × 20 cm) on an Owl P10DS Dual Gel System. Use the prep comb to make a single large well. Use 1× TBE in both chambers.

2. Pre-run the gel for 20–30 min by using constant power at 25 W supplied by EC Apparatus electrophoresis power supply EC 600.

3. Flush out the urea from well before loading samples.

4. Dissolve the synthesized DNA library in ddH$_2$O. Take out one-fifth of the DNA solution and mix well with an equal volume of Urea PAGE loading buffer. Load the DNA sample slowly into the well and assemble the lid.

5. Run the gel at constant power of 25 W until the bromophenol blue reaches the bottom of the gel. This may take 90 min.

6. Turn off the power supply and disassemble the gel system. Transfer the gel from the glass plates to Saran wrap on both sides. Visualize the DNA band by UV shadowing and cut it out with a fresh razor blade.

3.4.2. Electroelution

1. Assemble the Whatman Elutrap electroelution System.

2. Put the gel slice into the chamber, and mush it into small pieces with a pipette tip. Fill the chamber with 0.5× TBE buffer until the gel is covered.

3. Run the electroelution for 2 h at 300 V on a Bio-Rad PowerPac basic power supply.

4. At the end of running, switch the electrodes and run it backwards for 1 min at 300 V.

5. Remove the solution from the small chamber between the two membranes with a plastic dropper, and expel into a 1.6 mL tube.

6. Measure the volume of eluent and add 0.1 volume of 3 M NaOAc and 3 volumes of 100% ethanol. Mix well and freeze the tube at −20°C for 30 min.

7. Spin the tube at $16,000 \times g$ for 20 min at 4°C.

8. Discard the supernatant and add 500 μL of 70% ethanol.

9. Spin again at $16,000 \times g$ for 1 min.

10. Discard the supernatant and air-dry the pellet for 5–10 min at room temperature.

11. Dissolve the DNA in ddH$_2$O and measure the absorbance at 260_{nm} on a spectrophotometer. Calculate the concentration by using the online software Oligonucleotide Properties Calculator (http://www.unc.edu/~cail/biotool/oligo/index.htmL). Adjust the concentration to 50 μM and store the DNA at −20°C.

3.5. PCR and TOPO Cloning

3.5.1. PCR Amplification of Library DNA

1. Obtain a 5′ and 3′ primer for amplification of the library DNA. The 5′ primer sequence is: TAATACGACTCACTATAGG. The 3′ primer sequence is: CTAGCTACCTATAGCCGGTGG.

2. Resuspend the primers in ddH$_2$O and adjust the concentration to 10 μM.

3. Prepare a series of tenfold diluted stock of the library DNA from Subheading 3.4. The concentrations of the diluted stock DNA are 5, 0.5, and 0.05 μM.

4. Prepare a series of PCR reaction mixtures in 0.2 mL PCR tubes containing 5, 0.5, and 0.05 pmol of the library DNA (1 μL of the diluted stock DNA), 50 pmol of each primer (5 μL of 10 μM primer), 2 μL of 10 mM dNTP mix, 10 μL of 10× ThermoPol buffer, and add ddH$_2$O to 100 μL.

5. Place the tubes on a PCR machine and denature the template at 94°C for 2 min. Take the tubes out and sit on ice for 2 min. Add 1 μL Taq DNA polymerase (5 U/μL) to each tube. Program the PCR machine to run 30 cycles of PCR of 94°C for 15 s, 60°C for 30 s, and 72°C for 45 s and start the PCR.

6. While the reactions are cycling, prepare a 2% Agarose gel with 1× TBE. Add 2 μL ethidium bromide solution (10 mg/mL) per 50 mL gel.

7. Monitor the PCR cycling. When the cycles have reached number 10, pause the PCR cycling at 2 s before the end of the elongation step (72°C, 45 s total). Take 4 μL out of each PCR reaction and pipette into 4 μL 5× Agarose gel loading buffer with Orange G as the dye which was previously aliquoted into a 96-well plate on ice. Do this for cycles 12, 14, 16, etc., until the end of PCR.

8. Run the samples on the Agarose gel together with 5 μL 100 bp DNA ladder in an adjacent lane.

9. Observe the PCR product on a UV transilluminator.

10. Based upon the results of the gel, determine when the PCR product has reached a plateau in concentration. Note the number of cycles.

11. Repeat the PCR with the condition (dilution of library DNA and cycles of PCR) from above and proceed to TOPO TA cloning.

3.5.2. TOPO Cloning and Sequencing

1. Follow the instructions of the TOPO TA Cloning Kit from Invitrogen and ligate the PCR product directly with the TOPO TA cloning vector.

2. Transform the ligation mixture into Top 10 competent cells and plate onto LB plates with 50 μg/mL ampicillin.

3. Incubate the plate overnight at 37°C in an incubator chamber.

4. Sequence at least 20 clones to ensure that the library was synthesized as designed and there is no strong bias in the nucleotide composition. If there is a bias, consider resynthesizing the library DNA.

3.6. Transcription and Purification of mRNA

1. Set up the transcription reaction by combining the following reagents: 100 μL of 10× Transcription buffer, 8.33 μL 300 mM spermidine, 25 μL 1 M $MgCl_2$, 10 μL 1 M DTT, 50 μL each of the 100 mM NTPs, 40 μL extra 100 mM GTP, 5 μL RNase inhibitor (40 U/μL), 10 μL Inorganic pyrophosphatase, 50 μL T7 RNA polymerase (20 U/μL), and library DNA (add volume sufficient for 5–50 nM final concentration). Add ddH_2O to 1 mL and mix well.

2. Incubate overnight at 37°C in an incubator chamber.

3. DNase treatment: remove the transcription reaction from the incubator and add 50 μL Turbo DNase. Incubate for 15 min at 37°C. After incubation, add 435 mg solid urea so that the final concentration of urea is 8 M. Add 100 μL 0.5 M EDTA and 20 μL Urea PAGE loading buffer and mix well.

4. Heat the mixture at 90°C for 5 min and place on ice until loading.

5. Purify the mRNA as described in Subheading 3.6. Use 3 M KOAc, pH 5.5 and 100% ethanol to precipitate the mRNA after electroelution (see Note 16). Wash the pellet twice with 70% ethanol. Air-dry the pellet. Dissolve the mRNA in ddH$_2$O at a final concentration of 50 μM and store at −20°C.

3.7. Psoralen Photo-Crosslinking

1. In a 1.5 mL tube, add the following reagents: 8 μL 1 M HEPES-KOH, pH 7.6, 40 μL 1 M KCl, 16 μL 25 mM spermidine, 0.8 μL 0.5 M EDTA, 24 μL of 50 μM mRNA, 24 μL of 125 μM XL-PSO oligonucleotide, and 287 μL ddH$_2$O. Mix well and spin briefly to collect the liquid. Aliquot into four 0.2 mL PCR tubes with 100 μL in each tube.

2. Place the PCR tubes in a PCR machine, heat to 70°C for 5 min, then cool to 25°C over 5 min (0.1°C/s).

3. After this, transfer the mixtures to a crosslink plate, 100 μL per well.

4. Place the plate on a stable surface in a cold room. Put a 365-nm handheld UV lamp on top of the plate, making sure the window of the UV lamp is directly above the wells with crosslinking mixtures.

5. Turn on the UV lamp and irradiate the plate for 20 min at 4°C (see Note 17).

6. Transfer the mixture into a 2-mL tube and precipitate the mRNA with 3 volumes of ethanol and 0.1 volume of 3 M KOAc. Wash the pellet twice with 70% ethanol. Air-dry the pellet at room temperature.

3.8. Translation

3.8.1. Expression and Purification of AARS

1. Inoculate a 30-mL solution of LB media with appropriate antibiotics, grow overnight at 37°C, shaking at 200–250 rpm in a 250-mL flask.

2. Inoculate 500 mL of LB media with antibiotics with 25 mL of the starter culture from above (5% inoculation volume), and shake at 250 rpm at 37°C in a 2 L flask.

3. Grow until the OD$_{600}$ of 0.6 is reached (60–120 min).

4. Add 50 μL 1 M IPTG to a final concentration of 0.1 mM.

5. Grow for another 4–5 h at 37°C.

6. Harvest the cells by centrifugation at 4,000 × g for 25 min 4°C. The cell pellets can be stored at −20°C.

7. Take frozen cells from the freezer and resuspend in 25 mL B-PER phosphate buffer along with 100 μL His compatible protease inhibitor cocktail (Merck/EMD).

8. Transfer to a 50 mL Nalgene Oakridge centrifuge tube.

9. Shake gently for 15 min at room temperature on a rocking platform shaker.

10. Centrifuge the lysate at $16,000 \times g$ for 30 min at 4°C to pellet debris. Transfer the supernatant to a 50 mL Falcon tube.

11. Add 3 mL Ni-NTA Agarose slurry (mix well before using) to the lysate and shake in a rocking platform shaker at 4°C for 60 min.

12. Load all the solutions onto a Bio-Rad Econo-Pac column. Remove bottom cap and allow to drain by gravity. Wash the Falcon tube with the flow through to ensure all the Ni-NTA Agarose is transferred into the column.

13. Wash with 2×15 mL Ni-NTA wash buffer (I).

14. Elute with 6 mL Ni-NTA elution buffer (I).

15. Dialyze the eluents in a 3–12 mL volume Slide-A-Lyzer dialysis cassette overnight at 4°C against 500 mL enzyme dialysis buffer with glycerol. Change the dialysis buffer next morning and continue the dialysis for another 4–5 h.

16. Take out the protein solutions from the Slide-A-Lyzer with a needle and transfer into a 15-mL Falcon tube. Use the dialysis buffer as the blank and measure the OD_{280} on a spectrophotometer. Calculate the protein concentration according to each protein's calculated extinction coefficient. Aliquot into small amounts and flash freeze at −80°C (see Note 18).

3.8.2. Purification of Escherichia coli Ribosomes

Ribosomes are prepared at 4°C from *E. coli* strain A19 (19, 20).

1. Prepare a 150-mL LB media in a 500-mL flask and $4 \times 2,000$ mL of LB media in 4×8 L flasks. Adjust the pH to 7 if necessary. Autoclave for 20 min at 121°C. Cool down to room temperature.

2. Inoculate the 150-mL LB from A19 glycerol stock for overnight starter culture.

3. Inoculate $4 \times 2,000$ mL LB with 30 mL starter culture and shake at 250 rpm and 37°C.

4. Check the OD_{600} and stop the growth when $OD_{600} = 0.6$–0.8, about 2.5 h.

5. Pour the culture into 4×2.25 L centrifuge bottles and cool down on ice.

6. Pellet the cells at $4,000 \times g$ for 30 min at 4°C.

7. Resuspend and combine the cells with a total of 600 mL ribosome buffer A in a 1 L bottle and spin at $4,000 \times g$ for 30 min at 4°C.

8. Freeze the bottle with bacterial pellet at −20°C.

9. Resuspend the pellet in 14 mL of ribosome buffer A and transfer to the medium blender chamber of the BeadBeater. Rinse the centrifuge bottle with 3 mL ribosome buffer A and transfer to the chamber. Repeat the rinse once.

10. Add pre-cooled glass beads to fill the blender chamber. Assemble the BeadBeater according to manufacturer's instructions.
11. Induce cell lysis with 6×20 s pulses. Allow the chamber to cool down for 40 s between pulses.
12. Transfer the glass beads-lysate mixture into a centrifuge tube. Use 5 mL ribosome buffer A to rinse the blender chamber and transfer to the centrifuge tube. Repeat the rinse once.
13. Centrifuge at $16,000 \times g$ for 15 min at 4°C.
14. Carefully transfer the supernatant into a clean 50 mL Falcon tube and keep on ice. Centrifuge again if necessary.
15. Distribute 15 mL of cold ribosome buffer B with 30% sucrose into each of four ultracentrifuge tubes on ice.
16. Carefully layer 7.5 mL of the cleared lysate onto the sucrose in the four tubes.
17. Seal all the tubes by using the Quick-Seal Tube Rack.
18. Spin at $300,000 \times g$ for 2.5 h in Type Ti70.1 rotor at 4°C.
19. Cut off the top part of the tubes and discard the supernatant. Wash the pellet with ribosome buffer A, taking care to remove the loosely packed brown flocculent material on top of the clear ribosome pellet.
20. Add 1 mL ribosome buffer B and a magnetic stir bar to each tube. Place the tubes in beakers containing half-full ice water. Resuspend the ribosome pellet with gentle stirring. Make sure no water gets into the tubes.
21. Combine the ribosomes and repeat the ultracentrifugation once.
22. Repeat steps 19–20 (see Note 19), use 2 mL buffer C to resuspend the ribosome pellet.
23. Determine the concentration based on $OD_{260}=1,000$ being equal to 23 nmol of ribosomes. Aliquot the ribosomes in 20 µL portions on ice and flash freeze in liquid nitrogen (see Note 20). Store at −80°C.

3.8.3. Translation

1. Prepare a standard 3× polymix buffer.
2. Place a 15-mL falcon tube on ice. Add 3.33 mL of 3× polymix buffer supplemented with 1 mM DTT, 2 mM ATP, 2 mM GTP, 10 mM Creatine phosphate, and 30 µM (6R,S)-5,10-formyl-5,6,7,8-tetrahydrofolic acid. Add final concentrations of 4 µg/mL creatine kinase, 3 µg/mL myokinase, 1.1 µg/mL nucleotide diphosphate kinase, 1 µg/mL inorganic pyrophosphatase, 0.2 µM MTF, 1.0 µM IF1, 0.3 µM IF2, 0.7 µM IF3, 3.2 µM EF-Tu, 0.6 µM EF-Ts, 0.5 µM EF-G, 0.3 µM RF1, 0.4 µM RF3, 0.1 µM RRF, 0.5 µM ribosomes, and 0.05 OD_{260} U/L total tRNA. In addition, the reaction contains

natural amino acids (200 μM each), unnatural amino acids (400 μM to 6.6 mM) (see Note 21), and AARSs (0.1 μM MetRS, 0.3 μM LeuRS, 0.6 μM GluRS, 0.2 μM ProRS, 1.0 μM GlnRS, 1.0 μM HisRS, 0.25 μM PheRS A294G, 1.5 μM TrpRS, 0.2 μM SerRS, 0.2 μM IleRS, 0.4 μM ThrRS, 0.6 μM AsnRS, 0.6 μM AspRS, 0.5 μM TyrRS, 0.5 μM LysRS, 0.4 μM ArgRS, 0.2 μM ValRS, 0.2 μM AlaRS, 0.5 μM CysRS, and 0.06 μM GlyRS). 0.2 μM of ^{35}S-Met is added to isotopically label the peptides.

3. Start translation by addition of photo-crosslinked mRNA (see Subheading 3.7) to 1.0 μM followed by incubation for 1 h at 37°C.

4. At the end of incubation, add KCl and $MgCl_2$ to a final concentration of 550 and 50 mM, respectively (see Note 22). Mix by inverting the tube several times.

5. Incubate the tube at room temperature for 15 min.

6. Transfer the tube to a −20°C freezer and incubate overnight.

3.9. Oligo(dT) Purification and Cyclization

1. Take six 20 mL Bio-Rad Econo-Pac columns and rinse with ddH_2O.

2. Weigh 1 g of Oligo(dT)-cellulose in a 50 mL Falcon tube. Add 30 mL ddH_2O to swell the cellulose.

3. Transfer 5 mL Oligo(dT)-cellulose into each of six Econo-Pac column.

4. Allow the water flow through.

5. Add 10 mL of oligo(dT) binding buffer to the columns and drain. Repeat once.

6. Cap the bottom of the columns with a yellow end cap.

7. Take the translation reaction out of the −20°C freezer and vortex to resuspend any precipitates that might have formed. Remove 5 μL for scintillation counting to determine the total radioactivity of ^{35}S-Met added to the translation reaction.

8. Equally transfer the rest of the translation reaction to the columns, 2.6 mL for each column.

9. Rinse the translation reaction tube three times with 5 mL oligo(dT) Binding buffer and transfer to one of the columns.

10. Add more binding buffer to the columns so that each has a final volume of 20 mL. Cap the columns tightly on both ends.

11. Place the columns on a rocking platform shaker in 4°C cold room and shake for 30 min.

12. Place the columns on a column rack and carefully remove the bottom end caps. Drain the binding buffer from the columns into a disposable tube. Discard the liquid radioactive waste accordingly.

13. Add 25 mL oligo(dT) wash buffer and allow to drain. Repeat this once.
14. While the columns are draining, prepare the cyclization buffer.
15. Replace bottom end caps tightly to the columns. Add 6 mL of cyclization buffer to each column.
16. Shake the columns at room temperature for 30 min (see Note 23).
17. Drain the cyclization buffer, and wash the columns twice with 20 mL wash buffer. The first wash should contain 5 mM 2-mercaptoethanol and second wash should have 0.5 mM TCEP as the reducing agent (see Note 24).
18. After the columns have completely drained, elute the columns eight times with 1 mL ddH$_2$O with 0.5 mM TCEP.
19. Take 1 μL from each eluent and count in a scintillation counter.
20. Combine fractions with significant radioactivity and transfer into 6×50 mL Nalgene Oakridge centrifuge tubes and add 0.8 mL 3 M KOAc, pH 5.5, 32 mL 100% ethanol and 0.8 mL 5 mg/mL glycogen.
21. Freeze the tube in the −20°C freezer for 30 min.
22. Centrifuge at 16,000×g for 20 min at 4°C.
23. Discard the supernatant (see Note 25).
24. Resuspend the pellet in each tube with 500 μL 70% ice cold ethanol and transfer to a new microcentrifuge tube. Repeat twice. Combine all the washes in the microcentrifuge tube.
25. Centrifuge for 2 min at maximum speed on a bench-top centrifuge.
26. Discard the supernatant accordingly (see Note 25).
27. Wash the pellet with 500 μL 70% ethanol and centrifuge. Discard the supernatant accordingly. Spin again and completely remove residual liquid with a pipette and air-dry for 10 min at room temperature.
28. Calculate the pmol amount of the mRNA-peptide fusion based on the scintillation counts. Re-dissolve the pellet in ddH$_2$O so that the final concentration of the fusion is 100 nM (0.1 pmol/μL) (see Note 26).
29. If necessary, use centrifugal filter device to filter out residual Oligo(dT)-cellulose in the fusion solution.

3.10. Reverse Transcription and Ni-NTA Purification

1. In a 2-mL microcentrifuge tube, add the mRNA-peptide fusion from Subheading 3.9 and 0.5 mM dNTPs, 0.5 μM RT-primer TTTTTTTTTTTTTTGTGATGGTGATGGTGGCC TAAGC. Put the tube in a heating block and incubate for 5 min at 65°C. Immediately place the tube on ice and incubate for at least 1 min (see Note 27).

2. Add reagents to a final concentration of 5 mM $MgCl_2$, 1 mM DTT, 2 U/μL RNaseOUT, 5 U/μL Superscript III in a total volume of 2 mL (see Note 28). Mix well.

3. Incubate for 30 min at 55°C for elongation, then 15 min at 70°C to inactivate the Superscript III (see Note 29).

4. Transfer 1 mL Ni-NTA Agarose to an Econo-Pac column (see Note 30). Add 4 mL Ni-NTA denaturing binding buffer and drain the column. Cap the bottom end of the column tightly.

5. Transfer the RT reaction to the column. Rinse the tube with 1 mL Ni-NTA denaturing binding buffer and transfer to the column. Repeat the rinse two more times.

6. Add a total of 10 mL Ni-NTA denaturing binding buffer to the column.

7. Shake the column on a rocking platform shaker for 1 h at 4°C.

8. Drain the column. Use a 15-mL Falcon tube to collect the flow through.

9. Wash the column with 10 mL Ni-NTA wash buffer (II). Repeat the wash twice. Each time collect the wash in a new 15 mL falcon tube.

10. Elute six times with portions of 0.5 mL Ni-NTA elution buffer (II). Collect eluents in 1.5 mL microcentrifuge tubes.

11. For each elution, let the elution buffer sit on the beads for 5 min by capping the bottom end of the column.

12. Take 100 μL from the flow through and three washes, and 5 μL of each eluent and count in the scintillation counter.

13. Combine all eluents with high counts, and dialyze against 1 L pre-cooled selection buffer in a cold room overnight.

14. Precipitate the fusion with ethanol and NaOAc (see Note 31). Redissolve in 1 mL selection buffer. Place on ice before proceeding to selection.

3.11. Selection

1. Take two 1.5 mL tubes and add 200 μL of Pierce Magnetic Glutathione beads to each tube. Label one tube "GST" for negative selection and the other "GST fusion" for positive selection.

2. Add 1 mL GST Beads Wash buffer to each tube and mix the beads by tapping the tube gently. Place the tubes on the magnetic stand. Use a pipette to remove the supernatant while holding the magnetic stand. Be careful not to disturb the beads (see Note 32).

3. Repeat the wash two more times.

4. Add 1 mL GST beads wash buffer and 10 μM purified GST to the tube labeled "GST". Add 1 mL GST beads wash buffer

and 10 μM purified GST fusion of target protein to the tube labeled "GST fusion".

5. Incubate at 4°C for 1 h with rotation on a tube rotator.

6. At the end of incubation, briefly spin the tubes to collect the beads and liquid. Place the tubes on the magnetic stand and remove the supernatant.

7. Wash the beads twice with 1 mL GST beads wash buffer and once with 1 mL selection buffer.

8. Add dialyzed or precipitated mRNA-peptide fusions from Subheading 3.10 to the tube labeled "GST". Add BSA to a final concentration of 0.1 mg/mL. Place the tube labeled "GST fusion" on ice until use.

9. Incubate the tube labeled "GST" at 4°C for 1 h with rotation.

10. Separate the supernatant from the beads in the tube labeled "GST" using the magnetic stand and transfer to the tube labeled "GST fusion".

11. Incubate at 4°C for 1 h with rotation.

12. Remove the supernatant from the beads using the magnetic stand and save it in a new tube. The supernatant is the flow through.

13. Wash the beads three times with 400 μL selection buffer. Save the wash in a new tube each time.

14. Elute six times with portions of 100 μL GSH elution buffer.

15. At each elution step, let the elution buffer sit on beads for 5 min before separation on the magnetic stand.

16. Take 10 μL from the flow through and three washes, and 1 μL of each eluent, and count in the scintillation counter. Count all the magnetic beads in two labeled tubes.

17. Combine all eluents with high counts, and dialyze against 1 L pre-cooled dialysis buffer in the cold room overnight.

3.12. PCR Amplification of Selected Fusions

1. Set up a series of pilot PCR reactions using 10, 20 and 30 μL dialyzed eluent from Subheading 3.11 in 100 μL PCR reactions. Run the cycles and check PCR products on a 2% Agarose gel as described in Subheading 3.5. Analyze the intensity of DNA bands to find out the optimum conditions for PCR.

2. Set up a large-scale PCR using multiple tubes or PCR strips and use all the eluents from Subheading 3.11 to amplify the selected fusions.

3. After PCR, combine all the reaction mixtures and add equal volume phenol:chloroform:isoamyl alcohol (25:24:1). Vortex the tube vigorously and spin at maximum speed for 2 min to separate the organic and aqueous phases.

4. Transfer the upper phase to a new tube and precipitate with ethanol and NaOAc.

5. Redissolve the DNA in 500 μL ddH$_2$O. This is the template for the next round of selection (see Note 33).

3.13. Additional Rounds of Selection

Once the first round of selection is completed in Subheading 3.12, repeat the selection for additional 6–10 rounds. Use a 1–2 mL of translation reaction. Scale down all the subsequent steps accordingly (see Note 34). The progress of the selection can be monitored by measuring the percentage of ^{35}S peptides that are eluted from the GSH column for that round. Once this value plateaus, proceed to sequencing.

3.14. Analysis of Selected Sequences

1. Clone and sequence the library as described in Subheading 3.5. Usually one 96-well plate of clones is enough to find several families of unnatural peptide selection winners.

2. Once the individual sequences are obtained, they can be aligned manually or using web-based alignment programs. For sequence alignments, Clustal W2 (Jalview V. 2) is used (21). Unnatural amino acids were assigned based on the tRNA/AARS pairs responsible for the incorporation into peptides.

3. Select a few sequences from each family for further study.

4. Synthesize the peptide sequences using a solid phase peptide synthesizer.

4. Notes

1. Other gel elution methods such as crush and soak can be used. Electroelution with the Elutrap is relatively fast and efficient in nucleic acid recovery from PAGE gels.

2. One can also make a 10% APS stock solution and store at 4°C for up to a month. We prefer freshly made APS.

3. It is not necessary to filter the TBE buffer as the EDTA in the buffer is sufficient to inhibit DNase and RNase activity.

4. Standard PCR buffer can also be used. We prefer ThermoPol buffer for its optimized components for use with Taq polymerase and the inclusion of 0.1% Triton X-100 to reduce nonspecific binding.

5. Standard agarose gel loading buffer can also be used but the bromophenol blue in the standard loading buffer might interfere with the visualization of the PCR product. Orange G is smaller than bromophenol blue and would not block the visibility of smaller DNA molecules.

6. A stock solution of DTT can be made and aliquots can be stored at −20°C.

7. Inorganic pyrophosphatase can be stored as 1 mg/mL in enzyme dialysis buffer (50 mM HEPES-KOH, pH 7.6, 100 mM KCl, 10 mM $MgCl_2$, 7 mM 2-mercaptoethanol, 50% glycerol). Dilute the 1 mg/mL stock with the enzyme dialysis buffer to 0.1 mg/mL for routine use.

8. Purchase the XL-PSO oligonucleotide from a good resource such as Integrated DNA Technologies (IDT) or the Keck facilities at Yale. Dissolve the oligonucleotide in ddH_2O. Desalt on a NAP-25 column (GE Healthcare). Measure the OD_{260} on a spectrophotometer and adjust the concentration to 125 μM. Aliquot and cover with aluminum foil. Store at −20°C.

9. Other dialysis units or cassettes can be used, such as dialysis tubing or D-Tube dialyzer from Merck/EMD Chemicals. The molecular weight cut-offs should be smaller than the molecules that need to be dialyzed.

10. The protease inhibitor should not contain any EDTA or EGTA as these are chelating agents and will strip the nickel from Ni-NTA Agarose.

11. Some of the total tRNAs prepared from *E. coli* extracts are intrinsically acylated with their cognate natural amino acids. They need to be deacylated before use in PURE translation system. After deacylation and dialysis, the natural amino acids are removed from tRNAs, and the tNRAs can then be acylated with unnatural amino acids in PURE system.

12. Make a fresh stock of 0.5 M TCEP and dilute 1- to 1,000-fold into solution before use.

13. We chose GST fusion for its broad use in successful expression of soluble proteins in *E. coli*. We also use GST alone to perform the negative selection. In the negative selection, the peptides that bind to GST and the support matrix (agarose or other matrix) are removed from the library. The rest of the peptides are then selected on a GST fusion target. Other proteins with different tags can be used, but the negative selection needs to be adjusted accordingly.

14. Although we have not performed an exhaustive search it appears that peptides with a single cysteine form linear peptides containing a bis-thioether adduct between the cysteine and 2-mercaptoethanol.

15. One negative consequence of using the codon NNB is that it leads to more exaggerated biases between other amino acids. For example, in an NNB library, the ratio of serine to tryptophan is 5:1, vs. only 3:1 in an NNS or NNK library.

16. For mRNA that will be used in translation, use KOAc instead of NaOAc to precipitate. Residual Na$^+$ in the mRNA preparation will interfere with translation.
17. Longer exposure is not necessary. The efficiency of photo-crosslinking is usually around 50%.
18. It is important to aliquot the purified proteins in small volumes. Do not freeze and thaw the proteins more than twice.
19. If the ribosomes are to be used to study translation factors or their mutants, the trace factor activity that remains after second wash can be further reduced by washing the ribosomes a third or even fourth time.
20. It is important to aliquot the purified ribosomes in small amount. Do not freeze and thaw the ribosomes more than once. Discard the remaining ribosomes after thawing.
21. The concentrations of the amino acids need to be optimized for each new combination. The analogs are initially utilized at 400 μM and the concentration is increased if misincorporations are observed in translation reactions with FLAG or His-tagged templates. The process is outlined in ref. 12.
22. It is convenient to make a stock solution of KCl/MgCl$_2$. Mix 2.85 M KCl and 1 M MgCl$_2$ in a ratio of 11.9:3.1. Add 0.32 volume of this mixture to the translation reaction. The final concentrations of KCl and MgCl$_2$ in the reaction are approximately 500 and 50 mM, respectively.
23. It is necessary to release the pressure at least once during cyclization by opening the bottom cap while holding the column upside down, since the acetonitrile evaporates and pressure can be built inside the column. Use a paper towel to cover the cap when opening. Use caution to avoid any radioactive contamination.
24. The 2-mercaptoethanol in the first wash is used to neutralize the residual dibromo-*m*-xylene in the cyclization buffer.
25. Be sure to discard the supernatant properly since it contains residual ^{35}S.
26. The amount of mRNA-peptide fusion obtained after oligo(dT) purification varies from 1 to 100 pmol.
27. In the later rounds of selection, the volume needed for reverse transcription is usually less than 400 μL. Multiple PCR tubes and PCR machine can be used.
28. DTT is always added ahead of RNaseOUT.
29. After reverse transcription, do not denature the duplex.
30. In the later rounds of selection, the volume of Ni-NTA Agarose required is usually less than 400 μL. Centrifugal filter devices can be used to simplify the purification of mRNA-peptide fusion through Ni-NTA agarose.

31. In the later rounds of selection, the volume of mRNA-peptide fusion is small and no precipitation is required. The solution after dialysis can be directly used in selection.

32. The magnetic beads cannot withstand high speed centrifugation. Spin briefly if necessary.

33. Save half of the purified PCR template from the previous round in case anything goes wrong.

34. In the later rounds of selection, sometimes PCR amplification of selected mRNA templates can be difficult. Conditions for pilot PCR need to be explored and optimized. More eluent volume or cycles might be necessary for PCR. If PCR is successful, repeat the previous round of selection without the negative selection or use a larger volume of translation reaction.

Acknowledgments

This work was supported by the Concern Foundation and the Massey Cancer Center at Virginia Commonwealth University.

References

1. Forster, A. C., Weissbach, H., and Blacklow, S. C. (2001) A simplified reconstitution of mRNA-directed peptide synthesis: activity of the epsilon enhancer and an unnatural amino acid, *Anal. Biochem.* 297, 60–70.

2. Shimizu, Y., Inoue, A., Tomari, Y., Suzuki, T., Yokogawa, T., Nishikawa, K., and Ueda, T. (2001) Cell-free translation reconstituted with purified components, *Nat. Biotechnol.* 19, 751–755.

3. Hartman, M. C. T., Josephson, K., and Szostak, J. W. (2006) Enzymatic aminoacylation of tRNA with unnatural amino acids, *Proc. Natl. Acad. Sci. U. S. A.* 103, 4356–4361.

4. Murakami, H., Ohta, M., Ashigai, H., and Suga, H. (2006) A highly flexible tRNA acylation method for non-natural polypeptide synthesis, *Nat. Methods* 3, 357–359.

5. Merryman, C., and Green, R. (2004) Transformation of aminoacyl tRNAs for the *in vitro* selection of "drug-like" molecules, *Chem. Biol.* 11, 575–582.

6. Goto, Y., Ohta, A., Sako, Y., Yamagishi, Y., Murakami, H., and Suga, H. (2008) Reprogramming the translation initiation for the synthesis of physiologically stable cyclic peptides, *ACS Chem. Biol.* 3, 120–129.

7. Kawakami, T., Ohta, A., Ohuchi, M., Ashigai, H., Murakami, H., and Suga, H. (2009) Diverse backbone-cyclized peptides via codon reprogramming, *Nat. Chem. Biol.* 5, 888–890.

8. Sako, Y., Morimoto, J., Murakami, H., and Suga, H. (2008) Ribosomal synthesis of bicyclic peptides via two orthogonal inter-side-chain reactions, *J. Am. Chem. Soc.* 130, 7232–7234.

9. Yamagishi, Y., Ashigai, H., Goto, Y., Murakami, H., and Suga, H. (2009) Ribosomal synthesis of cyclic peptides with a fluorogenic oxidative coupling reaction, *Chembiochem* 10, 1469–1472.

10. Millward, S. W., Takahashi, T. T., and Roberts, R. W. (2005) A general route for post-translational cyclization of mRNA-displayed libraries, *J. Am. Chem. Soc.* 127, 14142–14143.

11. Josephson, K., Hartman, M. C. T., and Szostak, J. W. (2005) Ribosomal synthesis of unnatural peptides, *J. Am. Chem. Soc.* 127, 11727–11735.

12. Hartman, M. C. T., Josephson, K., Lin, C.-W., and Szostak, J. W. (2007) An expanded set of amino acid analogs for the ribosomal translation of unnatural peptides, *PLoS ONE* 2, e972.

13. Adessi, C., and Soto, C. (2002) Converting a peptide into a drug: strategies to improve stability and bioavailability, *Curr. Med. Chem.* 9, 963–978.
14. Roberts, R. W., and Szostak, J. W. (1997) RNA-peptide fusions for the *in vitro* selection of peptides and proteins, *Proc. Natl. Acad. Sci. U. S. A.* 94, 12297–12302.
15. Nemoto, N., Miyamoto-Sato, E., Husimi, Y., and Yanagawa, H. (1997) In vitro virus: bonding of mRNA bearing puromycin at the 3′-terminal end to the C-terminal end of its encoded protein on the ribosome in vitro, *FEBS Lett.* 414, 405–408.
16. Takahashi, T. T., and Roberts, R. W. (2009) In vitro selection of protein and peptide libraries using mRNA display, *Methods Mol. Biol.* 535, 293–314.
17. http://www.ambion.com/techlib/append/rbs_requirements.html. (Accessed 1 August 2010).
18. Kurz, M., Gu, K., and Lohse, P. A. (2000) Psoralen photo-crosslinked mRNA-puromycin conjugates: a novel template for the rapid and facile preparation of mRNA-protein fusions, *Nucleic Acids Res.* 28, E83.
19. http://www.daimi.au.dk/~g960801/Protocols/Riboso.pdf. (Accessed 1 August 2010).
20. Spedding, G. (1990) *Ribosomes and protein synthesis: a practical approach*, Oxford University Press, London.
21. Waterhouse, A. M., Procter, J. B., Martin, D. M., Clamp, M., and Barton, G. J. (2009) Jalview Version 2–a multiple sequence alignment editor and analysis workbench, *Bioinformatics (Oxford, England)* 25, 1189–1191.

Part VI

Case Studies

Chapter 22

Optimization of CAT-354, a Therapeutic Antibody Directed Against Interleukin-13, Using Ribosome Display

George Thom and Ralph Minter

Abstract

In this case study, we describe the use of in vitro protein evolution with ribosome display to improve the potency of a human interleukin-13-neutralising antibody by a factor of over 200-fold and derive a therapeutic candidate, CAT-354, for the treatment of asthma. A combination of directed and random mutagenesis enabled the identification of highly potent neutralising antibodies and highlighted the advantage of the ribosome display protein evolution approach in identifying beneficial mutations across the entire sequence space. This chapter describes in detail the process followed to achieve a successful in vitro affinity maturation outcome using ribosome display technology.

Key words: Ribosome display, Interleukin-13, Protein evolution, scFv, Antibody

1. Introduction

Technologies for the isolation of human monoclonal antibodies against disease-associated protein targets are now widely available and, as a result, many human antibody drugs are achieving success in the clinic (1). Achieving therapeutic efficacy remains a challenging task and so, to improve the chances of success, the emphasis has shifted in recent years towards optimising the potency, and other features, of antibodies. One of the key determinants of antibody potency is the affinity for the target antigen, and many different mutagenesis strategies have been used to identify antibody variants with higher target affinity than the original "parent" antibody. Oligonucleotide-directed mutagenesis is the favoured approach to target mutations to the complementarity-determining region (CDR) loops, where the chance of influencing antibody potency is expected to be high. Simultaneous saturation mutagenesis of all

CDR loop residues is made impossible by the sheer number of potential combinations of amino acids that must be explored. Despite the production and selection, by various display technologies, of large (10^9–10^{11}) variant antibody libraries becoming a more routine procedure, this allows the full exploration of only six or seven residues. As a result, the majority of affinity maturation approaches have concentrated on the region with the highest likelihood of yielding favourable affinity gains, the variable heavy (V_H) chain CDR3 loop (2–4). However, there is no guarantee that side chain replacements within the V_H CDR3 loop necessarily leads to significant gains in antibody affinity.

A more useful method to inform and direct antibody affinity maturation would be one that rapidly identifies positions within the whole antibody variable fragment (Fv) sequence, where change is tolerated and also associated with affinity gains. We have found that ribosome display selection (5), in conjunction with random mutagenesis by error-prone PCR, is a powerful method to scan parent antibody sequences for affinity-enhancing mutations. Multiple, iterative cycles of mutagenesis and selection can be applied to favour the accumulation of beneficial mutations in the pool of selected variants, and, by analysing the sequences of clones with improved potency, a map of hotspots can be derived that is effectively a functional scan of the Fv sequence. In vitro display technologies, such as ribosome display, offer two important advantages, the first being that large libraries can be made rapidly because there is no need to transform high copy numbers of mutant plasmids into a host. The second key advantage is that additional mutations can be introduced at every selection round because a PCR step is included in the amplification cycle between rounds.

Interleukin-13 (IL-13) is a cytokine which is known to be a key mediator of allergic respiratory and oesophageal inflammation and is strongly linked to the clinical development of asthma (6). In this study, V_H CDR3-targeted mutagenesis and error-prone mutagenesis were combined to engineer a 200-fold potency gain in a neutralising human antibody against IL-13, allowing this antibody to progress as a clinical candidate, CAT-354, for the treatment of asthma. Scanning of protein sequence space by using iterative cycles of random mutagenesis and in vitro selection was found to be a rapid way to gain knowledge about the protein interaction surface and simultaneously achieve higher antibody affinity.

2. Materials and Methods

Materials and methods relating to ribosome display are described in ref. 7. Other methods, not relating to ribosome display, are described elsewhere and references are provided.

3. Results

The affinity maturation of the IL-13-neutralising antibody BAK1 was performed in vitro by ribosome display. Three V_H CDR3 libraries, comprising, respectively, saturation mutagenesis of residues 94–99, 98–100C, and 100B–102, numbered according to Kabat (8), were prepared by oligonucleotide-directed mutagenesis using NNS triplets to diversify each target codon (3). More mutations were introduced by incorporating error-prone PCR amplification of the scFv sequences before the first and third rounds of selection. Ribosome display selections were performed stringently, as described in ref. 9, on decreasing concentrations of IL-13, reducing tenfold at each round, from 10 nM at round one to 100 pM at round three. The ribosome display affinity maturation strategy is summarised schematically in Fig. 1.

Unique variant scFvs were expressed and purified (10) before testing for improved potency over the parent BAK1 scFv in the IL-13-dependent TF-1 cell proliferation assay (4). Ribosome display yielded 25 variants with improved potency (BAK1.1–BAK1.25) as determined

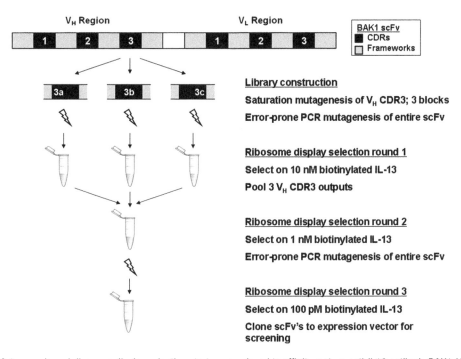

Fig. 1. Mutagenesis and ribosome display selection strategy employed to affinity mature anti-IL13 antibody BAK1. V_H CDR3 of the BAK1 scFv gene was targeted for saturation mutagenesis in three blocks, requiring the construction of libraries, 3a, 3b, and 3c. Error-prone PCR mutagenesis of each library added further sequence diversity to the entire scFv gene. Three rounds of ribosome display were performed, and affinity pressure was applied by reducing the concentration of the IL-13 target protein at each round. Additional error-prone PCR mutagenesis was included prior to the final round of selection and, subsequently, outputs were cloned to an expression vector for screening.

Table 1
Antibody sequences derived by ribosome display affinity maturation of BAK1

	VH CDR1	VH CDR2	VH CDR3	VL CDR1	VL CDR2	VL CDR3
BAK1	N Y G L S	W I S A N N G D T N Y G Q E F Q G	D S S S N W A R W F F D L	G G N N I G S K L V H	D D G D R P S	Q V W D T G S D P V V
BAK1.1	- - - - -	- - - - - - - - - - - - - - - - -	- - - S - - - - - - - - -	- I - - - - - - - - -	- - - - - - -	- - - - - - - - - - -
BAK1.2	- - - - -	- - - - - - - - - - - - - - - K -	- - - - - - - - - - - - -	- I - - - - - - - - -	- - - - - - -	- - - - - - - - - - -
BAK1.3	- - - - -	- - - - - - - D - - - - - - - - -	- - - S - - - - - - - - -	- - - - - - G - - - -	- - - - - - -	- - - - - - - - - - -
BAK1.4	- - - - -	- - - - - - - D - - - - - - - - -	- - - - - - - - - - - - -	- - - - - S - - - - R	- - - - - - -	- - - - - - - - - - -
BAK1.5	- - - - -	- - - - - - G - - - - - - - - - -	- - - N - - - - - - - - -	- - - - - - - - - - -	- - - - - - -	- - - - - - - - - - -
BAK1.6	- - - - -	- - - - - - - - - - - - - - - - -	- - - S - - - - - - - - -	- - - - - - G - - - -	- - - - - - -	- - - - - - - - - - -
BAK1.7	- - - - -	- - - - - - - - - - - - - - - - -	- - - - - - - - - - - - -	- - - D - - G - - - -	- - - - - - -	- - - - - - - - - - -
BAK1.8	- - - - -	- - - - - - - - R - - - - - - - -	- - - S - - - - - - - - -	- - - - - - G - - - -	- - - - - - -	- - - - - - - - - - -
BAK1.9	- - - - -	- - - - - - - - - - - - - - - - -	- - - N - - - - - - - - -	- - - - S - G - - - -	- - - - - - -	- - - - - - - - - - -
BAK1.10	- - - - -	- G - - - - - - - - - - - - - - -	- - - S - - - - - - - - -	- - - - S - - - - - -	- - - - - - -	- - - - - - - - - - -
BAK1.11	- - - - -	- - - - - - - - - - - - - R - - -	- - - - - - - - - - - - -	- - - - - - G - - - -	- - - - - - -	- - - - - - - - - - -
BAK1.12	D - - - -	- - - - - - - - - - - - - - - - -	- - - - - - - - - - - - -	- - - - - - G - - - -	- - - - - - -	- - - - - - - - - - -
BAK1.13	- - - - -	- - - - - - - - - - - - - - - - -	- - - S - - - - - - - - -	T - - - - - - - - - -	- - - - - - -	- - - - - - - - - - -
BAK1.14	- - - - -	- - - - - - - - - - - - - - - - -	- - - N - - - - - - - - -	- I - - - - - - - - -	- - - - - - -	- - - - - - - - - - -
BAK1.15	- - - - -	- - - - - - - - - - - - - - - - -	- - - - - - - - - - - - -	- - - - - - G - - - -	- - - - - - -	- - - - - - - - - - -
BAK1.16	- - - - -	- T - - - - - - - - - - - - - - -	- - - - - - - - - - - - -	- - - - - - - - - - -	- - - - - - T	- - - - - - - - - - -
BAK1.17	- - - - -	- - - - - - - - I R - - - - - - -	- - - - - - - - - - - - -	- I - - - - - - - - -	- - - - - - -	- - - - - - - - - - -
BAK1.18	- - - - -	- - - S - - - - - - - - - - - - -	- - - S - - - - - - - - -	- - - - - - - - - - -	- - - - - - -	- - - - - - - - - - -
BAK1.19	- - - - -	- - - - - - - K - - - - - - - - -	- - - S - - - - - - - - -	- - - - - - G - - - -	- - - - - - -	- - - - - - - - - I -
BAK1.20	- - - - -	- G - - - - - D - - - - - - - - -	- - - N - - - - - - - - -	- - - - - - - - - - -	- - - - - - -	- - - - - - - - - - -
BAK1.21	- - - - -	- - I - - - - - - - - - - - - - -	- - - - - - - - - - - - -	- - - - - - G - - - -	- - - - - - -	- - - - - - - - - - -
BAK1.22	- - - - -	- - - - - - - - - - - - - - - - -	- - - - - - - - - - - - -	T - - - - - G - - - -	- - - - - - -	- - - - - - - - - - -
BAK1.23	- - - - -	- - - - - - - - - R - - - - - - -	- - - - - - - - - - - - -	- - - - S - G - - - -	- - - - - - -	- - - - - - - - - - -
BAK1.24	- - - - -	- - - - - - - - - - - G - - - - -	- - - S - - - - - - - - -	- - - - S - - - - - -	- - - - - - -	- - - - - - - - - - -
BAK1.25	- - - - -	- - - - - - - - - R - - - - - - -	- - - S - - - - - - - - -	- - - - - - - - - - -	- - - - - - -	- - - - - - - - - - -

The full CDR sequences are shown for the parent antibody BAK1, and mutations found in variants BAK1.1–BAK1.25 are *highlighted*. Dashes indicate non-mutated residues

by their IC_{50} values of between 4.3 and 23 nM in the assay. In scFv format, these variants were therefore between two- and tenfold improved over BAK1 (IC_{50} of 44 nM). The sequences of the variants BAK1.1–BAK1.25 are summarised in Table 1. The V_H N99S mutation was strongly selected, as judged by its appearance in 10 of the 25 mutants shown. Despite targeting V_H CDR3 by saturation mutagenesis, relatively few beneficial changes were observed in this CDR3 loop, with only one other position (V_H S97N) mutated from wild type in the improved mutant population. It is worth noting that, typically, following saturation mutagenesis of six continuous CDR positions improved variants contain substitutions in four to six positions.

Sequence analysis of 42 BAK1 variants indicated that error-prone PCR added an average of 2.2 additional amino acid substitutions per variant, excluding changes in V_H CDR3. The V_L CDR1 loop, in particular, accumulated a disproportionately large number of changes, with 46 of 94 error-prone mutations (49.0%) found in V_L CDR1. This is despite the V_L CDR1 loop only comprising 11 of the 230 amino acids (4.8%) of the Fv region. The overall distribution of error-prone PCR mutations shows a very strong preference for CDR mutations over framework changes, with 77 out of 94 (81.9%) error-prone PCR mutations occurring in CDRs. This finding is illustrated in Fig. 2, which plots the mutation frequency observed in the ribosome display-selected V_H and V_L regions alongside the mutation frequency following somatic hypermutation, the natural mechanism of antibody affinity maturation in B lymphocytes in vivo.

Variant BAK1.1 was converted from scFv to IgG format (11) and tested for potency against the BAK1 parent IgG in the TF-1

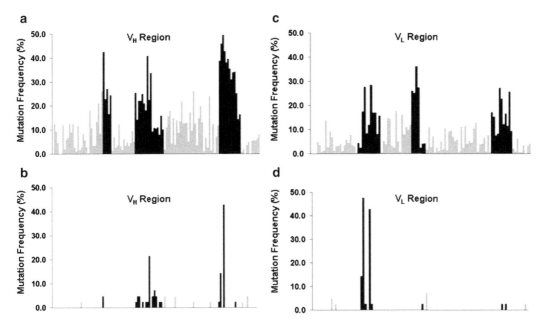

Fig. 2. Positional bias of mutations within V_H and V_L antibody regions following either somatic hypermutation in vivo or ribosome display protein evolution in vitro. The mutation frequency for each position in the V_H region is plotted following (**a**) somatic hypermutation ($n=2,526$) and (**b**) ribosome display evolution ($n=42$). Similarly, in the V_L region, the mutation frequency at each position is plotted following (**c**) somatic hypermutation ($n=3,904$) and (**d**) ribosome display evolution ($n=42$). Antibody framework positions are shown in *grey bars* and CDR positions in *black bars*.

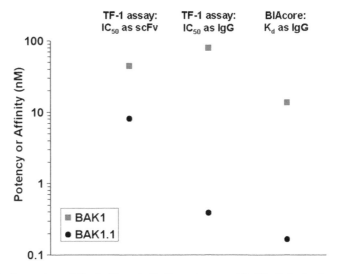

Fig. 3. Comparison of the IL-13 neutralisation potency and IL-13 affinity for the parent antibody BAK1 and affinity-matured variant BAK1.1. The potency, measured by IC_{50} in nM, of the BAK1 and BAK1.1 antibodies in both scFv and IgG formats is plotted alongside the affinity, also in nM, of both antibodies in IgG format.

proliferation assay, and the affinities were determined by kinetic analysis using surface plasmon resonance (12). Results are summarised in Fig. 3 and Table 2. The BAK1.1 IgG IC_{50} value of

Table 2
TF-1 IC$_{50}$ data are summarised, along with the binding kinetics of each IgG for IL-13, as measured by surface plasmon resonance

IgG	TF-1 IC$_{50}$, pM	TF-1 IC$_{50}$ BAK1/TF-1 IC$_{50}$ BAK1.1	k_{on}, per M/s	k_{off}, per second	K_d, pM	K_d BAK1/K_d BAK1.1
BAK1	78,900	–	5.49×10^5	7.41×10^{-3}	13,500	–
BAK1.1	388	203	2.49×10^6	4.09×10^{-4}	164	82

Potency and affinity improvements over BAK1 are also summarised. All TF-1 IC$_{50}$ figures were calculated from samples analysed in triplicate, and all values, including the kinetic values from surface plasmon resonance, had standard deviations of <10%

388 pM in the TF-1 assay and K_d of 164 pM corresponded to an overall improvement of 203-fold in potency and 82-fold in affinity relative to the parental BAK1 IgG. There was also a significant 21-fold potency gain upon conversion from scFv to IgG for BAK1.1, which may reflect an increase in stability conferred by the IgG format, given that the assay was performed at 37°C for 72 h.

4. Discussion

A validated method for antibody affinity maturation is to perform saturation mutagenesis of the V_H CDR3 loop. By following this approach for antibody BAK1, a substitution (V_H N99S) was discovered in the V_H CDR3 of several variant antibodies with improved potency. However, in all cases, these potent variants also harboured mutations in other CDR loops, most notably V_L CDR1 and V_H CDR2, which had derived from the random mutagenesis strategy. The crucial role of these additional CDR mutations in conferring the potency gain highlights a shortcoming in the use of directed mutagenesis alone to improve protein attributes and suggests there is benefit in the empirical exploration of antibody sequence space. We found that ribosome display was particularly well-suited to iterative cycles of random mutagenesis and affinity selection because the PCR amplification employed between rounds allows an opportunity for further sequence diversification without the burden of transforming the library into cells, as is required for other display technologies, such as phage or yeast display.

The random mutagenesis strategy used here resulted in a high frequency of mutations in CDR positions, with a distribution reminiscent of human antibodies that have undergone somatic hypermutation in vivo. The preference for CDR mutations is strong evidence that the majority of replacements were incorporated because of a beneficial effect on antigen binding, as intended through the use of affinity selections, rather than an effect on protein folding or any other characteristic. The CDR preference is particularly favourable when developing antibodies as therapeutic candidates because it is desirable for the framework regions to align exactly to the respective human germ-line sequence to reduce the potential for immunogenicity in patients. Although the scanning approach is comparable to somatic hypermutation, the advantage of the in vitro approach is that once mutational hotspots have been identified, those key positions can be revisited to introduce the full diversity of amino acid replacements. Both somatic hypermutation and error-prone PCR are limited by the fact that multiple (2–3) base substitutions within a codon are extremely rare. This rarity leads to an inherent bias towards conservative substitutions, and there is some evidence to suggest that non-conservative substitutions can be of significant benefit in the process of protein evolution (13). Furthermore, the pool of hotspots identified in the initial scan is expected to span several non-continuous regions of sequence space, which could then be randomised in a single library to take advantage of combinatorial effects on potency, where two or more mutations combine synergistically. The in vitro nature of ribosome display allows both the initial scan and the follow-up mutagenesis to be carried out rapidly, and, furthermore, the large library sizes available to ribosome display would enable as many as seven hotspots to be investigated in one library.

In conclusion, the approach of scanning a protein interaction surface for mutational hotspots is a rapid method to effectively exploit large regions of protein sequence space to identify positions amenable to optimisation. In this study, ribosome display in vitro evolution isolated an antibody of high affinity (a K_d of 164 pM) and high potency (an IC_{50} value of 388 pM in the IL-13-dependent TF-1 cell proliferation assay). The antibody resulting from this work, CAT-354, has been shown to exert an IL-13-neutralising effect in an in vivo model of human IL-13-induced inflammatory responses (14) and has consequently advanced into clinical studies for the treatment of asthma (15, 16).

References

1. Reichert, J. M. Antibodies to watch in 2010, *mAbs* 2, 84–100.
2. Thompson, J., Pope, T., Tung, J. S., Chan, C., Hollis, G., Mark, G., and Johnson, K. S. (1996) Affinity maturation of a high-affinity human monoclonal antibody against the third hypervariable loop of human immunodeficiency virus: use of phage display to improve affinity and broaden strain reactivity, *Journal of molecular biology* 256, 77–88.
3. Baker, K. P., Edwards, B. M., Main, S. H., Choi, G. H., Wager, R. E., Halpern, W. G., Lappin, P. B., Riccobene, T., Abramian, D., Sekut, L., Sturm, B., Poortman, C., Minter, R. R., Dobson, C. L., Williams, E., Carmen, S., Smith, R., Roschke, V., Hilbert, D. M., Vaughan, T. J., and Albert, V. R. (2003) Generation and characterization of LymphoStat-B, a human monoclonal antibody that antagonizes the bioactivities of B lymphocyte stimulator, *Arthritis and rheumatism* 48, 3253–3265.
4. Thom, G., Cockroft, A. C., Buchanan, A. G., Candotti, C. J., Cohen, E. S., Lowne, D., Monk, P., Shorrock-Hart, C. P., Jermutus, L., and Minter, R. R. (2006) Probing a protein-protein interaction by in vitro evolution, *Proceedings of the National Academy of Sciences of the United States of America* 103, 7619–7624.
5. Hanes, J., and Pluckthun, A. (1997) In vitro selection and evolution of functional proteins by using ribosome display, *Proceedings of the National Academy of Sciences of the United States of America* 94, 4937–4942.
6. Caramori, G., Ito, K., and Adcock, I. M. (2004) Targeting Th2 cells in asthmatic airways, *Current drug targets* 3, 243–255.
7. Lewis, L., and Lloyd, C. (2010) Optimisation of antibody affinity by ribosome display using site-directed or error-prone mutagenesis, in *Ribosome Display and Related Display Technologies: Methods and Protocols* (Jackson, R. H., and Douthwaite, J., Eds.), Humana Press, New York.
8. Kabat, E. A. *Sequences of proteins of immunological interest 5th ed*, Bethesda, MD : U.S. Dept. of Health and Human Services, Public Health Service, National Institutes of Health, 1991. NIH publication ; no. 91–3242.
9. Hawkins, R. E., Russell, S. J., and Winter, G. (1992) Selection of phage antibodies by binding affinity. Mimicking affinity maturation, *Journal of molecular biology* 226, 889–896.
10. Vaughan, T. J., Williams, A. J., Pritchard, K., Osbourn, J. K., Pope, A. R., Earnshaw, J. C., McCafferty, J., Hodits, R. A., Wilton, J., and Johnson, K. S. (1996) Human antibodies with sub-nanomolar affinities isolated from a large

non-immunized phage display library, *Nature biotechnology* 14, 309–314.
11. Persic, L., Roberts, A., Wilton, J., Cattaneo, A., Bradbury, A., and Hoogenboom, H. R. (1997) An integrated vector system for the eukaryotic expression of antibodies or their fragments after selection from phage display libraries, *Gene* 187, 9–18.
12. Karlsson, R., Michaelsson, A., and Mattsson, L. (1991) Kinetic analysis of monoclonal antibody-antigen interactions with a new biosensor based analytical system, *Journal of immunological methods* 145, 229–240.
13. Miyazaki, K., and Arnold, F. H. (1999) Exploring nonnatural evolutionary pathways by saturation mutagenesis: rapid improvement of protein function, *Journal of molecular evolution* 49, 716–720.
14. Blanchard, C., Mishra, A., Saito-Akei, H., Monk, P., Anderson, I., and Rothenberg, M. E. (2005) Inhibition of human interleukin-13-induced respiratory and oesophageal inflammation by anti-human-interleukin-13 antibody (CAT-354), *Clin Exp Allergy* 35, 1096–1103.
15. Oh, C. K., Faggioni, R., Jin, F., Roskos, L. K., Wang, B., Birrell, C., Wilson, R., and Molfino, N. A. An open-label, single-dose bioavailability study of the pharmacokinetics of CAT-354 after subcutaneous and intravenous administration in healthy males, *British journal of clinical pharmacology* 69, 645–655.
16. Singh, D., Kane, B., Molfino, N. A., Faggioni, R., Roskos, L., and Woodcock, A. A phase 1 study evaluating the pharmacokinetics, safety and tolerability of repeat dosing with a human IL-13 antibody (CAT-354) in subjects with asthma, *BMC pulmonary medicine* 10, 3.

Chapter 23

Affinity Maturation and Functional Dissection of a Humanised Anti-RAGE Monoclonal Antibody by Ribosome Display

Simon E. Hufton

Abstract

The pursuit of more potent, safe, and cost-effective drugs has placed a greater emphasis on antibody optimisation within the drug discovery process. Technologies to rapidly improve antibody drug performance, such as phage display, ribosome display, and yeast display, are playing a key role in this effort. Among these ribosome display is a particularly powerful technology and has recently been applied to the affinity optimisation of a humanised anti-receptor for advanced glycation end products (anti-RAGE) antibody (Finlay et al., J Mol Biol 388:541–558, 2009). By using a combination of error-prone PCR with ribosome display each amino acid position within this humanised antibody was scanned for both its functional importance and its capacity to increase affinity resulting in both affinity-matured antibody variants and a functional map of the antibody paratope.

Key words: Affinity maturation, Ribosome display, Mutational scanning, Antibody humanization

1. Introduction

Monoclonal antibodies are now firmly established as a highly successful class of drug. Their therapeutic and commercial success owes much too several major advances in antibody engineering which have led to the generation of highly specific, high affinity and safe biotherapeutics to treat a range of important diseases. As growth in antibody drug discovery continues, there is an increasing emphasis on optimisation of therapeutic leads prior to clinical development. This is driven principally by the need for even better antibodies, often to already well-validated disease targets, that are more potent, have a better safety profile and are ultimately more cost-effective treatments. Antibodies are frequently optimised for affinity using in vitro molecular evolution technologies; however, other properties like specificity, stability, solubility, pharmacokinetics, effector function, and immunogenicity can all be engineered using such technologies (1). Selection of antibodies from genetically diverse libraries using display technologies, such as phage display (2, 3), ribosome display (4, 5), and yeast display (6), are frequently used for in vitro antibody affinity maturation. The expectation being that increased affinity may improve pharmacokinetic and safety profiles and reduce dosing, toxicity, and cost of therapy. There is a growing number of candidate therapeutic antibodies that have all been successfully affinity matured using display technologies, for example anti-$\alpha v \beta 3$ integrin (7), anti-RSV (8, 9), anti-VEGF receptor 2 (10), anti-CEA (6), and anti-RAGE (11).

A key step in optimising a given antibody is deciding where to introduce sequence diversity. In the case of affinity maturation, there are several mutagenesis techniques which can be used and these can be broadly divided into targeted and non-targeted strategies. Many targeted strategies have successfully focussed on the CDR loops, primarily the VH-CDR3 loop as this is naturally the most variable loop and its central position within the antigen-combining site supports its crucial role in antigen recognition (12). In other cases, mutagenesis strategies have been implemented to evaluate the potential for affinity improvements outside of the CDR3 loops using non-targeted strategies like error-prone PCR (4, 11, 13). A particularly powerful approach is to couple error-prone PCR with ribosome display. This allows the rapid construction of very large libraries in vitro without the need to transform mutant plasmids into a host cell. Also additional mutations can be introduced at each round of selection as PCR is integral to the selection cycle (4, 11, 13).

By way of example, we describe the methodology applied to a recent affinity maturation study of a candidate therapeutic antibody specific for the human receptor for advanced glycation end products (RAGE) (11) which had previously been shown to antagonise RAGE interaction with multiple ligands and have a potent protective effect in the murine cecal ligation and puncture model of polymicrobial sepsis (14). This antibody being humanised in origin comprised a "non-natural" combination of rat CDRs and human framework regions and as such was predicted to have a greater scope for improvements in antigen binding outside of the normal CDR loops than there would be for a fully human antibody which had already gone through a degree of optimisation in vivo. Furthermore, as is the case for many antibodies there was no crystal structure and little knowledge of the antibody antigen interface available. In addition to applying phage display in combination with CDR-targeted mutagenesis, we used a second strategy of ribosome display in combination with mutational scanning by error-prone PCR. This approach allowed the introduction of diversity over the whole length of the antibody sequence and subsequently identified key mutations at distal sites in the antibody sequence that correlated with increased affinity and further humanisation (11). Conventional humanisation and affinity maturation are often performed as discrete steps; however, this may overlook beneficial combinations of framework and CDR mutations. Using this approach, both affinity maturation and humanisation can be performed simultaneously resulting in antibodies with murine sequence limited to only those residues essential for determining antibody affinity and specificity.

2. Materials

Precautions must be taken to avoid DNA and nuclease contamination. All solutions, tubes, and tips must be sterilised and nuclease-free.

2.1. Ribosome Display Library Construction of M4-scFv Antibody Using Error-Prone PCR

1. Humanised antibody (e.g. XT-M4) reformatted to an scFv antibody fragment (GeneArt) and cloned into a suitable ribosome display vector (e.g. pWRIL-3) (see Fig. 1) (11).
2. Nuclease-free water.
3. GeneMorph II™ Random Mutagenesis kit (Agilent).
4. FwSD primer: AGACCACAACGGTTTCCCTCTAGAAATAATTTTGTTTAACTTTAAGAAGGAGATATATCCATGGACTACAAAGA.
5. M4-overlap primer: AACCAGAACCGCCGCCCTCGGCCCCTGAGGCCTGATCCGAGGACA
6. TAE buffer (Bio-Rad).

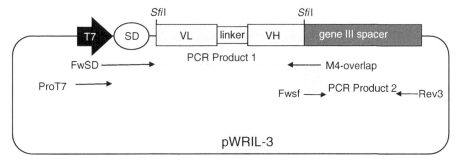

Fig. 1. The vector pWRIL-3 (11) contains all the elements required to carry out ribosome display as described in (15). The scFv antibody construct is ligated into the ribosome display vector as an Sfi1 fragment thereby genetically fusing it in frame to the gene 3 spacer sequence and providing a T7 promotor (T7) and a Shine–Delgarno sequence (SD) at the 5′ end of the construct. The primers used to amplify and assemble a ribosome display library are shown. The scFv error-prone PCR product 1 is amplified using primers FwSD and M4-overlap. The gene III spacer sequence (PCR product 2) is amplified using primers Fwsf and Rev3.

7. 1.5% (w/v) agarose, 1× TAE, 0.5 μg/ml ethidium bromide gel.
8. QIAquick™ gel extraction kit (Qiagen).
9. Rev3 primer: CCGCACACCAGTAAGGTGTGCGGTATCACCAGTAGCACC.
10. ProT7 primer: ATACGAAATTAATACGACTCACTATAGGGAGACCACAACGG.
11. Fwsf primer: GCCTCAGGGGCCGAGGGCGGCGGTT.
12. Phusion™ master mix kit (New England Biolabs).
13. Dimethyl sulphoxide (DMSO) (New England Biolabs).
14. *Sfi*1 restriction enzyme, supplied with reaction buffer (New England Biolabs).
15. Phage display vector (e.g. pWRIL-1) or a suitable alternative (11).
16. T4 DNA ligase, supplied with T4 DNA ligase buffer (New England Biolabs).
17. TG1 electrocompetent cells (Agilent).

2.2. In Vitro Transcription of Error-Prone PCR Library and Purification of RNA

1. RNase Zap™ (Sigma).
2. DNA Away™ (Molecular Bioproducts).
3. T7 RiboMAX™ Express RNA production kit (Promega).
4. Illustra™ ProbeQuant G-50 Micro Columns Purification kit (GE Healthcare).
5. UltraPure™ distilled water, DNase, RNase free (Invitrogen).
6. RNAsin ribonuclease inhibitor (Promega).
7. TAE buffer (Bio-Rad).
8. 2% (w/v) agarose, 1× TAE, 10 μg/ml ethidium bromide gel.
9. RNA sample loading buffer (Sigma).
10. RNA markers: 0.1–1 kb and 0.2–10 kb (Sigma).

2.3. In Vitro Translation of Purified RNA

1. UltraPure™ distilled water, DNase, RNase free (Invitrogen).
2. 1× phosphate-buffered saline (PBS) prepared by diluting 10× PBS (Invitrogen) with Ultrapure™ nuclease-free water (Promega).
3. PureSystem™ S-S kit (Wako).
4. 0.1 μg/μl protein disulfide isomerase (PDI) diluted from a 1 mg/ml (w/v) stock made in nuclease-free 1× PBS.
5. Wash buffer: 50 mM Tris–acetate, pH 7.5, 150 mM NaCl, 50 mM magnesium acetate, 0.1% (v/v) Tween 20, 25 mg/ml heparin sodium salt, and 0.5 μg/ml (w/v) bovine serum albumin (from 10 mg/ml BSA; New England Biolabs).

2.4. Affinity Selection of Ribosomal Complexes in Solution

1. UltraPure distilled water, DNase, RNase free (Invitrogen).
2. 4% (w/v) Sterilised milk (Merck) in water.
3. PBS, see Subheading 2.3, item 2.
4. Wash buffer, see Subheading 2.3, item 5.
5. Dynabeads, M-280 Streptavidin (Invitrogen).
6. Biotinylated antigen, e.g. human RAGE-Fc.
7. Elution buffer: 50 mM Tris–acetate, pH 7.5, 150 mM NaCl, 50 mM EDTA, 20 μg/ml *Saccharomyces cerevisiae* RNA (Sigma).
8. High pure RNA isolation kit (Roche).

2.5. Conversion of Eluted RNA to cDNA and PCR Amplification for Subsequent Cloning and Selection

1. UltraPure distilled water, DNase, RNase free (Invitrogen).
2. Superscript III™ reverse transcriptase (Invitrogen), provided with 5× first strand reaction buffer and 0.1 M DTT.
3. RNAsin™ ribonuclease inhibitor (Promega).
4. Rev3 primer, see Subheading 2.1, item 9.
5. Phusion™ master mix kit (New England Biolabs).
6. 10 mM Nucleotide triphosphate mix (Promega).
7. DMSO.
8. FwSD primer, see Subheading 2.1, item 4.
9. TAE buffer (Bio-Rad).
10. 1% (w/v) and 1.5% (w/v) agarose, 1× TAE, 0.5 μg/ml ethidium bromide gels.
11. *Sfi*1 restriction enzyme, see Subheading 2.1, item 14.
12. Expression vector, e.g. pWRIL-1 or equivalent.
13. 5,000 U/ml antarctic phosphatase.
14. QIAquick™ gel extraction kit (Qiagen).
15. T4 DNA ligase (New England Biolabs).
16. TG1 electro-competent cells (Agilent).
17. SOC medium (Sigma).
18. 2YT medium agar 22-cm bioassay plates containing 100 μg/ml carbenicillin and 2% (w/v) glucose.

2.6. ScFv Periplasmic Expression of Selected Antibody Variants

1. 2YT-CG medium: 2YT medium containing 100 μg/ml carbenicillin and 2% (w/v) glucose.
2. 1 M isopropyl β-D-1 thiogalactopyranoside (IPTG).
3. Periplasmic buffer: 50 mM HEPES, 0.5 mM EDTA, 20% (w/v) sucrose, pH 7.5.
4. 60% (v/v) Glycerol.

5. 2YT-CG medium: 2YT medium containing 100 μg/ml carbenicillin and 0.1% (w/v) glucose.
6. 96-Deep well plates (Axygen or alternative supplier).
7. 1 M $MgCl_2$.
8. HIS-Select™ filter plates (Sigma) or equivalent.

2.7. ELISA and Homogeneous Time-Resolved Fluorescence Screening Assays

1. Biotinylated antigen.
2. Maxisorb™ immunoplates (Nunc).
3. PBS, see Subheading 2.3, item 2.
4. PBS containing 0.05% (v/v) Tween 20.
5. PBS containing 3% (w/v) dried milk protein, 1% (w/v) BSA.
6. Anti-*c*-myc horseradish peroxidase (HRP) conjugate (Roche) diluted in PBS containing 3% (w/v) dried milk protein.
7. UltraTMB™ (Pierce).
8. 0.18 M phosphoric acid.
9. Parental antibody (e.g. anti-RAGE XT-M4) for cryptate labelling.
10. Europium cryptate labelling kit (Cisbio).
11. 1:1,600 dilution of streptavidin-XL665.
12. 1× assay buffer: 50 mM sodium phosphate pH 7.5, 400 mM potassium fluoride, 0.1% (w/v) BSA.

3. Methods

In order to affinity mature the humanised monoclonal anti-RAGE antibody XT-M4, both phage display and ribosome display strategies were used in combination with CDR-targeted and error-prone PCR, respectively (11). This chapter focuses on a description of the methods used for the affinity maturation and functional dissection of this antibody using ribosome display. The first step was to reformat XT-M4 into an scFv antibody fragment and confirm the retention of antigen-binding activity by ELISA. The reformatted M4-scFv antibody was cloned into the ribosome display vector pWRIL-3 (see Fig. 1) which contains all the expression machinery necessary for ribosome display (15) and was used as the template for the construction of an M4-scFv library using error-prone PCR. The M4-scFv library PCR product was then transcribed to RNA and in vitro translated to generate the RNA-ribosome-M4-scFv antibody complexes which were used for selection with biotin-labelled human RAGE-Fc (see Fig. 2a). The RNA from complexes bound to antigen was then recovered and converted to cDNA using RT-PCR, and was subsequently used for further rounds of selection

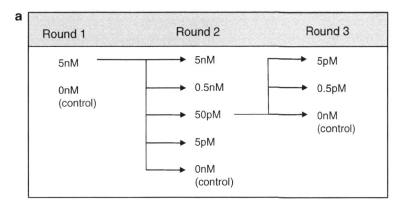

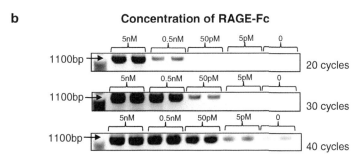

Fig. 2. (**a**) An example selection strategy used for anti-RAGE XT-M4 (11). For affinity-driven selection antigen concentrations are reduced in successive rounds of selection with the objective of preferentially selecting higher affinity clones. A "no antigen" control is included at each round of selection to ensure that antigen-specific enrichment of antibody product is occurring. The starting concentration for the first round of selection is normally below the dissociation constant (K_d) of the parental antibody and then reduced incrementally during subsequent rounds. In the case of the anti-RAGE antibody M4, the starting antigen concentration chosen was 5 nM. (**b**) Example semi-quantitative PCR analysis from ribosome display selection. The amount of RNA and consequently PCR product recovered at different concentrations of biotinylated RAGE-Fc is monitored at 20, 30, and 40 PCR cycles. A control selection containing "no antigen" is included. The ~1,100 bp scFv antibody gene III spacer PCR product is indicated.

or cloned into a suitable *Escherichia coli* expression vector (e.g. pWRIL-1) (11). After cloning into a suitable expression vector, the output variants were sequenced and screened for antigen binding. Correlation of antigen-binding activity with specific mutations allowed both the identification of affinity matured antibodies and the generation of a functional map of the antibody paratope (see Figs. 3 and 4).

3.1. Ribosome Display Library Construction of M4-scFv Using Error-Prone PCR

1. The scFv antibody to be optimised cloned into a suitable ribosome display vector (e.g. pWRIL-3) is the starting template for construction of the error-prone PCR library. Set up the error-prone PCR using the GeneMorph II™ random mutagenesis kit as follows (see Note 1):

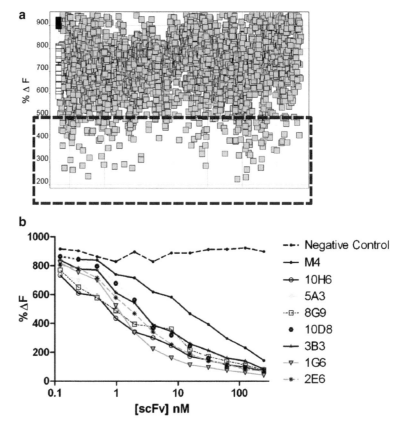

Fig. 3. (a) An example high-throughput HTRF screen on randomly picked clones from ribosome display selections on human RAGE – Fc (11). Output fluorescence for mutated clones (*grey squares*) was compared with parental M4 (*white squares*) and a negative control (*black squares*). Clones in the boxed population were purified and compared using titration HTRF. (b) A titration HTRF is shown comparing the parental antibody (M4) with 7 mutant variants with improved IC50 values (11). Reprinted from Finlay et al. (11).

Water	41.5 μl
10× Mutazyme II™ reaction buffer	5 μl
40 mM dNTP mix (200 μM)	1 μl
FwSD primer (100 μM)	0.25 μl
M4-overlap primer (100 μM)	0.25 μl
Mutazyme II™ DNA polymerase (2.5 U/μl)	1 μl
M4-scFv target DNA template (0.5 μg/μl)	1 μl
Total	50 μl

2. Amplify using the PCR parameters, 95°C for 2 min, followed by 30 cycles of 95°C for 30 s, 68°C for 1 min, 72°C for 1 min, followed by 72°C for 10 min, and cooling to 4°C. The PCR

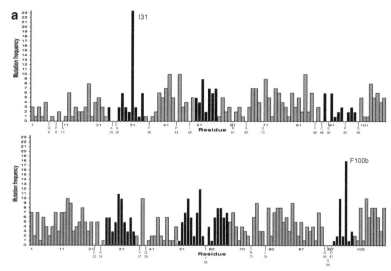

Fig. 4. (a) A mutational scanning map for the anti-RAGE antibody M4 is shown (11) as the frequency of amino acid substitutions in the VL (a) and the VH (b) from a total population of 384 functional-binding mutants. The CDRs are marked by *black bars*. Key mutational hotspots of I31 and F100b are shown (c) Mutational hotspots and sites intolerant of mutation (mutational cold spots) are indicated. Reprinted from Finlay et al. (11).

V region	VL domain		VH Domain	
	Hot spots	Cold spots	Hot spots	Cold spots
FR1	R18	Q6,P8,S10	P14,N30	C22,A24
CDR1	I31	A25,S26		
FR2	R45	F36,P44,I48		V37,Q39
CDR2	R50		N52a,D55,S62	Y58,N73,L79
FR3	F71,E81,F83	R61,S65,D70,Y86,C88,	F67,A93	
CDR3		E90,P95	F100b	G95,G96,D97
FR4	K103	G99	S113	

parameters should be optimised for each primer set but general guidelines are that the annealing temperature should be 5°C below the primer melting temperature. This is PCR product 1 in Fig. 1.

3. Separate the PCR product by electrophoresis on a 1.5% (w/v) agarose, 1× TAE, 0.5 μg/ml ethidium bromide gel and purify the band corresponding to the scFv antibody plus the Shine–Dalgarno sequence using a gel extraction kit. Quantify the PCR product by OD_{260} measurement using a conversion factor of OD_{260} of 1 is equivalent to a DNA concentration of 50 μg/ml.

4. Perform a standard PCR using the primers Fwsf and Rev3 to amplify the gene 3 spacer – 3′stem loop sequence from the ribosome display vector pWRIL-3 (11,15) as follows. This is PCR product 2 in Fig. 1:

Phusion™ master mix	25 µl
DMSO	2.5 µl
Fwsf primer (10 µM)	3 µl
Rev3 primer (10 µM)	3 µl
Plasmid template (0.5 µg/µl)	1 µl
Water	x µl
Total	50 µl

5. Gel purify this PCR product as described in step 3.

6. Link the scFv fragment (PCR product 1) to the gene 3 spacer sequence (PCR product 2) using an assembly PCR as follows:

Phusion™ master mix	25 µl
DMSO	2.5 µl
PCR product 1 (50 ng)	x µl
PCR product 2 (50 ng)	y µl
Water	z µl
Total	44 µl

7. Perform the PCR with an initial annealing step of 98°C for 3 min, and then 10 cycles of 98°C for 10 s, 70°C for 20 s, 72°C for 20 s. Then, immediately after the annealing step add 3 µl of 10 µM ProT7 primer and 3 µl of 10 µM Rev3 primer to each reaction. Continue the PCR using the parameters 98°C for 1 min and 20 cycles of 98°C for 10 s, 66.8°C for 20 s, 72°C for 1 min 30 s followed by 72°C for 10 min followed by cooling to 4°C (see Note 2).

8. Gel purify and quantify the final assembled product corresponding to T7 promotor, Shine–Dalgarno sequence, scFv library, and gene3 spacer as described in step 3. This product is now ready for use for RNA transcription in Subheading 3.2.

9. To evaluate individual antibody sequences and determine mutation frequency of the library, digest a sample of this DNA with *Sfi*1 restriction enzyme and clone into pWRIL-1 phage display vector (see Subheading 3.5, steps 8–16), or any other suitable vector for DNA sequencing (see Note 3).

3.2. In Vitro Transcription of the Error-Prone PCR Library and Purification of RNA

It is crucial that all materials and consumables be RNase and DNase free. All work surfaces and pipettes should be treated with RNase Zap™ and DNA Away™.

1. Perform transcription using the T7 RiboMAX™ express RNA production kit using a total input scFv DNA library of 5 μg purified DNA corresponding to approximately 10^{12} molecules. Prepare a 25 mM rNTP mix using the 100 mM rNTP solutions provided in the kit. Set up the reaction at room temperature and add the components in the order shown (see Note 4):

5× T7 Transcription buffer	20 μl
25 mM rNTP mix	30 μl
scFv DNA library template (5 μg total input)	xμl
Nuclease-free water	yμl
T7 Enzyme Mix	10 μl
Total volume	100 μl

2. Mix gently by pipetting the reaction up and down and incubate at 37°C in a thermocycler for 2.5 h.

3. Add RNase-free DNase to a concentration of 1 U/μg of starting scFv library template and incubate at 37°C for 15 min.

4. Purify the transcribed RNA using an Illustra™ Probe Quant G-50 micro column. Use two separate columns per 100 μl transcription reaction. Resuspend the resin in the column by vortexing briefly. Snap off the closure at the bottom of the column and place the column in a collection tube for support.

5. Pre-spin the columns 1 min exactly at $735 \times g$ and used immediately after preparation to avoid the purification resin drying out.

6. Place the columns in 1.5-ml nuclease-free tubes and apply 50 μl of the transcribed RNA sample slowly to the centre of the resin (see Note 5).

7. Centrifuge the column at $735 \times g$ for 2 min to collect the purified RNA sample in the bottom of the tube.

8. Estimate RNA quantity using OD_{260}/OD_{280} measurements (see Note 6).

9. Estimate RNA quality by mixing a 2-μl sample of purified RNA with 10 μl of RNA sample loading buffer in a 1.5-ml nuclease-free tube. Denature at 70°C for 10 min and then place immediately on ice for 2 min. Run the sample on a 2% (w/v) agarose, 1× TAE, 10 μg/ml ethidium bromide gel (see Note 6).

10. To reduce the risk of RNA degradation, add 5 μl of RNAsin™ inhibitor to the RNA samples and use immediately. For storage, dispense the RNA into 10 μl aliquots and store at −70°C.

3.3. In Vitro Translation of PURIFIED RNA

1. Perform cell-free translation using the PureSystem™ S-S kit. Thaw the translation reagents on ice and thaw the template RNA sample at room temperature. Set up the appropriate number of reactions by adding the components into 0.5-ml nuclease-free PCR tubes in the order shown below to a final volume of 50 µl:

Nuclease-free water	x µl
Solution A (PURE system™)	25 µl
Solution B (PURE system™)	10 µl
Protein disulphide isomerise, PDI (0.1 µg/µl)	4.5 µl
Template RNA (1 µg/ml)	y µl
Total	50 µl

2. Mix the reaction gently by pipetting up and down and then collect at the bottom of the tube by brief centrifugation.

3. Incubate at 37°C for 30 min using a thermocycler and stop the reaction by placing the tubes on ice.

4. Stabilise the ternary ribosomal complexes by the addition of ten times (500 µl) the initial volume of ice-cold wash buffer to the reaction and mix gently. The ribosomal complexes are now ready for selection.

3.4. Affinity Selection of Ribosomal Complexes in Solution

The selection procedure up to the elution of RNA is carried out in a cold room at 4°C. All tubes, magnetic racks, pipette tips and solutions are kept ice-cold throughout the procedure.

1. Block nuclease-free 1.5-ml tubes with 4% (w/v) sterilised milk solution in water for 1 h with end-over-end rotation at room temperature. Wash the tubes three times with sterile, ice-cold PBS and three times with ice-cold wash buffer. Add 1 ml of wash buffer to each tube and equilibrate on ice for at least 20 min before using for selection.

2. Wash 100 µl of streptavidin magnetic beads per selection four times with ice-cold wash buffer and resuspend in 100 µl of ice-cold wash buffer in ice-cold tubes.

3. Add biotinylated antigen to the translation mixture at the required concentration (see Note 7). Duplicates of each antigen concentration plus a duplicate negative control containing no antigen are routinely used for each round of selection.

4. Wrap the 1.5-ml tubes with parafilm and place within a 50-ml Falcon tube filled with ice and rotate end-over-end for 1 h in a cold room.

5. After this incubation period, add the selection mix to the tube containing the streptavidin magnetic beads and rotate end-over-end on ice for 15 min in a cold room.

6. Capture the streptavidin magnetic beads using a magnetic rack and remove the supernatant containing the unbound ribosomal complexes. Wash the beads with bound ribosomal complexes twice with 1 ml of ice-cold wash buffer and once with 1 ml of ice-cold nuclease-free water to remove the non-specific binding complexes.

7. To elute bound ribosomal complexes, add 200 µl of ice-cold elution buffer to each tube and resuspend the magnetic beads. Rotate the tubes end-over-end in a cold room for 10 min.

8. Capture the magnetic beads again using the magnetic rack and transfer the eluted RNA solution to a 1.5-ml nuclease-free ice-cold tube. Purify the RNA using the High Pure RNA Isolation Kit as follows;

9. Transfer the sample to a filter column inserted into a collection tube and spin for 15 s at $8,000 \times g$.

10. Remove the filter tube from the collection tube and discard the flow through liquid. Re-insert the filter column into collection tube.

11. Wash the column according to the manufacturer's instructions and elute the RNA with 50 µl of nuclease-free water. Immediately denature the eluted RNA at 70°C for 10 min, followed by chilling on ice for 2 min. It is preferable to convert the eluted RNA to cDNA immediately to avoid the risk of degradation.

3.5. Conversion of Eluted mRNA to cDNA and PCR Amplification for Subsequent Cloning and Selection

Selected RNA is subjected to RT-PCR and the resulting cDNA is used for RNA transcription for subsequent rounds of selection, or can be cloned into a suitable *E. coli* expression vector for analysis of individual scFv antibody clones.

1. Prepare a master mix on ice corresponding to the number of reactions being performed and containing all the components listed below except eluted RNA:

Rev3 primer (100 µM)	3 µl
dNTP mix (10 mM)	5 µl
Sterile nuclease-free water	11 µl
5× Reverse transcriptase buffer	20 µl
Dithiothreitol (DTT) (0.1 M)	5 µl
RNasin™ ribonuclease inhibitor (50 U)	1 µl
Superscript reverse transcriptase III	5 µl
Eluted RNA	50 µl
Total	100 µl

2. Aliquot 50 μl of this master mix into 0.5-ml nuclease-free PCR tubes and then add 50 μl of the eluted RNA. Mix the contents by gently pipetting up and down and incubate at 55°C for 30–60 min, 70°C for 15 min followed by cooling to 4°C using a thermocycler. This first strand cDNA synthesis product can either be used immediately for PCR or stored at −20°C.

3. Perform PCR to analyse the selection outputs at different antigen concentrations and to determine the optimum number of PCR cycles to give sufficient product to proceed to the next round of selection or to clone for monoclonal analysis (see Note 8). Typically, semi-quantitative PCR is performed although real-time PCR can also be used. Analyse samples after 20, 30, and 40 PCR cycles. Therefore, three PCRs are set up for each cDNA sample. The following procedure is for PCR using 0.2-ml PCR tubes in strips of eight.

4. Prepare a reaction master mix corresponding to the number of reactions required. Add the reaction components listed below into a 1.5-ml nuclease-free tube in the order shown, except for the cDNA sample:

Water	8.2 μl
Phusion™ master mix	15 μl
DMSO	1.5 μl
Primer FwSD (100 μμM)	0.15 μl
Primer Rev3 (100 μM)	0.15 μl
cDNA sample	5 μl
Total	30 μl

5. Add 5 μl of each cDNA sample to a 0.2-ml PCR tube on ice and then add 25 μl of the above master mix to each tube. Mix gently by tapping the PCR strips followed by a brief centrifugation.

6. PCR amplify using the conditions 98°C for 3 min, and 20, 30, and 40 cycles of 98°C for 10 s, 50°C for 20 s, 72°C for 1 min 30 s, followed by 72°C for 10 min. This can be done in a single PCR machine by preparing separate strips containing eight different cDNA samples, and removing one strip each at 20 cycles, 30 cycles, and 40 cycles.

7. Analyse a 10-μl sample from each PCR on a 1.5% (w/v) agarose, 1× TAE, 0.5 μg/ml ethidium bromide gel (see Fig. 2). Choose the optimum selection output and the optimum number of PCR cycles, then scale up the PCR to generate a larger amount of PCR product. The purified PCR product can be transcribed for subsequent rounds of selection or cloned into a suitable expression vector.

8. Generally, outputs from sequential rounds of selection are amplified and cloned into a suitable expression vector (e.g. pWRIL-1) (11). This allows both sequence analysis and binding assays to be performed. For this, digest 2 μg of purified PCR product and 5 μg of vector (e.g. pWRIL-1) with *Sfi*1 at 50°C for 3 h as follows:

Nuclease-free water	y μl
PCR product or vector	x μl
Buffer 4	5 μl
BSA	0.5 μl
*Sfi*1 (40 U)	2 μl
Total	50 μl

9. To reduce re-ligation of *Sfi*1 digested vector, treat with antarctic phosphatase (5,000 U/ml) according to the manufacturer's protocol by incubation at 37°C for 60 min followed by heat inactivation at 65°C for 5 min.

10. Separate the digested vector on a 1% (w/v) agarose, 1× TAE, 0.5 μg/ml ethidium bromide gel. Cut the linearised vector band from the gel and purify using a QIAquick™ gel purification kit. Quantify A_{260} measurement.

11. Separate the *Sfi*1 digested PCR fragments on a 1.5% (w/v) agarose, 1× TAE, 0.5 μg/ml ethidium bromide gel. Cut out the ~750 bp band corresponding to the scFv antibody DNA and purify using the QIAquick™ gel extraction kit. Quantify by A_{260} measurement.

12. Ligate 140 ng of digested pWRIL-1 vector and 70 ng of digested scFv insert per reaction as follows. Include a self-ligation control using the same quantity of cut vector but no scFv insert:

Digested pWRIL-1 vector (140 ng)	x μl
Digested scFv insert (70 ng)	y μl
5× Ligase buffer	5 μl
DNA ligase (1 U/μl)	1 μl
Water	20 μl

13. Incubate the ligation reactions overnight at 16°C.

14. Use 1 μl of each ligation reaction to transform 40 μl of electrocompetent TG1 cells according to manufacturer's instructions.

15. Remove cells from the electroporation cuvettes using 1 ml of SOC medium pre-warmed to 37°C and add to an appropriate culture tube followed by incubation at 37°C for 1 h to allow expression of antibiotic resistance.

16. Spread the library onto 2YT-CG agar 22 cm bioassay trays and incubate overnight at 37°C (see Note 9).

3.6. ScFv Periplasmic Expression of Selected Antibody Variants

1. Pick individual colonies from the bioassay trays into standard sterile 96-well plates containing 2YT-CG medium. Ideally use an automated colony picking device for this. Incubate plates overnight at 37°C and 80% humidity (if available) with shaking.

2. Add glycerol to the plates to give a final concentration of 30% (v/v) and store plates at −80°C, or use immediately to inoculate 96-deep well plates containing 900 μl of 2YT-CG medium.

3. Grow the cultures in deep well plates at 37°C, 80% humidity and shaking at 600 rpm for 5–6 h until an OD_{600} of ~0.8 is reached. Induce expression by the addition of IPTG to a final concentration of 0.02 mM and incubate at 30°C overnight.

4. Pellet cells by centrifugation at $1,260 \times g$ and resuspend in 150 μl of ice-cold periplasmic buffer.

5. Place the samples on ice for 30 min and then centrifuge at $3,220 \times g$ for 20 min. Recover the supernatant comprising the periplasmic fraction and containing individual expressed scFv antibodies.

6. Screen the crude periplasmic extracts for antigen binding using a suitable ELISA or homogeneous time-resolved fluorescence (HTRF) assay using single point analysis.

7. For more detailed HTRF titration analysis, small-scale single-step scFv purification can be performed as follows.

8. Inoculate *E. coli* clones containing scFv antibodies of interest are into 10 ml of 2YT medium supplemented with 100 μg/ml (w/v) carbenicillin and 0.1% (w/v) glucose in a 50-ml Falcon tube.

9. Grow the cultures overnight at 37°C with shaking at 250 rpm until an OD_{600} of 0.6 is reached. Induce antibody expression by adding IPTG to a concentration of 0.02 mM, followed by further incubation overnight at 30°C.

10. Centrifuge the cultures for 10 min discard the supernatants. Resuspend the bacterial pellet in 1.5 ml of ice-cold periplasmic buffer and then add 1.5 ml of a 1 in 5 dilution of ice-cold periplasmic buffer in water.

11. Incubate on ice for 30 min and centrifuge the samples at $3,220 \times g$ for 10 min. Collect the supernatant and add $MgCl_2$ to a final concentration of 10 mM. Purify the scFv antibodies from periplasmic preps using metal chelate chromatography, for example by using HIS-Select™ filter plates.

3.7. ELISA and HTRF Screening Assays

1. For high-throughput analysis of crude periplasmic extracts, perform a single-point, antigen-binding ELISA. Coat 96-well Maxisorp™ plates with 75 µl per well of 1–10 µg/ml antigen solution in PBS overnight at 4°C.

2. Wash the plates three times with 300 µl of PBS containing 0.05% (v/v) Tween 20, preferably using a liquid handling robot. Block plates with 200 µl of PBS, 3% (w/v) dried milk protein, and 1% (w/v) BSA per well for 1 h at room temperature.

3. Dilute the crude periplasmic extract 1 in 4 in PBS, 3% (w/v) dried milk protein, 1% (w/v) BSA and add to the blocked ELISA plate and incubate for 1 h.

4. Wash five times with 300 µl of PBS containing 0.05% (v/v) Tween 20.

5. Detect binging by the addition of 75 µl per well of an HRP-conjugated anti-*c*-myc antibody diluted 1:2,500 in PBS, 3% (w/v) dried milk protein, and incubate at room temperature for 1 h.

6. Wash the plates seven times with PBS containing 0.05% (v/v) Tween 20.

7. Develop the HRP reaction by the addition of 75 µl per well of UltraTMB™ and stop by the addition of 75 µl per well of 0.18 M phosphoric acid. Read the plate at 450 nm.

8. In cases where screening for both antigen binding and the retention of a functional epitope is desired, a competition HTRF assay is preferred (see Note 10) and is performed as follows. The assay can be run in either a high-throughput mode using single-point analysis or in a more quantitative mode using a concentration range of purified scFv (see Fig. 3a, b for example data).

9. Label the parental antibody (e.g. anti-RAGE XT-M4) with europium–cryptate using a cryptate labelling kit according to the manufacturer's instructions.

10. Add the following reagents sequentially into a 384-well low volume black plate; 1.5-nM biotinylated antigen, a 1:1,600 dilution of streptavidin-XL665, a 1:1,000 dilution of the europium–crypate-labelled parental antibody, and 0.5% (v/v) periplasmic extract containing the scFv antibody of interest in a total reaction volume of 20 µl in 1× assay buffer.

11. Allow the reaction to proceed for 3 h at room temperature and read the plates on a suitable plate reader with excitation at 340 nm and two emission readings at 615 nm (measuring input donor fluorescence from europium–cryptate labelled parental antibody) and 665 nm (measuring output acceptor fluorescence from streptavidin-XL665).

12. Express all readings as a percentage change in fluorescence (%ΔF), where %$\Delta F = [(F_{665}$ sample$/F_{615}$ sample$) - (F_{665}$ control$/F_{615}$ control$)]/(F_{665}$ control$/F_{615}$ control$) \times 100$. "Control" represents the background fluorescence energy transfer in wells containing 1:1,000 labelled XT-M4, in assay buffer, alone.

13. Correlating functional binding data with sequence information allows the identification of both affinity-matured antibodies and the generation of a functional map of the antibody paratope (see Note 11). Example data from a study of a humanised anti-RAGE antibody is shown in Fig. 4.

4. Notes

1. The amount of target DNA template is determined by the mutation frequency required. The lower the amount of input DNA the greater the mutation frequency as a greater number of PCR cycles are required to complete amplification process. An intermediate mutation frequency is about 4.5–9 mutations/kb which uses between 100 and 500 ng of target DNA template.

2. The PCR conditions need to be optimised for the particular antibody primers used, the PCR machine, annealing temperature, and magnesium concentration.

3. Sequencing of random clones can be performed to ensure a broad spread of mutations over the length of the scFv antibody at the predicted frequency. In the case of the M4-scFv library, an average of 2.5 mutations per antibody was obtained (11).

4. This reaction can be scaled up or down to suit your template requirements. A 100-μl reaction will typically produce between 200 and 500 μg of RNA in 2–4 h.

5. The columns are used immediately after preparation to avoid the resin drying out. After the pre-spin, the resin will come away from the tube slightly to form a column within the tube. It is essential that the RNA sample being purified is added slowly to the centre of the column without touching the resin bed and ensuring that the solution does not run into the sides of the column.

6. The concentration of RNA samples should be between 2,300 and 3,500 μg/ml with an OD_{260}/OD_{280} ratio of between 1.9 and 2.1. A single band on an agarose gel indicates RNA is of sufficient quality to proceed to the in vitro translation reaction. There may be some smearing but as long as the RNA band is of the expected size and is >95% of the total then the RNA is of sufficient quality to proceed. Up to 5 μg of RNA is used per translation reaction and this corresponds to approximately 10^{12} molecules.

7. The initial antigen concentration used for selection is generally lower than the starting K_d of the parental antibody antigen interaction and is lowered progressively in successive rounds. In the case of the M4 scFv antibody, we predetermined an optimum starting antigen concentration by empirically testing the minimum antigen concentration at which we were able to recover a strong band compared to "no antigen" control selection. A series of model selections were performed at several antigen concentrations [500 nM, 50 nM, 5 nM, 500 pM, 0 nM (no antigen control)], and the minimum antigen concentration where M4-scFv product was recovered in an RT-PCR was taken as the starting point for each selection step. In the case of the M4 antibody under study, we chose 5 nM as the starting antigen concentration antibody library selections and this was progressively lowered in subsequent rounds (see the examples given in Fig. 2). Output RNA samples from consecutive selection rounds are converted to cDNA, PCR amplified and cloned into a suitable expression vector for screening for antigen binding and sequence analysis.

8. Between rounds of selection, further diversity can be introduced by using error-prone PCR or low fidelity PCR to amplify cDNA after each selection step.

9. Titration of transformed cultures can be performed to assess transformation and cloning efficiency; however, generally a successful cloning is indicated by a greater number of colonies being present on the test plates compared to the corresponding titration plates prepared for the self-ligation control.

10. In cases where it is critical to maintain the functional epitope of the parental antibody, then a competition HTRF assay is preferred. The HTRF assay (11) format described here was specifically applied to the affinity maturation of an anti-RAGE antibody for which the epitope was already correlated with efficacy in an animal model of severe sepsis and systemic infection (14). The assay measures the decrease in fluorescence observed upon binding of a europium–cryptate labelled parental antibody to antigen in the presence of competing scFv test antibodies and can be performed as either single point analysis or in a more quantitative form with increasing concentration of purified scFv antibodies.

11. ScFv variants with improved activity in an HTRF assay can be reformatted to full IgG molecules for affinity determination by BIAcore. Individual mutations correlated with improved HTRF binding activity can also be combined in a single IgG molecule to assess potential synergistic mutations and further affinity improvement.

Acknowledgments

Xuemei Liu, Alfredo Darmanin-Sheehan and former colleagues at Wyeth Research Ireland are gratefully acknowledged.

References

1. Carter, P. J. (2006) Potent antibody therapeutics by design, *Nat. Rev. Immunol.* 6, 343–357.
2. Schier R, McCall A, Adams GP et al (1996) Isolation of picomolar affinity anti-c-erbB-2 single-chain Fv by molecular evolution of the complementarity determining regions in the center of the antibody binding site, *J. Mol. Biol.* 263, 551–567.
3. Schier R, Bye J, Apell G et al (1996) Isolation of high-affinity monomeric human anti-c-erbB-2 single chain Fv using affinity-driven selection, *J. Mol. Biol.* 255, 28–43.
4. Thom G, Cockroft AC, Buchanan AG et al (2006) Probing a protein-protein interaction by in vitro evolution, *Proc. Natl. Acad. Sci. U. S. A* 103, 7619–7624.
5. Luginbuhl B, Kanyo Z, Jones RM et al (2006) Directed evolution of an anti-prion protein scFv fragment to an affinity of 1 pM and its structural interpretation, *J. Mol. Biol.* 363, 75–97.
6. Graff CP, Chester K, Begent R et al (2004) Directed evolution of an anti-carcinoembryonic antigen scFv with a 4-day monovalent dissociation half-time at 37 degrees C, *Protein Eng Des Sel* 17, 293–304.
7. Wu H, Beuerlein G, Nie Y et al (1998) Stepwise in vitro affinity maturation of Vitaxin, an alpha v beta3-specific humanized mAb, *Proc. Natl. Acad. Sci. U. S. A* 95, 6037–6042.
8. Wu H, Pfarr DS, Tang Y et al (2005) Ultra-potent antibodies against respiratory syncytial virus: effects of binding kinetics and binding valence on viral neutralization, *J. Mol. Biol.* 350, 126–144.
9. Wu H, Pfarr DS, Johnson S et al (2007) Development of motavizumab, an ultra-potent antibody for the prevention of respiratory syncytial virus infection in the upper and lower respiratory tract, *J. Mol. Biol.* 368, 652–665.
10. Lu D, Shen J, Vil MD et al (2003) Tailoring in vitro selection for a picomolar affinity human antibody directed against vascular endothelial growth factor receptor 2 for enhanced neutralizing activity, *J. Biol. Chem.* 278, 43496–43507.
11. Finlay WJ, Cunningham O, Lambert MA et al (2009) Affinity maturation of a humanized rat antibody for anti-RAGE therapy: comprehensive mutagenesis reveals a high level of mutational plasticity both inside and outside the complementarity-determining regions, *J. Mol. Biol.* 388, 541–558.
12. Barbas CF III, Bain JD, Hoekstra DM et al (1992) Semisynthetic combinatorial antibody libraries: a chemical solution to the diversity problem, *Proc. Natl. Acad. Sci. U. S. A* 89, 4457–4461.
13. Jermutus L, Honegger A, Schwesinger F et al (2001) Tailoring in vitro evolution for protein affinity or stability, *Proc. Natl. Acad. Sci. U. S. A* 98, 75–80.
14. Lutterloh EC, Opal SM, Pittman, DD et al (2007) Inhibition of the RAGE products increases survival in experimental models of severe sepsis and systemic infection, *Crit Care* 11, R122.
15. Hanes J, Jermutus L, Pluckthun A. (2000) Selecting and evolving functional proteins in vitro by ribosome display, *Methods Enzymol.* 328, 404–430.

INDEX

A

Affinity maturation 13, 15, 20, 140, 145, 163–189, 261, 263, 275–277, 394–396, 399, 403–421

Antibody
    fragment 13, 56, 76, 79, 82, 140, 163, 164, 187, 214, 263, 281, 315, 405, 408
    humanization 405
    library 6, 13, 81, 421
    technology 349, 350
    therapeutics 393–400, 404

C

cDNA-protein fusion 114, 119, 121–123, 128–132, 134, 237, 238, 240, 241, 245–247
Cell-free expression 38, 76, 192
Cell-free protein synthesis 251, 252
Complementarity determining region (CDR) 141, 393
Conditional protein-protein interaction 288, 289
Covalent coupling of mRNA and protein 87–99
Cyclic thioether peptides 344

D

Designed Ankyrin Repeat Proteins (DARPins) 14
Directed mutagenesis 139–160, 393, 395, 399
Disulfide bridge 315, 316
Disulfide-rich peptide 237–250
Dithiothreitol 48, 116, 144, 192, 196, 197, 254, 258, 415
DNA display 113–134, 146, 237–250, 307

E

Emulsion 8, 18–19, 22, 102, 103, 105, 106, 108–110, 349, 356
Epitope mapping 253
Escherichia coli 5, 6, 10, 17, 59, 65, 191, 195, 230, 252, 299, 336, 357, 370, 380–381, 409
Eukaryotic ribosome display 9, 45–57, 75–84
Eukaryotic ribosome display, mRNA elution from 46

G

Genetic code reprogramming 335
Genotype-phenotype conjugation 87, 287
Genotype-phenotype linkage 87, 287

H

High diversity 88, 261, 280, 287
Hydrophobic interaction chromatography 192, 193, 199

I

In situ RT-PCR 76, 81, 82
Interleukin–13 393–401
In vitro compartmentalisation 101
In vitro protein selection 299
In vitro selection 59–73, 88, 113, 114, 121, 123, 127, 152, 209, 238, 242, 253–255, 263, 287, 288, 315, 358, 361, 367–389, 394
In vitro transcription translation 31–40, 101
In vitro virus 113, 237

L

Library construction
    error prone 158, 201–203, 210
    oligonucleotide directed 328
    random peptide 60–61, 66

M

Model selection 56, 102, 105–110, 160, 184, 210, 211, 421
mRNA display 16–19, 87–99, 102, 113–134, 237, 287–296, 349, 350, 356, 367–389
mRNA-peptide fusion 368, 374, 383, 385, 388, 389
mRNA secondary structure 292
MSp-Cv interaction 60
Mutational scanning 404, 411

N

Naive ribosome display library 214
Nanofitin 316, 326, 327
Natural proteome library 88, 288, 289, 291–292

Julie A. Douthwaite and Ronald H. Jackson (eds.), *Ribosome Display and Related Technologies: Methods and Protocols*,
Methods in Molecular Biology, vol. 805, DOI 10.1007/978-1-61779-379-0, © Springer Science+Business Media, LLC 2012

Index

O
Off-rate 20, 21, 184, 261–263, 271, 276–278, 281–284

P
Panning 17–19, 55, 102, 105–107, 209
Pathogen.. 300–311
Pathogen genomic DNA library....................................... 300
Peptide aptamer.. 237–250
Peptide library 60–61, 71, 87, 88, 127, 238, 242, 249, 257, 258, 289–291, 294, 344, 374
Peptide screening.. 251–259
Peptidomimetic 352–354, 356, 362
Phage ELISA 167, 178–179, 186, 187, 189
Protein folding..11–12, 191, 399
Protein scaffold..13, 18, 263, 288
PURE translation system 351, 368, 387
Puromycin 16, 17, 88–91, 93–95, 98, 113, 114, 116–117, 121–129, 131, 133, 237, 238, 240, 243, 245, 291, 292, 356, 368, 376

R
Rabbit reticulocyte lysate...................... 45–57, 71, 76, 78, 81, 83, 89, 91, 94, 118, 128, 131, 249, 292
Radiolabelling...35, 91, 94, 95, 98
Random mutagenesis.......................... 6, 12, 23, 49, 141, 142, 148, 156, 165, 182, 194, 196, 199, 201, 316, 394, 399, 405, 409
Real-time PCR.................................. 84, 145, 148, 152, 155, 157, 160, 165, 166, 171, 175, 181, 185, 197, 207, 209, 211, 345, 347, 416
Ribosome binding site........................... 7, 9–11, 16, 47, 195, 257, 262, 268, 311, 320, 375
Ribosome display..3–23, 31, 45–57, 59–73, 75–84, 87, 102, 139–160, 163–189, 191–211, 213–235, 251–259, 261–284, 299–313, 315–330, 336, 349, 393–400, 403–421
Ribosome display construct 9, 47, 200, 320
Ribozyme 335–347, 353, 355, 356

S
Sac7d.. 315–330
ScFv... 5, 49, 76, 140, 163, 192, 213, 251, 263, 395, 405
ScFv library 158, 169, 170, 182, 222–224, 412, 413
Selection stringency...50, 200, 283, 287, 292, 330
S30 extract............................... 11, 19, 20, 151, 157, 158, 172, 173, 204, 209, 210, 230, 252, 262, 266, 267, 270, 275–276, 302, 307, 308, 313, 318, 324, 327, 329
Single-chain antibody... 76, 79, 82
SNAP-tag......................................101–103, 105, 108, 109
Stability selection..................... 192, 193, 196–200, 203–205
Stabilized ribosome display 59–73
Stem loop 9–11, 46, 47, 56, 146, 147, 168, 170, 180, 262, 268, 274, 311, 312, 320, 412
Stop template 143, 149, 156
Synthetic combinatorial peptide library............. 88, 289–291

T
Thioether... 335–347, 387
T7 promoter 7, 10, 54, 65, 78, 79, 82, 91, 120, 126, 146, 180, 183, 253, 257, 262, 291, 320, 375, 376
tRNA 8, 31, 88, 109, 144, 166, 196, 252, 262, 291, 302, 318, 335–347, 350, 367

U
Unnatural amino acid 102, 349–363, 367–369, 374, 375, 382, 386, 387

V
Vaccine genes... 299–313